ELASTICITY

TENSOR, DYADIC, and ENGINEERING APPROACHES

PEI CHI CHOU

Billings Professor of Mechanical Engineering
Drexel University
Philadelphia, Pennsylvania

NICHOLAS J. PAGANO

Air Force Senior Scientist
Wright Laboratory
Wright-Patterson Air Force Base, Ohio

DOVER PUBLICATIONS, INC., NEW YORK

Published in Canada by General Publishing Company, Ltd., 30 Lesmill Road, Don Mills, Toronto, Ontario.
Published in the United Kingdom by Constable and Company, Ltd., 3 The Lanchesters, 162–164 Fulham Palace Road, London W6 9ER.

This Dover edition, first published in 1992, is an unabridged, corrected republication of the work first published by the D. Van Nostrand Company, Inc., Princeton, New Jersey, 1967.

Manufactured in the United States of America
Dover Publications, Inc., 31 East 2nd Street, Mineola, N.Y. 11501

Library of Congress Cataloging-in-Publication Data

Chou, Pei Chi, 1924–
Elasticity—tensor, dyadic, and engineering approaches / by Pei Chi Chou and Nicholas J. Pagano.
p. cm.
Originally published: Princeton, N.J. : Van Nostrand, 1967.
Includes bibliographical references and index.
ISBN 0-486-66958-0 (pbk.)
1. Elasticity. I. Pagano, Nicholas J. II. Title.
[QA931.C5 1992]
620.1'1232—dc20 91-31041
CIP

TO Rosalind and Marianne

Preface

This textbook is designed for advanced undergraduate and beginning graduate engineering courses in elasticity. It is suitable for students who have already taken basic engineering courses including statics and strength of materials. The mathematical background required consists of calculus and vector analysis but not tensor analysis. A chapter on Cartesian tensors is included in the book just prior to the chapters where tensor notation is introduced. The principles of dyadic notation are also presented when they are needed.

The first part of this book treats the theory of elasticity by the most elementary approach, emphasizing physical significance, and using engineering notations. Most of the treatments in the first few chapters follow the inductive approach rather than the deductive approach. Equations and principles are first explained in one-dimensional or two-dimensional cases, then extended to the general three-dimensional cases. The main purpose of this part of the book is to give the engineering students a clear, basic understanding of linear elasticity. Conciseness in expression of equations and coverage of more general cases are not the objectives in these chapters.

The latter part of the book, after the Cartesian tensor and dyadic notations are introduced, gives a more general treatment of elasticity. Most of the equations in the earlier chapters are repeated in Cartesian tensor notation, and again in vector-dyadic notation. The duplication is deemed desirable from the pedagogic point of view. By combining this threefold presentation in one book, we hope the beginning students will have the benefit of cross referencing, and thus make the learning process easier.

In the last chapter, several representations of the general solution of the elasticity equations are presented, and the deductive approach is used to show how these representations can be specialized for classes of problems of interest.

For the advanced undergraduate or beginning graduate engineering students, elasticity is considered here not only as a course in formulating and solving problems in elasticity, but also as an introduction to the boundary value problems in partial differential equations, to the concept of tensor qualities, and to Cartesian tensor and dyadic notations. The practical application of boundary value problems may be explained by using any branch of engineering sciences, such as heat transfer, fluid mechanics, or electric fields. The choice of elasticity is first of all our personal preference; but it is also felt that in dealing with a solid free body, stress, and strain, it is perhaps easier for the engineering student to grasp. The same argument can also be applied to the explanation of the concept of a tensor quantity.

Although Cartesian tensor notation is widely used today in elasticity, fluid mechanics, as well as all branches of continuum mechanics, it is felt that there is still a need of an elementary disposition of this topic for engineering students. Many students have been subjected to the ordeal of learning a new subject topic (e.g., elasticity) and a strange new notation (tensor notation) at the same time. In our experience, most students by their junior year are well drilled in studies of vectors and have a basic understanding of the vector nature of quantities, such as velocity and force; but very few have the correct concept of the tensor nature of stress and strain. In this text, we ask the student first to firmly grasp the tensor *nature* of stress and strain, using familiar engineering notation. When Cartesian tensor notation is introduced, the student may concentrate on learning the new notation, as applied to familiar equations.

The question as to the merits of symbolic notations (vector, dyadic) versus tensor notation has always been controversial. In 1950, Nadai [1] maintained that, "The author questions the value of this new symbolism [tensor notation] in the theories of elasticity and plasticity and from the engineer's point of view for a number of reasons. Space prohibits enumerating them here, but the primary reason is that it does not introduce any new concepts apart from its apparent brevity, alleged by a few of its adherents, . . ." At the present time, there are still some people who agree with Nadai, even though tensor notation, especially Cartesian tensor, is very popular. The symbolic [vector and dyadic] notation has been subjected to even more severe attacks. Jeffreys and Jeffreys [2] commented, "In elasticity and the dynamics of viscous fluids the suffix notation [Cartesian tensor notation] adapts itself far more naturally, and in the theory of relativity vector notation breaks down completely because the parallelogram law fails for velocities. Consequently, some mathematical physicists hold that vector notation is pure waste of time and delays the acquisition of familiarity with the more generally useful method. . . ." Prager [3] presented an objective evaluation and indicated

[1] A. Nadai, *Theory of Flow and Fracture of Solids,* McGraw-Hill Book Company, Inc., New York, 1950, Chapter 14, p. 171.

[2] Harold and Bertha S. Jeffreys, *Methods of Mathematical Physics,* 3rd ed., Cambridge University Press, New York, 1962, p. 64.

[3] William Prager, *Introduction to Mechanics of Continua,* Ginn and Company, Boston, 1961.

that, in treating vector quantities, the symbolic notation [vector notation] is most suitable; in dealing with second-order tensor quantities, both symbolic and component [tensor] notations could be used; but for higher-order tensors, symbolic notation becomes progressingly more cumbersome as the order of the tensors increases and is not practical.

One reason the symbolic notation is not widely accepted today in the linear theory of elasticity is probably due to the special symbols and operators that must be defined. In contrast with this, in using Cartesian tensor notation, only a few conventions and special tensors need be introduced, and the conventional algebra and calculus remain applicable without any modification.

There are, however, some major drawbacks in using Cartesian tensor notation. Equations derived with Cartesian tensor notation cannot be changed readily into curvilinear coordinates. They must be converted to general tensor notation, or to symbolic notation, and then specialized into particular curvilinear coordinates. Otherwise, one is forced to derive all the equations involving spacial derivatives in each particular curvilinear coordinate system separately, which is obviously very tedious. Therefore, in treating equations in curvilinear coordinates, only general tensor and symbolic approaches are practical.

In linear elasticity, most of the quantities involved are either scalars, vectors, or second-order tensors (except the coefficients in generalized Hooke's law). Thus, for the purpose of treating linear elasticity in curvilinear coordinates, we feel that dyadic notation is more expedient than general tensor notation.

Another drawback in using Cartesian tensor notation is that the idea it conveys is sometimes not as complete as that implied by the symbolic notation. For instance, it is well known that, in the theory of linear elasticity, the components of displacement in Cartesian coordinates are biharmonic. In Cartesian tensor notation, this is represented by the expression

$$u_{i,\,jjkk} = 0 \qquad (\text{or} \qquad \nabla^2\nabla^2\, u_i = 0)$$

Some authors caution that this equation is true only in Cartesian coordinates and is not true in other coordinates. The fact that the displacement *vector* is biharmonic in any coordinate system is somewhat obscured and seldom mentioned in textbooks. In symbolic (vector) representation, this equation is

$$\nabla^2\nabla^2\, \mathbf{u} = 0$$

which leaves little room for misunderstanding.

On the other hand, it would be very cumbersome to discuss generalized Hooke's law, which involves a fourth-order tensor coefficient, with symbolic notation. Cartesian tensor notation is probably best suited in this case.

We have managed in this text to introduce only a few dyadic symbols in the symbolic treatment. Any student familiar with vector analysis should be able to read the chapter on dyadics with a minimum expenditure of effort and time.

Aside from the relative merits of each, both Cartesian tensor and symbolic notations are being used in the literature today; therefore, a student should be at home with both.

The instructor using this book may omit certain chapters depending on the nature of his course. For instance, it is possible to present a course using the first ten chapters and not considering vectors and dyadics at all. Advanced students may skip the first seven chapters and delve right into the tensor and/or dyadic approaches.

ACKNOWLEDGMENTS

We are indebted to Professor Y. C. Fung of the University of California for reading and commenting on the manuscript, and to Professor E. J. Brunelle of Rensselaer Polytechnic Institute for his constructive criticism of the manuscript. Many of our students have contributed in different ways toward the preparation of the manuscript. In particular we would like to thank Mr. Herbert A. Koenig and Mr. Alan Benson, who have read the final manuscript and presented many helpful suggestions and corrections. We would also like to give special credit to Messrs. Bruce MacDonald, George Nice, and Richard Perry, who worked out most of the problems and helped in clarifying some of the statements in the manuscript. Our sincere appreciation goes to Miss Karen Walis for her expertness in deciphering and typing the seemingly endless volumes of corrections and revisions to the manuscript.

PEI CHI CHOU
NICHOLAS J. PAGANO

Contents

Introduction

The theory of elasticity deals with the systematic study of the stress, strain and displacement in an elastic body under the influence of external forces. It differs from the traditional undergraduate strength of materials in that the latter is more elementary in theory with more emphasis on convenient formulas for practical application. Elementary strength of materials treats each problem separately; for example, a beam and a shaft are analyzed as separate problems. Although they are of practical importance, the formulas derived apply under very restrictive conditions. Thus, they are only approximately correct in many cases and they frequently violate conditions which are brought to light by the more refined investigations of the theory of elasticity. On the other hand, the theory of elasticity deals with general equations which must be satisfied by an elastic body in equilibrium under any external force system. Simplifying assumptions are also used in elasticity, but usually with a better knowledge about the approximations involved.

It will be shown that the response of a solid body to external forces is influenced by the geometric configuration of the body as well as the mechanical properties of the material. We shall restrict our discussion to elastic materials, i.e., materials in which the deformation and stress disappear with the removal of external forces.

The condition of equilibrium in a continuous medium is expressed in terms of stress components; in Chapter 1, accordingly, we shall define the stress components and study their properties at a point and the equations governing their variation in space. We must satisfy, in addition to equilibrium conditions ($\Sigma F = 0$, $\Sigma M = 0$), certain conditions such that the deformed medium is continuous. In Chapter 2 we shall define the strain components, which are measures of unit relative displacement, and study their properties. The constitutive equations for elastic materials, which relate the stress components to the strain components (generalized Hooke's law), will be discussed in Chapter 3. In Chapter 4, all the governing equations introduced in the first three chapters will be grouped together and the formulation of elasticity problems will be presented.

Chapter 5 presents the plane-stress and plane-strain problems, and Chapter 6, torsion problems. The main purpose of those two chapters is to demonstrate how the general governing equations in elasticity can be simplified in certain practical problems and some of the methods of solving these equations. Chapter 7 contains the discussion of strain energy and energy methods.

The concept of tensor quantities and the use of Cartesian tensor notation, which will be used in later chapters, are presented in Chapter 8, the order of presentation here being strongly influenced by Chapter 1 of the book, *Cartesian Tensors,* by Harold Jeffreys, Cambridge University Press, 1961. Chapter 8 is included here mainly because many of the undergraduate engineering curricula do not treat this topic. Chapters 9 and 10 cover the three-dimensional formulation of elasticity problems using Cartesian tensor notation. In Chapter 11, the use of vector and dyadic notations in elasticity is introduced. The derivations in this chapter normally follow from consideration of the Cartesian components of vectors and dyadics, while orthogonal curvilinear coordinates are presented in Chapter 12. In this chapter, the form of the various vector and dyadic operators will be expressed in an orthogonal curvilinear coordinate system, thus making it possible to write the governing equations in any such coordinate system.

The last chapter gives a unified treatment of the governing equations of elasticity in terms of displacement and stress functions. Here it is demonstrated that the solutions of specific problems discussed earlier in the text can be derived from certain general functions.

1

Analysis of Stress

1.1. Introduction

In this chapter we shall define stress and discuss two important aspects of stress, namely, the state of stress at a point, which leads to the transformation of stress equations, and the equations governing the variation of the stress components in space. Emphasis will be placed on the physical nature of stress in contrast to the material of Chapter 9, where we are more concerned with the mathematical structure of tensor quantities. The two-dimensional case is used for the detailed derivation of equations, and most of the three-dimensional equations are presented without derivation. A graphical representation of the transformation of stress components is given in addition to the analytical equations. The equations of equilibrium are derived from the engineering point of view by taking a small element as a free body.

Most of the material in this chapter, which deals essentially with two-dimensional problems, is usually covered in elementary courses, such as strength of materials. It is included here and discussed in detail in order to clarify the basic nature of the stress tensor. The notations and sign conventions adopted, where possible, are consistent with those used in the general approach in Chapter 9, and, in general, they follow the most accepted ones in engineering literature.

The theory of stress presented in this chapter is applicable to any continuum, e.g., elastic or plastic solids, viscous fluids, regardless of the mechanical properties of the material.

1.2. Body Forces, Surface Forces, and Stresses

When a body is subjected to an applied load system, internal forces are induced in the body. The behavior of the body, i.e., the changes in its dimensions (its deformation) or in some cases, its eventual failure, is mainly a function of the internal force distribution, which in turn depends upon the external force system. The response of a body to an external force system is conveniently studied by grouping forces into two categories, body forces and surface forces. Body forces are associated with the mass of the body and are distributed throughout the volume of a body; they do not result from direct contact with other bodies. Gravitational, magnetic, and inertia forces are all body forces. They are specified in terms of force per unit volume, i.e., body force intensity. The x, y, and z components of body force intensity are given the symbols F_x, F_y, and F_z. Many authors use the term "body force," with units of force per unit volume, to mean "body force intensity." The exact meaning of the term is always clear from the context in which the term is used. Surface forces result from physical contact between two bodies, or more subtly, they may represent the force which an imaginary surface within a body exerts on the adjacent surface.

If an imaginary cutting plane is assumed to pass through a body as shown in Fig. 1.1 and part I is analyzed as a free body, we observe that

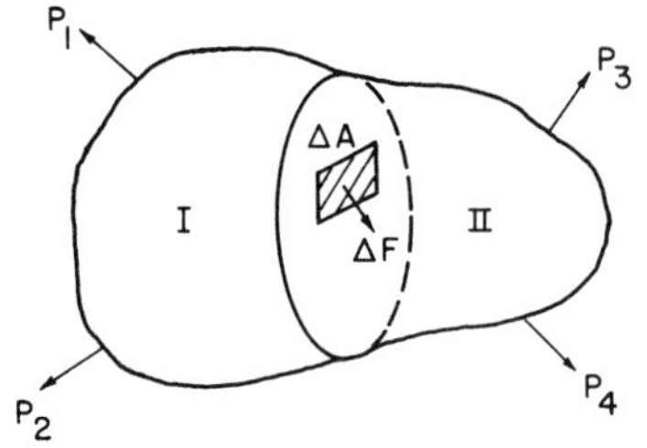

Fig. 1.1 External Surface Forces and Internal Forces

surface forces P_1 and P_2 are held in equilibrium (assuming the body is in equilibrium) by the force exerted on part I by part II. This force, however, is distributed over the entire plane; that is, any elementary area ΔA is subjected to a force ΔF; consequently, the average force per unit area is

$$p_{\text{ave}} = \frac{\Delta F}{\Delta A}$$

The stress at a point in ΔA is defined as the limiting value of average force per unit area as the area ΔA approaches zero, or

$$p = \lim_{\Delta A \to 0} \frac{\Delta F}{\Delta A} = \frac{dF}{dA}$$

Notice that the force dF (or stress p) is not necessarily in a direction normal to or tangential to the surface on which it acts. If we consider the equilibrium of a free body, where stresses in general act on all of the external surfaces, we must determine the *force* resulting from the stress acting on each surface (as well as the resultant force caused by the distributed body forces) in order to express the conditions of equilibrium, since statics demands that the resultant *force* must vanish. It should be noted that the stress on an element of area dA is a vector acting in the direction of the force vector dF. In other words, the stress on a *given plane* is a vector, the stress vector. If several stresses act on the *same plane*, their resultant is simply found by vector addition. A stress vector may also be decomposed into components which we shall call the stress components.

In order to study the nature of stress in more detail we consider the stressed element of Fig. 1.2. Now it is not proper to ask for the stress

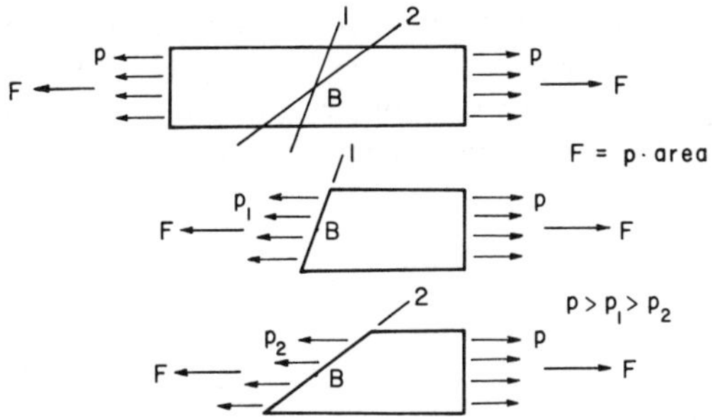

Fig. 1.2 Force (a Vector) and Stress (Not a Vector)

at a point, such as B, since we must define the direction of the plane through B on which we wish to determine the stress. Suppose, however, that we wish to determine the stress on any arbitrary plane through point B. Since there is a different (stress) vector associated with each plane (or each direction in space), stress does not behave as a vector. That is, a vector quantity associates only a scalar (its component) with each direction in space, whereas stress does not. Thus, unless we are referring to a specific plane, stress is not a vector quantity, but, as will be shown in Chapter 9, a second-order tensor. It is proper, however, to ask for the *state of stress* at B, where in this case we imply that our description makes it possible to determine the stress on every plane passing through this point. In Section 1.3 it will be shown that the specification of the state of stress is, in general, complete if we know the stress on *two* given planes passing through the point (in the two-dimensional case). In other words, *two* stress vectors (on two different planes) must be given in order that the stress vector on any plane through a given point be known. Furthermore, we observe that the forces acting on the two planes shown in Fig. 1.2 are the same, but that the stresses on the two planes are different since their areas are not equal. Thus, in order to completely define a stress vector, we must specify its magnitude and direction, as well as the plane on which it acts. The complete designation of a stress vector may be accomplished by using two indices for its components, e.g., τ_{xy}, $\tau_{r\theta}$, as well as a sign to indicate the direction of a particular component. However, when the plane under consideration is understood, we may eliminate one of the indices. The term stress is customarily used quite loosely to represent stress vector, stress component, or stress tensor. The precise meaning of the term is usually clear from the context in which it is used.

There are two convenient ways which are used to designate the components of a stress vector; the x, y, and z components, and the normal and tangential (shearing) components. The x, y, and z components of an *external* stress, i.e., the force per unit area acting on the boundary of a body, will be designated as T_x^μ, T_y^μ, T_z^μ (Fig. 1.3). The subscript gives the *direction* of the component, and the superscript μ defines the plane, i.e., the outward normal of the plane is in the μ direction. For

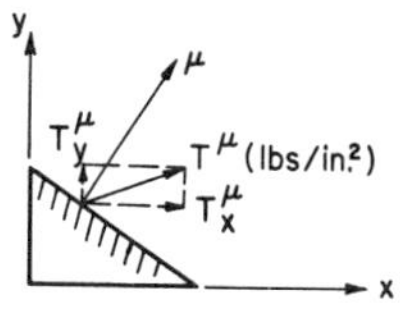

Fig. 1.3 External Stress Vector and Its Components

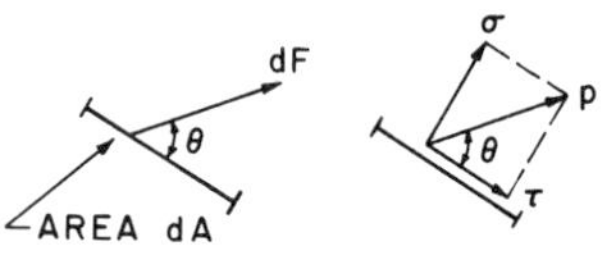

Fig. 1.4 Normal and Shearing Stress Components

internal stresses on imaginary cutting planes, the x, y, and z components are designated as p_x, p_y, p_z. Since these are most frequently utilized as an intermediate step in derivations, and the plane is understood, we use only a subscript.

Normal and shearing components of both external and internal stresses are more significant in some problems. We let σ represent the normal stress, i.e., the component of stress *perpendicular* to the plane on which it acts. The shear stress is denoted by the symbol τ. The latter is the component of stress which lies *in* the plane. Thus, referring to Fig. 1.4, where $p = dF/dA$, we have

$$p^2 = \sigma^2 + \tau^2$$

Let us next consider a three-dimensional state of stress acting on an element as shown in Fig. 1.5. Body forces are omitted in the figure, since we are merely interested in establishing a nomenclature for stress at this time. In order to develop the terminology necessary to specify a stress component, we must define the plane on which the component

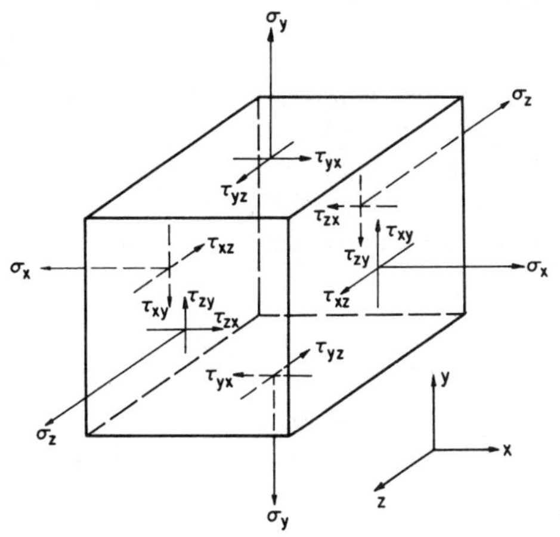

Fig. 1.5 Three-Dimensional State of Stress

acts. Unless otherwise specified, a plane is designated by the direction of its outward normal. In order to determine the direction of the outward normal, we must first establish a free body. The outward normal of any surface, then, points *away* from the interior of the free body. For example, the right-hand face of the element in Fig. 1.5 is called the x plane, since its outward normal points in the positive x direction. The left-hand face is termed the negative x plane.

The stress acting on the x plane may be resolved into normal and shearing components. However, the shear stress may act in any direction in its plane so that it is further resolved into components acting in the y and z directions, τ_{xy} and τ_{xz}. The first subscript in the notation refers to the plane on which the shear stress acts; the second gives the direction of the shear stress. The normal stress on the x surface is denoted simply by σ_x. Only one subscript is necessary for this component, since the normal stress must necessarily act in the direction of the normal to the plane.

We adopt the following sign convention for the various stress components. On a positive surface (the outward normal to a positive surface points in the positive coordinate direction), all stresses which act in positive coordinate directions are considered as positive. On a negative surface, all stresses acting in negative coordinate directions are positive. This convention, when applied to normal stress, follows the customary rule that tension is positive, compression negative. We observe that all stress components acting on the element in Fig. 1.5 are positive.

1.3. Uniform State of Stress (Two-Dimensional)

If the state of stress in a body is the same at all points, this body is said to be in a uniform state of stress. In this case, the stress vectors and stress components are independent of the space coordinates, although they still depend on the inclination of the plane on which they act. In a uniform state of stress the body force must necessarily be zero; otherwise the stresses must vary in space in order to maintain equilibrium.

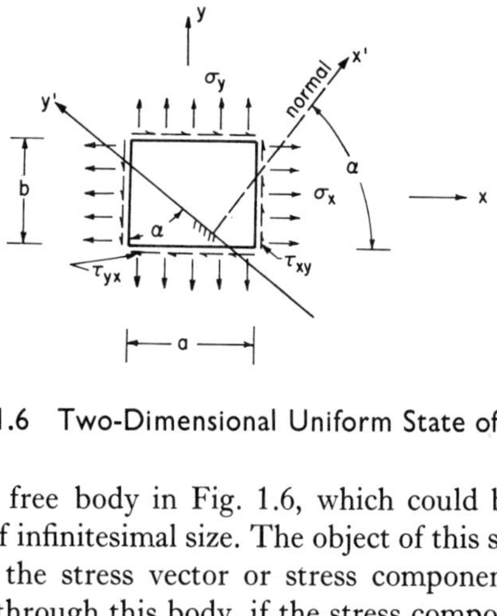

Fig. 1.6 Two-Dimensional Uniform State of Stress

Consider the free body in Fig. 1.6, which could be either of finite dimensions or of infinitesimal size. The object of this section is to derive expressions for the stress vector or stress components acting on any arbitrary plane through this body, if the stress components on the x, y, and z planes are given. Furthermore, we wish to determine the orientation of certain significant planes, which contain, for example, the maximum normal stress or the maximum shear stress.

We shall first consider a two-dimensional case. That is, we are interested in the x and y components of the stresses acting on planes parallel to the z axis (normal to the z plane). Furthermore, we assume that[1] $\tau_{xz} = \tau_{zx} = \tau_{yz} = \tau_{zy} = 0$. In this case, it is only necessary to write the

[1] Two specific classes of problems satisfying these relations will be discussed in Chapters 4 and 5. These are called *plane stress* and *plane strain* problems.

equilibrium expressions in the x and y directions; the stress σ_z need not be considered here, although it may be different from zero.

Right-handed coordinate systems will be employed throughout this text. For two-dimensional systems, the y axis will always be taken 90° counterclockwise from x, and similarly, y' is 90° counterclockwise from x', as shown in Fig. 1.6. The angle from x to x' is called α, and it is measured in the counterclockwise direction.

Considering the free body in Fig. 1.6 to be of unit length in the z direction and taking moments about the lower left corner, we see that

$$(\sigma_x b)(b/2) - (\sigma_x b)(b/2) + (\sigma_y a)(a/2) - (\sigma_y a)(a/2) - (\tau_{xy} b)a + (\tau_{yx} a)b = 0$$

and since ab is not zero, the result is

$$\tau_{xy} = \tau_{yx} \tag{1.1}$$

Thus the stress component on the x plane acting in the y direction is equal to the stress component on the y plane in the x direction. This result is quite general for any two perpendicular planes, i.e., $\tau_{x'y'} = \tau_{y'x'}$.

Next we shall consider the summation of forces in the x direction on the free body of Fig. 1.7(a). Letting the x and y components of stress

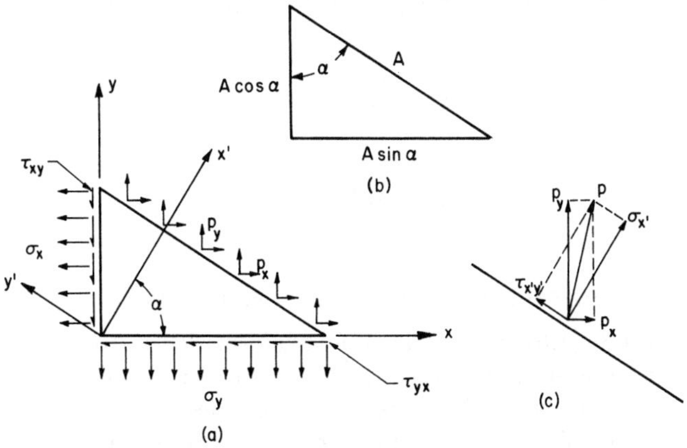

Fig. 1.7 Free Body Showing the Stress Components on the x' Plane

on the x' plane be p_x and p_y, the inclined surface area be A, and using Eq. (1.1), we find by statics

$$\sum F_x = 0$$

$$p_x A - \sigma_x A \cos\alpha - \tau_{yx} A \sin\alpha = 0$$

$$p_x = \sigma_x \cos\alpha + \tau_{xy} \sin\alpha \tag{1.2}$$

Similarly, from the equilibrium condition in the y direction we have

$$p_y = \sigma_y \sin \alpha + \tau_{xy} \cos \alpha \tag{1.3}$$

The normal stress on the x' plane is found by projecting p_x and p_y in the x' direction. Thus

$$\sigma_{x'} = p_x \cos \alpha + p_y \sin \alpha \tag{1.4}$$

Using Eqs. (1.2) and (1.3), this takes the form

$$\sigma_{x'} = \sigma_x \cos^2 \alpha + \sigma_y \sin^2 \alpha + 2\tau_{xy} \sin \alpha \cos \alpha \tag{1.5}$$

Similarly,

$$\tau_{x'y'} = p_y \cos \alpha - p_x \sin \alpha$$

or

$$\tau_{x'y'} = (\sigma_y - \sigma_x) \sin \alpha \cos \alpha + \tau_{xy}(\cos^2 \alpha - \sin^2 \alpha) \tag{1.6}$$

The stress $\sigma_{y'}$ may now be found by exposing the y' surface on a free body, or more simply, by substituting $(\alpha + \pi/2)$ for α in Eq. (1.5). This yields the relation

$$\sigma_{y'} = \sigma_x \cos^2(\alpha + \pi/2) + \sigma_y \sin^2(\alpha + \pi/2) + 2\tau_{xy} \sin(\alpha + \pi/2) \cos(\alpha + \pi/2)$$

and since

$$\sin(\alpha + \pi/2) = \cos \alpha$$

and

$$\cos(\alpha + \pi/2) = - \sin \alpha$$

we obtain

$$\sigma_{y'} = \sigma_x \sin^2\alpha + \sigma_y \cos^2\alpha - 2\tau_{xy} \sin \alpha \cos \alpha \tag{1.7}$$

Equations (1.5), (1.6), and (1.7) are the transformation of stress equations, which can be used to determine the stresses $\sigma_{x'}$, $\sigma_{y'}$, and $\tau_{x'y'}$, if σ_x, σ_y, τ_{xy}, and α are known. There are seven variables, σ_x, σ_y, τ_{xy}, $\sigma_{x'}$, $\sigma_{y'}$, $\tau_{x'y'}$, and α, involved in these three equations. In general, if any four of these seven variables are given, the other three may be determined from the three transformation of stress equations.[2] Thus, in the two-dimensional case, the state of stress is completely defined if the three stress components on two orthogonal planes are given, since the stress components on any other plane can be found once α is specified. In the three transformation of stress equations, any four of the seven variables can be prescribed arbitrarily without violating the equilibrium equations.

[2] See Problem 1-7 for the exceptional case where specification of four of these variables does not fix the others.

If more than four are prescribed arbitrarily, no solution of Eqs. (1.5), (1.6), and (1.7) can, in general, be obtained and the corresponding state of stress is impossible.

We also note from Eqs. (1.5) and (1.7) that

$$\sigma_{x'} + \sigma_{y'} = \sigma_x + \sigma_y = \text{constant}$$

Thus, the sum of the normal stresses on two perpendicular planes is invariant, i.e., independent of α.

1.4. Principal Stresses

Before determining the extreme values of normal and shear stress, it is convenient to write Eqs. (1.5), (1.6), and (1.7) in terms of 2α. Using the trigonometric identities

$$\sin 2\alpha = 2 \sin \alpha \cos \alpha; \quad \sin^2 \alpha = \tfrac{1}{2}(1 - \cos 2\alpha); \quad \cos^2 \alpha = \tfrac{1}{2}(1 + \cos 2\alpha) \tag{1.8}$$

we find that

$$\sigma_{x'} = \frac{\sigma_x + \sigma_y}{2} + \frac{\sigma_x - \sigma_y}{2} \cos 2\alpha + \tau_{xy} \sin 2\alpha \tag{1.9a}$$

$$\sigma_{y'} = \frac{\sigma_x + \sigma_y}{2} - \frac{\sigma_x - \sigma_y}{2} \cos 2\alpha - \tau_{xy} \sin 2\alpha \tag{1.9b}$$

and

$$\tau_{x'y'} = \frac{\sigma_y - \sigma_x}{2} \sin 2\alpha + \tau_{xy} \cos 2\alpha \tag{1.10}$$

To determine the orientation of the planes of maximum and minimum normal stress, Eq. (1.9a)[3] is differentiated with respect to α and the derivative equated to zero, so that

$$d\sigma_{x'}/d\alpha = -(\sigma_x - \sigma_y)\sin 2\alpha + 2\tau_{xy} \cos 2\alpha = 0 \tag{1.11}$$

or

$$\tan 2\alpha = \frac{2\tau_{xy}}{\sigma_x - \sigma_y} \tag{1.12}$$

This equation has two roots. These two roots of 2α are 180° apart; therefore the two values of α differ by 90° and the two planes of stationary (maximum, minimum, or possibly a point of inflection) normal stress are perpendicular. By comparing Eqs. (1.10) and (1.11) we observe that the shear stress on these planes is zero. The planes on which the shear stress vanishes are called the principal planes; the normal stresses on these planes are called the principal stresses.

[3] The same result can be obtained by using Eq. (1.9b).

The sine and cosine of the angles 2α defined by eq. (1.12) are

$$\sin 2\alpha = \pm \frac{2\tau_{xy}}{\sqrt{4\tau_{xy}^2 + (\sigma_x - \sigma_y)^2}}$$

$$\cos 2\alpha = \pm \frac{\sigma_x - \sigma_y}{\sqrt{4\tau_{xy}^2 + (\sigma_x - \sigma_y)^2}}$$

By substituting these expressions into Eq. (1.9a), and noting that the sine and cosine are either both plus or both minus, we find

$$\begin{aligned} \sigma_{\max} = \sigma_1 &= \frac{\sigma_x + \sigma_y}{2} + \sqrt{\left(\frac{\sigma_x - \sigma_y}{2}\right)^2 + \tau_{xy}^2} \\ \sigma_{\min} = \sigma_2 &= \frac{\sigma_x + \sigma_y}{2} - \sqrt{\left(\frac{\sigma_x - \sigma_y}{2}\right)^2 + \tau_{xy}^2} \end{aligned} \tag{1.13}$$

The fact that σ_1 and σ_2 are the maximum and minimum values of σ may be seen by comparing Eqs. (1.13) and (1.9a) or by considering the second derivative of Eq. (1.9a) evaluated at the two roots of Eq. (1.12). Although the values of the angle α corresponding to the planes of σ_1 and σ_2 differ by 90°, the values of tan 2α from Eq. (1.12) are the same for both. In order to determine the particular value of 2α corresponding to either σ_1 or σ_2, it is necessary to consider the signs of the numerator and denominator of $2\tau_{xy}/(\sigma_x - \sigma_y)$.

The value of 2α corresponding to the direction of σ_1 (Fig. 1.8) is between 0 and $\pi/2$ if both $2\tau_{xy}$ and $(\sigma_x - \sigma_y)$ are positive, i.e.,

$$0 < 2\alpha < \pi/2, \qquad \text{if} \quad \tau_{xy} > 0 \quad \text{and} \quad (\sigma_x - \sigma_y) > 0$$

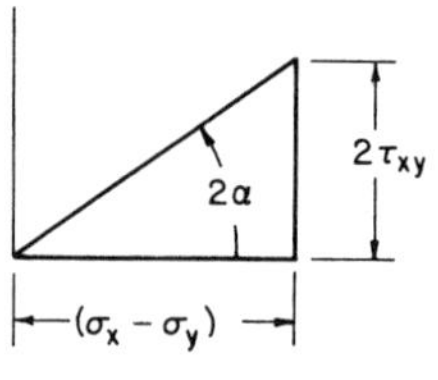

Fig. 1.8 Direction of the Maximum Principal Stress

The other cases are

$$\begin{aligned} \pi/2 < 2\alpha < \pi, &\qquad \text{if} \quad \tau_{xy} > 0 \quad \text{and} \quad (\sigma_x - \sigma_y) < 0 \\ \pi < 2\alpha < (3/2)\pi, &\qquad \text{if} \quad \tau_{xy} < 0 \quad \text{and} \quad (\sigma_x - \sigma_y) < 0 \\ (3/2)\pi < 2\alpha < 2\pi, &\qquad \text{if} \quad \tau_{xy} < 0 \quad \text{and} \quad (\sigma_x - \sigma_y) > 0 \end{aligned}$$

where, in each instance, α is the angle between the x axis and the direction of σ_1 ($\sigma_{\max}$), and 2α, of course, is double this angle.

Next we seek to determine the planes of maximum shear stress. Here we differentiate the expression for $\tau_{x'y'}$ and set this equal to zero. Thus

$$d\tau_{x'y'}/d\alpha = (\sigma_y - \sigma_x)\cos 2\alpha - 2\tau_{xy}\sin 2\alpha = 0$$

or

$$\tan 2\alpha = -(\sigma_x - \sigma_y)/(2\tau_{xy}) \qquad (1.14)$$

The two roots of 2α from Eq. (1.14) also define a set of perpendicular planes, so that the shear stresses on these two planes are equal. We also observe that the values of tan 2α given by Eqs. (1.12) and (1.14) are negative reciprocals of each other, so that the roots for 2α from these two equations are 90° apart and the corresponding planes are 45° apart.

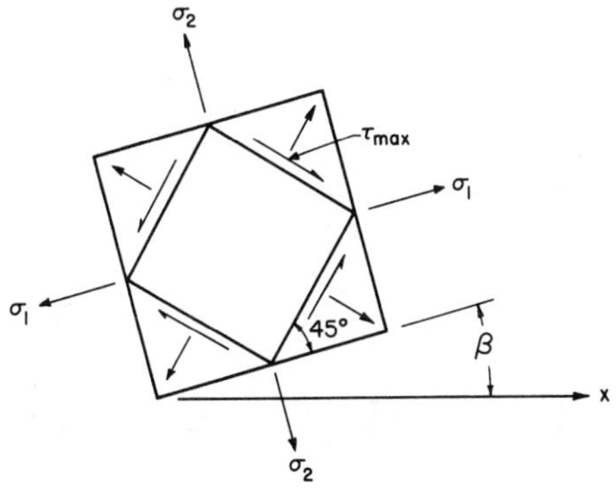

Fig. 1.9 Planes of Maximum Stress

That is, the planes of maximum shear stress are 45° from the principal planes. These results are illustrated in Fig. 1.9. Now from Eqs. (1.14) and (1.10) we find

$$\tau_{x'y'\max} = \pm\sqrt{\left(\frac{\sigma_x - \sigma_y}{2}\right)^2 + \tau_{xy}^2} \qquad (1.15)$$

If x' is taken as the plane for which $\alpha < \pi/2$, where α is the counterclockwise angle between x and x', the negative sign in Eq. (1.15) is chosen if $\sigma_x > \sigma_y$ and the positive sign if $\sigma_y > \sigma_x$. If $\sigma_y = \sigma_x$, then $\alpha = 0$ and $\tau_{x'y'} = \tau_{xy}$.

1.5. Mohr's Circle of Stress

The basic equations of two-dimensional transformation of stress (Eqs. (1.5), (1.6), (1.7), or (1.9), (1.10)), as well as the relations governing the principal stresses and maximum shear stress (Eqs. (1.12), (1.13),

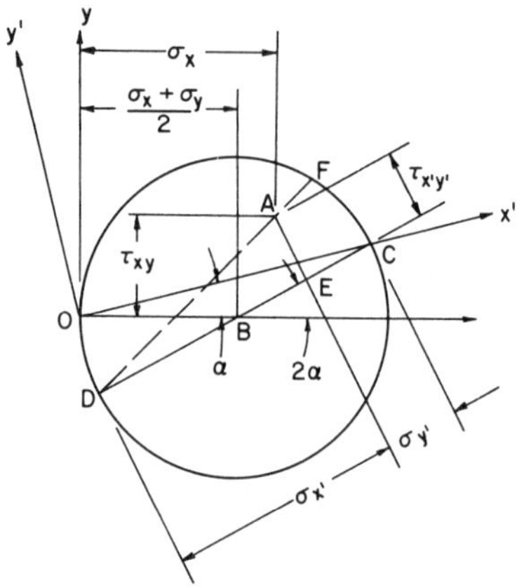

Fig. 1.10 The Dyadic Circle

(1.14), and (1.15)), may be conveniently represented graphically. The best known graphical representation is the Mohr circle.[4]

Rewriting Eqs. (1.9) and (1.10), we get

$$\sigma_{x'} - \frac{\sigma_x + \sigma_y}{2} = \frac{\sigma_x - \sigma_y}{2} \cos 2\alpha + \tau_{xy} \sin 2\alpha$$

$$\sigma_{y'} - \frac{\sigma_x + \sigma_y}{2} = -\frac{\sigma_x - \sigma_y}{2} \cos 2\alpha - \tau_{xy} \sin 2\alpha$$

$$\tau_{x'y'} = \frac{\sigma_y - \sigma_x}{2} \sin 2\alpha + \tau_{xy} \cos 2\alpha$$

Squaring the first and third of these expressions[5] and adding, we obtain

$$\left(\sigma - \frac{\sigma_x + \sigma_y}{2}\right)^2 + \tau^2 = \left(\frac{\sigma_x - \sigma_y}{2}\right)^2 + \tau_{xy}^2 \tag{1.16}$$

[4] Another representation, also originated by Mohr, is the *dyadic circle*. The dyadic circle is constructed as follows: Select an origin O and lay out the x and y axes. Also lay out the x' and y' directions by measuring the angle α as shown in Fig. 1.10. Locate point A with abscissa σ_x and ordinate τ_{xy}, and also point $B[(\sigma_x + \sigma_y)/2, 0]$. Draw a circle centered at B with radius OB. From point C, the point of intersection of the x' axis and the circle, construct the diameter CBD and line AE normal to this diameter. The stresses $\sigma_{x'}$, $\sigma_{y'}$, and $\tau_{x'y'}$ are then given by the lengths DE, EC, and EA, respectively. The stress vector on the x' plane has the magnitude given by length DA and it acts in the direction OF. Furthermore, the edge view of the x' plane lies in the direction of OD. See A. Durelli, Phillips, and Tsao, *Introduction to the Theoretical and Experimental Analysis of Stress and Strain*, McGraw-Hill Book Company, Inc., New York, 1958, pp. 160–162, or H. M. Westergaard, *Theory of Elasticity and Plasticity*, John Wiley & Sons, Inc., New York, 1952, p. 51.

[5] Using the second and third of these expressions, with $\sigma_{y'} = \sigma$, we get the same result.

where we have put $\sigma_{x'} = \sigma$ and $\tau_{x'y'} = \tau$. This is the equation of a circle in the (σ, τ) coordinates with its center on the σ axis, translated $(\sigma_x + \sigma_y)/2$ units to the right of the origin, and with a radius of

$$\sqrt{\left(\frac{\sigma_x - \sigma_y}{2}\right)^2 + \tau_{xy}^2}$$

as shown in Fig. 1.11. This circle is known as Mohr's circle of stress. It represents the relationship between σ and τ on any plane, i.e., it states the relation between the normal stress and the shear stress on any given plane through an element with the given values of σ_x, σ_y, and

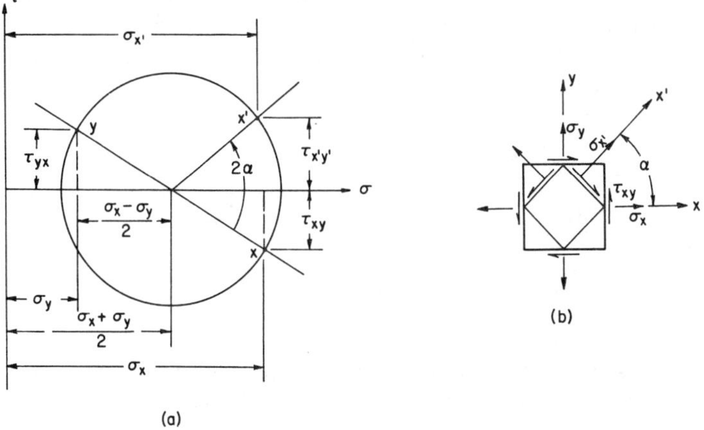

Fig. 1.11 Mohr's Circle of Stress

τ_{xy} acting on it. Also, as we will see, the angle 2α can be determined conveniently. The Mohr circle for any state of stress (two-dimensional) is easily plotted by locating the two points corresponding to the stress conditions on the x and y planes. To do this we must, however, adopt a different convention to represent the shear stress, to be used *only in drawing* and *interpreting Mohr's circle*. This convention is that shear stress causing clockwise moment about any point in the element is plotted above the horizontal axis of the Mohr circle. For example, the shear stress τ_{xy} in Fig. 1.11 causes a counterclockwise moment, thus the point on the Mohr circle which gives the stress on the x plane is plotted below the horizontal axis. Similarly, the point on the Mohr circle representing the stress on the y plane is plotted above the horizontal axis.

For a given state of stress, the *direction* of the shear stress on any plane is fixed, but the *sign* of the shear stress as defined in Section 1.2 (Fig. 1.5) depends on the coordinate system. For instance, the shear stress $\tau_{x'y'}$ as shown in Fig. 1.7 is positive, but the same shear stress component will become negative if the y' axis is relabeled as x'. A point on the Mohr circle gives the magnitude and direction of the normal and shear stresses on a given plane in the element, regardless of the coordinate system used.

In any analytical representation the coordinate axes are always specified; therefore, the use of the standard sign convention given in Section 1.2 does not introduce any ambiguity.

In order to illustrate how the angle 2α is found from the Mohr circle, thus obtaining the maximum benefit from this procedure, we assume that the element is oriented along the principal directions, i.e., we assume the principal stresses σ_1 and σ_2 are known. We then ask how to locate the point on the Mohr's circle which represents the stresses on a plane inclined at an angle α with respect to the plane of σ_1. Substituting $\sigma_x = \sigma_1$, $\sigma_y = \sigma_2$, and $\tau_{xy} = 0$ into the transformation of stress equations (1.9) and (1.10), we have

$$\sigma_{x'} = \frac{\sigma_1 + \sigma_2}{2} + \frac{\sigma_1 - \sigma_2}{2} \cos 2\alpha$$

$$\sigma_{y'} = \frac{\sigma_1 + \sigma_2}{2} - \frac{\sigma_1 - \sigma_2}{2} \cos 2\alpha$$

$$\tau_{x'y'} = \frac{\sigma_2 - \sigma_1}{2} \sin 2\alpha$$

On Mohr's circle, these are precisely the stresses found by rotating diameter AB through a counterclockwise angle 2α as shown in Fig. 1.12.

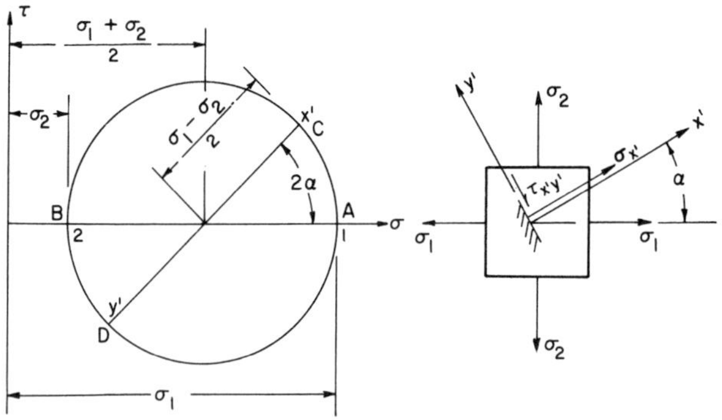

Fig. 1.12 Stress Components on x' and y' Determined from Mohr's Circle

Point C gives the stress components on the x' plane and D gives the stress condition on the y' plane after correct interpretation of the direction of the shear stress.

In general, then, every point on the Mohr circle corresponds to a plane in the body under stress which has the stresses given by the point. Thus the point A in Fig. 1.12 is denoted by 1, B by 2, C by x', D by y'. Then the angle between the radii drawn to two points on the Mohr circle, for example 1 and x', is twice the angle in space between the corresponding planes. Furthermore, the angle in space is measured in the same direction as the double angle on the Mohr circle.

The general procedure to follow in constructing the Mohr circle corresponding to a given stress condition other than the principal stresses is shown in Fig. 1.11. We plot the two points corresponding to the stresses on the x and y planes, keeping the convention for the representation of shear stress in mind. The two points are connected by a straight line, which is the diameter of the circle, and the intersection of this line with the σ axis is the center of the circle. The two points just plotted are labeled x and y, respectively. Then the stress components on any plane are found by measuring twice the angle between x or y and the normal to the plane, in the same direction, on the Mohr circle and evaluating the ordinate and abscissa at this point. The reader should now verify Eqs. (1.9), (1.10), (1.12), (1.13), (1.14), and (1.15), as well as the double angle relationship for all known planes, by using the Mohr circle construction.

1.6. State of Stress at a Point

In Sections 1.3 through 1.5, the transformation of stress equations were examined for a body under a uniform state of stress. We shall now show that these same relations, i.e., Eqs. (1.1) through (1.16), also apply at each *point* in a body under a nonuniform stress distribution, including the effects of body forces.

Consider the state of stress at point O in Fig. 1.13. We denote the stresses at O by σ_x, τ_{xy}, σ_y, the stress components on a plane parallel to AB through O by p_x, p_y, and the body force components at O by F_x, F_y (which may include inertia forces). The dimensions of the element, Δx, Δy, Δs, are small quantities. The normal stress on plane OB at point B, due to the nonuniform nature of stress distribution, is[6] $\sigma_x + (\partial\sigma_x/\partial y)(\Delta y)$; thus the average stress on OB is $\sigma_x + \frac{1}{2}(\partial\sigma_x/\partial y)(\Delta y)$. The average values for σ_y and τ_{xy} are represented in similar manner. Since F_x and F_y are the body forces at point O, the average body forces in the element are different and will be written as $F_x + \Delta F_x$ and $F_y + \Delta F_y$,

[6] If we denote the stress component σ_x at O by $\sigma_x(x, y)$, then the corresponding stress component at B is $\sigma_x(x, y + \Delta y)$. Thus, assuming the stress varies continuously with x and y, and if Δy is small, the difference between σ_x at O and B is given by $(\partial\sigma_x/\partial y)\Delta y$.

respectively. The term ΔF_x represents the deviation of the average from that at point O. In the limit, as Δx and Δy approach zero in such a way that $\Delta y/\Delta x = \text{constant}$, plane AB passes through point O. Assum-

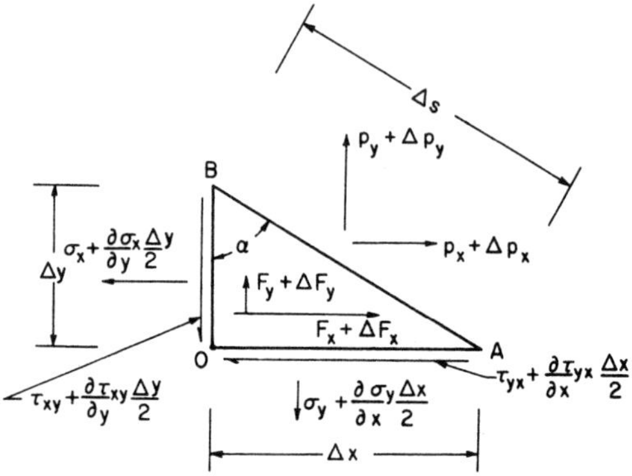

Fig. 1.13 Free Body of an Element Under a Nonuniform State of Stress

ing the depth of the element is unity, and summing forces in the x direction, we have

$$(p_x + \Delta p_x)\Delta s = \left(\sigma_x + \frac{\partial \sigma_x}{\partial y}\frac{\Delta y}{2}\right)\Delta y + \left(\tau_{yx} + \frac{\partial \tau_{yx}}{\partial x}\frac{\Delta x}{2}\right)\Delta x - (F_x + \Delta F_x)\frac{\Delta x \Delta y}{2}$$

Dividing by Δs gives

$$p_x + \Delta p_x = \left(\sigma_x + \frac{\partial \sigma_x}{\partial y}\frac{\Delta y}{2}\right)\cos\alpha + \left(\tau_{yx} + \frac{\partial \tau_{yx}}{\partial x}\frac{\Delta x}{2}\right)\sin\alpha - (F_x + \Delta F_x)\frac{\Delta x \cos\alpha}{2} \tag{1.17}$$

and as Δx and Δy approach zero, Δp_x and ΔF_x become vanishingly small; therefore

$$p_x = \sigma_x \cos\alpha + \tau_{yx} \sin\alpha$$

Similarly, by summing forces in the y direction,

$$p_y = \sigma_y \sin\alpha + \tau_{xy} \cos\alpha$$

The expressions for p_x and p_y are therefore identical to Eqs. (1.2) and (1.3). It will be shown in the next section that the relation

$$\tau_{xy} = \tau_{yx}$$

is also valid for a variable stress distribution; it follows that Eqs. (1.1) through (1.16) apply in the analysis of the state of stress at a point. It should be noted that these relations cannot be used to determine the stress distribution throughout a body. Once the stress distribution in a body is known, however, the transformation of stress equations can be used to define the stress components on any plane at any given point in the body.

1.7. Differential Equations of Equilibrium

Thus far we have considered stress relations only under a *uniform* stress condition (or at a point). In general, however, the stress components vary from point to point in a stressed body, these variations being governed by the equilibrium conditions of statics. The resulting expressions relate the space derivatives of the various components of stress and are known as the differential equations of equilibrium.

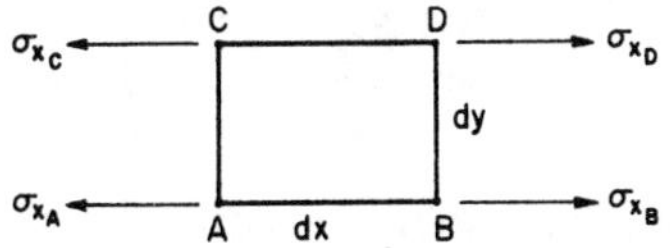

Fig. 1.14 Stress Variation on an Element

Consider, for example, the variation of one of the stress components, say σ_x, from point to point in the body. If the stress at A in Fig. 1.14 is σ_x, the stress at B is increased by an amount $(\partial\sigma_x/\partial x)\,dx$, where $\partial\sigma_x/\partial x$ is the rate of change of σ_x with respect to x and dx is the distance in the x direction. The partial derivative sign is required because σ_x is also a function of y (and z in the three-dimensional case). We are assuming that the stress components and their first derivatives are continuous. The stress at B then is given by

$$\sigma_{x_B} = \sigma_x + \frac{\partial\sigma_x}{\partial x}\,dx \tag{1.18}$$

Similarly, the stress at C and that at D are

$$\sigma_{x_C} = \sigma_x + \frac{\partial\sigma_x}{\partial y}\,dy \qquad \sigma_{x_D} = \sigma_{x_B} + \frac{\partial\sigma_{x_B}}{\partial y}\,dy$$

since x is constant from B to D. Using Eq. (1.18), this becomes

$$\sigma_{x_D} = \sigma_x + \frac{\partial \sigma_x}{\partial x} dx + \frac{\partial}{\partial y}\left(\sigma_x + \frac{\partial \sigma_x}{\partial x} dx\right) dy$$

or

$$\sigma_{x_D} = \sigma_x + \frac{\partial \sigma_x}{\partial x} dx + \frac{\partial \sigma_x}{\partial y} dy \tag{1.19}$$

where the second-order term (product of dx and dy) is neglected since it is small as compared with the first-order terms (terms containing only one differential).

Neglecting small quantities of higher order[7] then, the stresses on a surface of the infinitesimal element vary linearly. Referring to

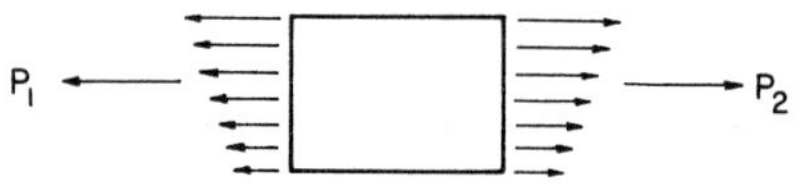

Fig. 1.15 Resultant Force on an Element

Fig. 1.15 then, the force on the left face of the element is

$$P_1 = \left(\frac{\sigma_x + \sigma_x + \dfrac{\partial \sigma_x}{\partial y} dy}{2}\right) dy$$

assuming the depth of the prism in the z direction is unity. Simplifying, we get

$$P_1 = \sigma_x dy + \frac{1}{2}\frac{\partial \sigma_x}{\partial y} dy^2$$

Similarly, the force on the right-hand face is

$$P_2 = \left(\frac{\sigma_x + \dfrac{\partial \sigma_x}{\partial x} dx + \sigma_x + \dfrac{\partial \sigma_x}{\partial x} dx + \dfrac{\partial \sigma_x}{\partial y} dy}{2}\right) dy$$

[7] Small quantities of higher order (higher-order terms) contain the product of more small quantities than another term. For example, if f and g are small quantities, the product fg is a higher-order term compared to f or g and is negligible compared to either f or g. This is certainly valid if f and g are of infinitesimal magnitude. It should be noted, however, that the omission of higher-order terms depends upon the final expression which is written. The omission of higher-order terms must be postponed until the final equation is obtained so that the relative magnitudes of the final terms can be evaluated, since lower-order terms might be cancelled out by a subtraction. For example, in Eq. (1.17) the terms Δp_x, $\Delta x, \ldots$ are omitted because it is the final equation. In Eq. (1.20) the terms containing $dxdy$ are not neglected because terms containing one differential length dx will cancel out.

or

$$P_2 = \sigma_x\, dy + \frac{\partial \sigma_x}{\partial x}\, dx\, dy + \frac{1}{2}\frac{\partial \sigma_x}{\partial y}\, dy^2$$

Therefore, the resultant force on the element is

$$P_2 - P_1 = \frac{\partial \sigma_x}{\partial x}\, dx\, dy$$

If we assume that the mean stresses are σ_x and $\sigma_x + (\partial\sigma_x/\partial x)\, dx$ acting at the centers of the left and right faces, respectively, the resultant force will be the same. The same result holds for the moment effect of the two systems. Therefore, in deriving the equilibrium equations, we will use a simplified stress system consisting of a uniform stress distribution on each face, which we shall represent by a single vector applied at the center of each face.

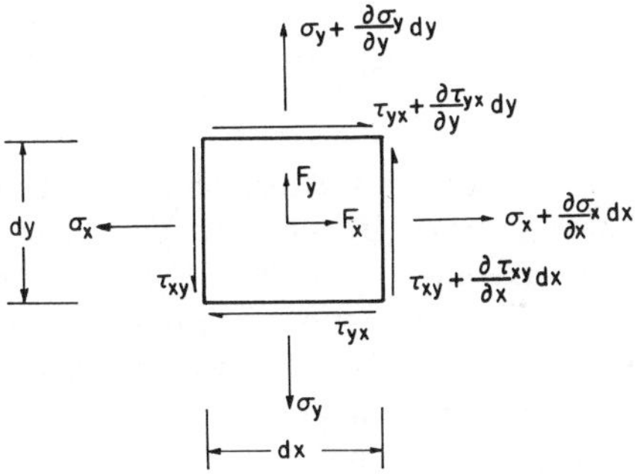

Fig. 1.16 Average Values of Stress on an Element

Assuming $\sigma_z = \tau_{xz} = \tau_{yz} = \tau_{zx} = \tau_{zy} = F_z = 0$, we have the conditions shown in Fig. 1.16. We also assume that σ_x, σ_y, τ_{xy}, and τ_{yx} are independent of z, as are the body force intensities F_x and F_y. The situation satisfying all of these conditions is known as plane stress. By writing $\sum x$-Forces $= 0$ we obtain, assuming a unit depth,

$$F_x\, dx\, dy + \left[\sigma_x + \frac{\partial \sigma_x}{\partial x}\, dx\right] dy - \sigma_x\, dy + \left[\tau_{yx} + \frac{\partial \tau_{yx}}{\partial y}\, dy\right] dx - \tau_{yx}\, dx = 0 \tag{1.20}$$

which when simplified becomes

$$\left[\frac{\partial \sigma_x}{\partial x} + \frac{\partial \tau_{yx}}{\partial y} + F_x\right] dx\, dy = 0$$

and since $dx\, dy$ is not zero, the quantity in parenthesis must vanish.

Thus we obtain

$$\frac{\partial \sigma_x}{\partial x} + \frac{\partial \tau_{yx}}{\partial y} + F_x = 0$$

Similarly, the force summation in the y direction yields the relation

$$\frac{\partial \sigma_y}{\partial y} + \frac{\partial \tau_{xy}}{\partial x} + F_y = 0$$

The equilibrium equations may be generalized by considering the three-dimensional counterpart of Fig. 1.16 with the following result:

$$\begin{aligned}
&\frac{\partial \sigma_x}{\partial x} + \frac{\partial \tau_{yx}}{\partial y} + \frac{\partial \tau_{zx}}{\partial z} + F_x = 0 \\
&\frac{\partial \sigma_y}{\partial y} + \frac{\partial \tau_{xy}}{\partial x} + \frac{\partial \tau_{zy}}{\partial z} + F_y = 0 \\
&\frac{\partial \sigma_z}{\partial z} + \frac{\partial \tau_{xz}}{\partial x} + \frac{\partial \tau_{yz}}{\partial y} + F_z = 0
\end{aligned} \tag{1.21}$$

For a body in equilibrium then, the manner in which the stresses vary from point to point is governed by the above equations of equilibrium.[8]

Notice that the second and third of Eqs. (1.21) can be obtained from the first through a cyclic permutation, i.e., by changing x to y, y to z and z to x. Most of the basic equations which will be developed later follow this rule. The replacements of x by y, etc., simply correspond to a renaming of the coordinate axes.

We may also apply a third equation of statics to the stresses in Fig. 1.16, i.e., $\Sigma M = 0$. Taking moments about the lower left corner, we find that

$$\left(\frac{\partial \sigma_y}{\partial y}\, dy\, dx\right)\frac{dx}{2} - \left(\frac{\partial \sigma_x}{\partial x}\, dx\, dy\right)\frac{dy}{2} + \left(\tau_{xy} + \frac{\partial \tau_{xy}}{\partial x}\, dx\right) dy\, dx$$
$$- \left(\tau_{yx} + \frac{\partial \tau_{yx}}{\partial y}\, dy\right) dx\, dy + F_y\, dx\, dy\, \frac{dx}{2} - F_x\, dx\, dy\, \frac{dy}{2} = 0$$

Neglecting terms containing triple products of dx and dy, we find that

$$\tau_{xy} = \tau_{yx}$$

[8] If the body is not in static equilibrium, the inertia force terms

$$-\rho a_x, \quad -\rho a_y, \quad -\rho a_z$$

where a_x, a_y, and a_z are the components of acceleration, must be included in the body force components F_x, F_y, and F_z, respectively, in Eqs. (1.21).

By considering the three-dimensional case and writing $\Sigma M = 0$ about x, y, and z axes, we would discover that

$$\tau_{xy} = \tau_{yx}, \qquad \tau_{xz} = \tau_{zx}, \qquad \tau_{yz} = \tau_{zy} \tag{1.22}$$

Therefore only six of the nine stress components at a point are independent.

Since the three moment equations have been utilized to relate the shear stress components, they cannot be used again as governing equations in the theory of elasticity. Only the three force equilibrium equations (1.21) remain to be satisfied by the stress components. We also observe that we have three equations relating the six variables, so that additional equations are required for a complete solution of the stress distribution throughout a body. As will be shown later, these additional relations are in the form of strain-displacement equations and, for elastic bodies, the generalized Hooke's law equations. It is quite natural to expect that we would not obtain a complete solution at this stage since we have made no mention of the material properties and the manner in which the body is deformed. These considerations are certainly required in order to define the stress distribution in a body under load. Thus the stress distribution in a body, in the terminology of strength of materials, is always statically indeterminate.

1.8. Three-Dimensional State of Stress at a Point

In Section 1.6 we considered the state of stress at a point only in dealing with the equilibrium in the x and y directions, and therefore only four stress components (three independent ones) were involved. In the three-dimensional case, there are nine stress components, six of which are independent, i.e., σ_x, σ_y, σ_z, τ_{xy}, τ_{yz}, and τ_{zx}. The equations governing the transformation of stress and the principal stresses in the three-dimensional case may be obtained by approaches similar to those applied to the two-dimensional state of stress. In this section we shall briefly discuss the three-dimensional state of stress, emphasizing physical understanding and application. In Chapter 9 a complete development for a general state of stress will be presented.

A limiting process similar to that carried out in Section 1.6 reveals that it is sufficient to consider the stress components without the space derivative terms as shown in Fig. 1.17 in order to determine the stress components on another plane through the given element. Similar to the two-dimensional case, our objective now is to determine the components of stress on an inclined x' plane (or plane ABC in Fig. 1.17), if the stresses on x, y, and z planes, as well as the direction of x', are given.

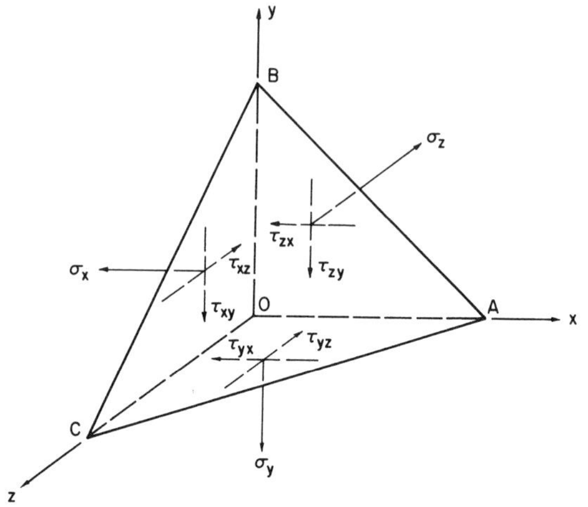

Fig. 1.17 Stress Components on a Tetrahedron

The direction of plane ABC is defined by the angles its normal makes with x, y, and z. We let the cosines of those angles be a_{11}, a_{21}, and a_{31}, respectively (see Table 1.1). It can be shown[9] that the area of triangle AOC in Fig. 1.18 in the y plane is related to the area of ABC by

$$A_{AOC} = A_{ABC} \cos(y, x') = Aa_{21}$$

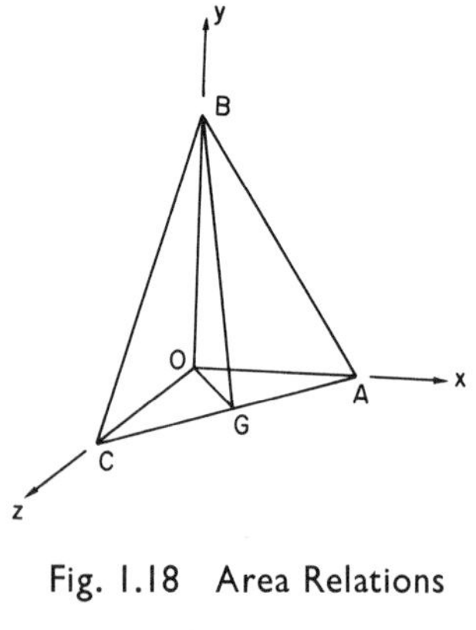

Fig. 1.18 Area Relations

[9] To determine the relation between A_{AOC} and A_{ABC}, we construct a vertical plane perpendicular to AC. This plane intersects ABC along BG, and AOC along OG. Thus $(OG) = (BG) \cos(y, x') = (BG)a_{21}$, and $A_{ABC} = (1/2)(BG) \times (AC)$, $A_{AOC} = (1/2)(OG) \times (AC)$, so that $A_{AOC} = A_{ABC}\, a_{21}$.

and similarly

$$A_{AOB} = Aa_{31} \qquad A_{BOC} = Aa_{11} \tag{1.23}$$

where A is the area of triangle ABC and the notation (y, x') stands for the angle between y and x'. In general, then, the ratio of the projected area on any plane to the area ABC itself is equal to the cosine of the angle between the planes or between their normals.

The given stress components on x, y, and z planes are shown in Fig. 1.17. The stress vector p acting on surface ABC together with its components in the x, y, and z directions, p_x, p_y, and p_z, are shown in Fig. 1.19.

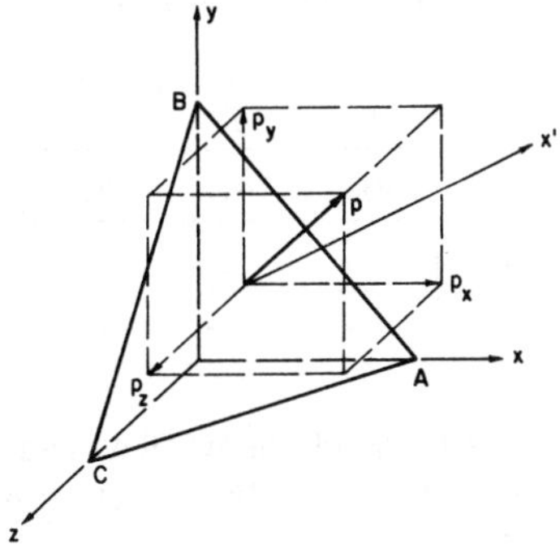

Fig. 1.19 Components of the Stress Vector on the x' Plane

From the force equilibrium condition and Eqs. (1.23) we get

$$\sum F_x = \sum F_y = \sum F_z = 0$$

$$\begin{aligned} p_x &= \sigma_x a_{11} + \tau_{yx} a_{21} + \tau_{zx} a_{31} \\ p_y &= \tau_{xy} a_{11} + \sigma_y a_{21} + \tau_{zy} a_{31} \\ p_z &= \tau_{xz} a_{11} + \tau_{yz} a_{21} + \sigma_z a_{31} \end{aligned} \tag{1.24}$$

Recalling that the stress on a given surface is a vector, we may obtain its component in any direction by summing up the projections of its x, y, and z components in this direction. The normal stress $\sigma_{x'}$ may then be found by projecting p_x, p_y, and p_z in the x' direction and adding. With Eqs. (1.24), then, we find

$$\sigma_{x'} = \sigma_x a_{11}^2 + \sigma_y a_{21}^2 + \sigma_z a_{31}^2 + 2\tau_{xy} a_{11} a_{21} + 2\tau_{yz} a_{21} a_{31} + 2\tau_{zx} a_{31} a_{11} \tag{1.25}$$

In order to determine the shear stress τ on this plane, we note that

$$\tau^2 = p^2 - \sigma_{x'}{}^2 \tag{1.26}$$

where

$$p^2 = p_x^2 + p_y^2 + p_z^2$$

Equation (1.26) gives the magnitude of the shear stress on the x' plane. If the direction of τ is required, it is necessary to introduce an orthogonal coordinate system x', y', z'. Since x' is normal to plane ABC, y' and z' must lie in this plane. We define the directions of x', y', and z' by the cosines given in the following table,[10]

Table 1.1
NOTATIONS FOR
DIRECTION COSINES

	x'	y'	z'
x	a_{11}	a_{12}	a_{13}
y	a_{21}	a_{22}	a_{23}
z	a_{31}	a_{32}	a_{33}

where a_{11} is the cosine of the angle between x' and x, a_{23} is the cosine of the angle between z' and y, etc. By projecting p_x, p_y, and p_z in the y' direction, we find

$$\begin{aligned}\tau_{x'y'} = {} & \sigma_x a_{11} a_{12} + \sigma_y a_{21} a_{22} + \sigma_z a_{31} a_{32} \\ & + \tau_{xy}(a_{11} a_{22} + a_{21} a_{12}) \\ & + \tau_{yz}(a_{21} a_{32} + a_{31} a_{22}) \\ & + \tau_{zx}(a_{31} a_{12} + a_{11} a_{32})\end{aligned} \tag{1.27}$$

Similarly,

$$\begin{aligned}\tau_{x'z'} = {} & \sigma_x a_{11} a_{13} + \sigma_y a_{21} a_{23} + \sigma_z a_{31} a_{33} \\ & + \tau_{xy}(a_{11} a_{23} + a_{21} a_{13}) \\ & + \tau_{yz}(a_{21} a_{33} + a_{31} a_{23}) \\ & + \tau_{zx}(a_{31} a_{13} + a_{11} a_{33})\end{aligned} \tag{1.28}$$

The other stress components acting on planes oriented normal to y' or z' can be found by aligning the normal of plane ABC along y' or z', or by cyclic permutation of Eqs. (1.25), (1.27), and (1.28). It must be noted that the nine direction cosines are not independent because x', y', and z' are orthogonal, and also because the three direction cosines

[10] Note that the first subscript in the direction cosines is associated with the unprimed axes, where 1, 2, and 3 refer to x, y, and z, respectively. The second subscript is associated with the primed axes, where 1, 2, and 3 refer to x', y', and z', respectively.

of a particular direction in space are not independent but interrelated. It will be shown in Chapter 8 that the various direction cosines must satisfy the following relations:

$$\begin{gathered} a_{11}^2 + a_{21}^2 + a_{31}^2 = 1 \\ a_{12}^2 + a_{22}^2 + a_{32}^2 = 1 \\ a_{13}^2 + a_{23}^2 + a_{33}^2 = 1 \\ a_{11}a_{12} + a_{21}a_{22} + a_{31}a_{32} = 0 \\ a_{12}a_{13} + a_{22}a_{23} + a_{32}a_{33} = 0 \\ a_{11}a_{13} + a_{21}a_{23} + a_{31}a_{33} = 0 \end{gathered} \tag{1.29}$$

Thus from Table 1.1 we see that the sum of the squares of the cosines in any column equals unity, and that the sum of the products of the adjacent cosines in any two columns is zero. The same rules apply for the rows in Table 1.1, but these additional relations are not independent of Eq. (1.29).

It will be shown in Chapter 9 that, in a three-dimensional state of stress at a point, there always exist three mutually perpendicular principal planes on which the shear stress vanishes and the normal stresses assume stationary values. In order to determine the orientation of the principal planes, we differentiate Eq. (1.25) with respect to a_{11} and a_{21} (see Problem 21), noting that a_{11}, a_{21}, and a_{31} are not independent since $a_{11}^2 + a_{21}^2 + a_{31}^2 = 1$, or $a_{31}^2 = 1 - a_{11}^2 - a_{21}^2$. Setting the resulting expressions equal to zero, we get the relations

$$\begin{aligned} \frac{a_{11}\sigma_x + a_{21}\tau_{xy} + a_{31}\tau_{xz}}{a_{11}} &= \frac{a_{11}\tau_{xy} + a_{21}\sigma_y + a_{31}\tau_{yz}}{a_{21}} \\ &= \frac{a_{11}\tau_{xz} + a_{21}\tau_{yz} + a_{31}\sigma_z}{a_{31}} \end{aligned} \tag{1.30}$$

Using Eqs. (1.24), we note that Eq. (1.30) becomes

$$\frac{p_x}{a_{11}} = \frac{p_y}{a_{21}} = \frac{p_z}{a_{31}} \tag{1.31}$$

which shows that, on a plane where the normal stress assumes a stationary value, the shear stress vanishes since the stress vector and the normal to the plane are parallel. Equation (1.30) may be written as

$$\frac{p_x}{a_{11}} = \sigma_p = \frac{p_y}{a_{21}} = \frac{p_z}{a_{31}}$$

where σ_p represents the stationary values of the normal stress $\sigma_{x'}$. Thus we have,

$$\begin{aligned} a_{11}(\sigma_x - \sigma_p) + a_{21}\tau_{xy} + a_{31}\tau_{xz} &= 0 \\ a_{11}\tau_{xy} + a_{21}(\sigma_y - \sigma_p) + a_{31}\tau_{yz} &= 0 \\ a_{11}\tau_{xz} + a_{21}\tau_{yz} + a_{31}(\sigma_z - \sigma_p) &= 0 \end{aligned} \tag{1.32}$$

Since Eqs. (1.32) are three homogeneous, linear equations in a_{11}, a_{21}, and a_{31}, and the latter quantities cannot all vanish, the only nontrivial solution results if the determinant of the coefficients is set equal to zero. This gives the relation

$$\begin{aligned} &\sigma_p^{\ 3} - (\sigma_x + \sigma_y + \sigma_z)\sigma_p^{\ 2} \\ &\quad + (\sigma_x\sigma_y + \sigma_y\sigma_z + \sigma_z\sigma_x - \tau_{xy}^2 - \tau_{yz}^2 - \tau_{zx}^2)\sigma_p \\ &\quad - (\sigma_x\sigma_y\sigma_z + 2\tau_{xy}\tau_{yz}\tau_{zx} - \sigma_x\tau_{yz}^2 - \sigma_y\tau_{xz}^2 - \sigma_z\tau_{xy}^2) = 0 \end{aligned} \tag{1.33}$$

There are always three real roots of Eq. (1.33), and the shear stresses on the corresponding planes vanish. These stationary values of normal stress are called principal stresses and are commonly represented by σ_1, σ_2, and σ_3, where $\sigma_1 > \sigma_2 > \sigma_3$.

Once the principal stresses are determined, the direction cosines of the principal planes may be found by substitution into Eqs. (1.32) (which are only two independent equations, since the determinant of their coefficients vanishes) and by using the relation

$$a_{11}^2 + a_{21}^2 + a_{31}^2 = 1$$

for each value of σ_p. It should also be noted that Eq. (1.33) defines the three values of σ_p regardless of the orientation of the *xyz* Cartesian coordinate system. Thus the quantities in parentheses in this equation are invariant with respect to Cartesian coordinate systems, e.g., $\sigma_x + \sigma_y + \sigma_z = \sigma_{x'} + \sigma_{y'} + \sigma_{z'}$ etc.

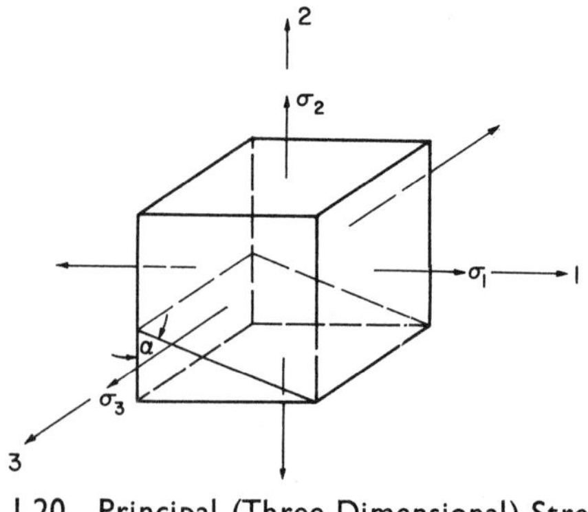

Fig. 1.20 Principal (Three-Dimensional) Stresses

Now let an element be oriented along the principal directions as shown in Fig. 1.20. Suppose that we write the equations for σ and τ on any plane normal to the 3-plane as shown. Clearly, σ_3 has no effect on the stress condition on this plane, so that the expressions for σ and τ in terms of σ_1 and σ_2 are the same as in the two-dimensional case. Thus σ and τ are related by a Mohr circle (Fig. 1.21). The stress condition on planes normal to the 1-plane is also given by a Mohr circle: the same holds true for planes normal to the 2-plane. It can be shown[11] that the stress conditions on all planes must lie in the shaded region between the circles in Fig. 1.21. From the circles in Fig. 1.21, then, it is evident

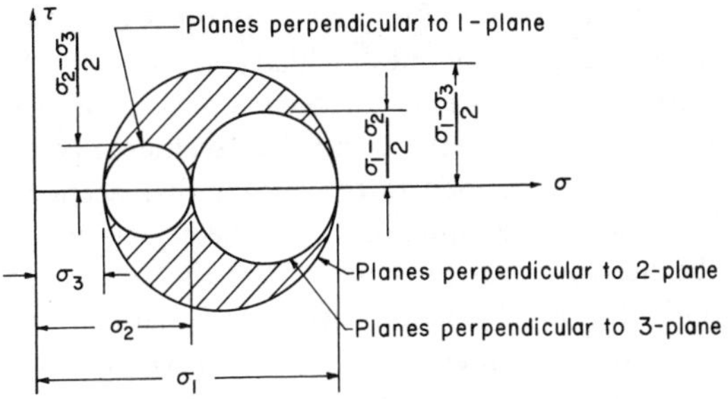

Fig. 1.21 Mohr's Circle for a Three-Dimensional State of Stress

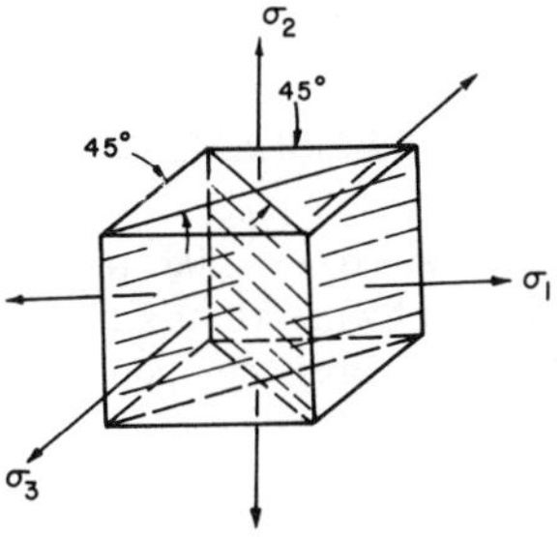

Fig. 1.22 Planes of Maximum Shear Stress

that the maximum shear stress $\tau_{max} = (\sigma_1 - \sigma_3)/2$ and acts on the planes bisecting the planes of maximum and minimum principal stress as shown in Fig. 1.22. The construction of Fig. 1.21 may be used to determine the stress on any plane graphically,[12] but this method will not be presented

[11] A. Nadai, *Theory of Flow and Fracture of Solids*, Vol. 1, Second Edition, McGraw-Hill Book Company, Inc., 1950, pp. 96–98.

[12] A. Durelli, Phillips, and Tsao, *Introduction to the Theoretical and Experimental Analysis of Stress and Strain*, McGraw-Hill Book Company, Inc., New York, 1958, pp. 78–79.

in this text. The planes of maximum shear stress may also be found by differentiating Eq. (1.26) with respect to a_{11} and a_{21} and setting the resulting expressions equal to zero. (See Problem 1-22.)

1.9. Summary

In summary, the most important features of Chapter 1 are
(a) the definitions of stress and stress components,
(b) the equations of equilibrium (1.21), and
(c) the transformation of stress equations (1.25), (1.27), (1.28).

Stress is defined as the limiting value of force per unit area as the area approaches zero. The state of stress at a point is characterized by nine components (six independent ones), which constitute the "stress tensor."

The solution of an elasticity problem consists of determining the distribution of stress (also strain and displacement) components in an elastic body under prescribed forces and/or displacements. We have, thus far, developed three equations, i.e., the equations of equilibrium, which can be used for this purpose. Once the stress distribution in a body is found, the transformation of stress equations define the stress components on arbitrary planes through any given point. A useful tool for visualizing and solving these relations, especially when $\tau_{xz} = \tau_{yz} = 0$, is the Mohr circle.

PROBLEMS

1-1 Given $\sigma_x = -14{,}000$ psi, $\sigma_y = 6{,}000$ psi, and $\tau_{xy} = -17{,}320$ psi, determine both by formulas and by the Mohr's circle, (a) the principal stresses and their directions and (b) the stress components on the x' and y' planes when $\alpha = 45°$.

1-2 Given a uniform state of stress with

$$\sigma_x = 10{,}000 \text{ psi} \qquad \tau_{xy} = 5{,}000 \text{ psi}$$
$$\sigma_y = 5{,}000 \text{ psi} \qquad \alpha = 30°$$

find $\sigma_{x'}$, $\sigma_{y'}$, and $\tau_{x'y'}$.

1-3 A rectangular block is under a uniformly distributed load as shown in the figure. Find the stress components on the plane $A - A$.

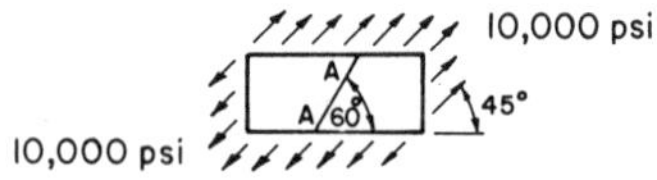

Problem 1-3

1-4 A thin plate is under a uniform state of stress as shown in the figure.

(a) Among all planes that are normal to the plane of the figure, find the one on which the maximum shear stress acts and the magnitude of this maximum shear stress.

(b) Find the greatest of all shearing stresses on any plane (three-dimensional).

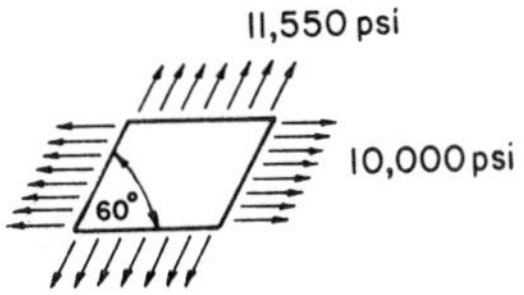

Problem 1–4

1-5 A pulley which is two feet in diameter is mounted on a 2-inch circular shaft, three feet from one of the bearings (and between them) which are eight feet apart. Power is transmitted through the 3-foot portion of the shaft such that the taut side of the belt has a tension of 300 lbs. and the slack side 100 lbs. Use the load stress relations from strength of materials and calculate the maximum tensile, compressive and shearing stresses in the shaft.

1-6 Calculate the values of principal stresses and the angles between the principal axes and the x axis for the element shown in the figure. Assume the following stresses are known:

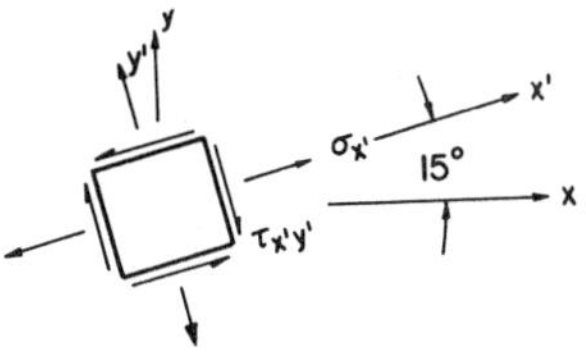

Problem 1–6

$$\sigma_{x'} = 5{,}000 \text{ psi}$$
$$\tau_{x'y'} = -2500 \text{ psi}$$
$$\sigma_{y'} = 0$$

1-7 (a) Show that the stresses

$$\sigma_x = \sigma_{x'} = \sigma_0$$
$$\tau_{xy} = -\tau_{x'y'} = \tau_0 \neq 0$$

do not define the state of stress at a point even though four of the seven variables in Eqs. (1.5) to (1.7) are specified. Hint: Show that the angle α between x and x' cannot be determined from the transformation of stress equations under the given stresses.

(b) Establish the result of part (a) geometrically by attempting to draw a Mohr circle.

(c) Show that the stresses given in part (a) define the state of stress at a point if, in addition, the angle α is prescribed.

1-8 A thin plate is under a uniform state of stress as shown in the figure. Find the principal stresses and their directions

(a) by using Mohr's circle only and

(b) by formulas.

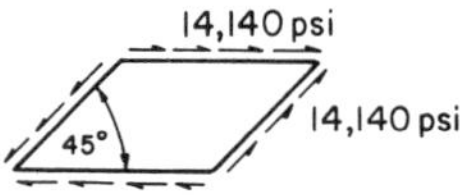

Problem 1–8

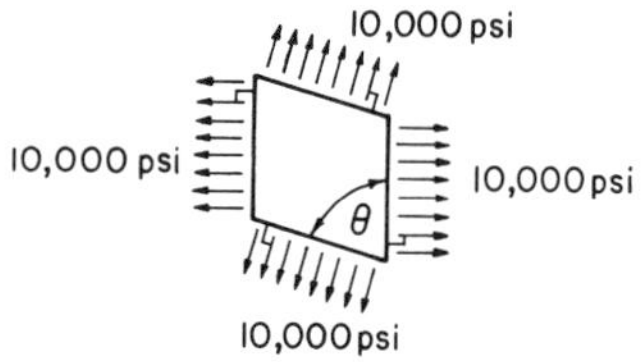

Problem 1–9

1-9 Solve the following problem both by using Mohr's circle and by formulas. A thin plate is under a uniform state of stress as shown in the figure. Determine the principal stresses.

1-10 A cylindrical bar of cross-sectional area A is subjected to a uniform tensile stress σ. Determine the maximum shearing stress and the plane it acts upon.

1-11 For a circular beam, the usual engineering equations for the stresses due to bending are

$$\sigma_y = \sigma_z = 0 \qquad \tau_{xz} = \tau_{yz} = 0$$

$$\sigma_x = \frac{My}{I} \qquad \tau_{xy} = \frac{V(R^2 - y^2)}{3I} \qquad I = \frac{\pi R^4}{4}$$

Do these equations satisfy the equilibrium equations?

1-12 (a) Verify Eqs. (1.9), (1.10), (1.12), (1.13), (1.14) and (1.15) by using Mohr's circle of stress and trigonometric formulas. These equations can also be verified by using Mohr's circle and geometrical theorems.

(b) Show by use of Mohr's circle that if $\tau_{xy} < 0$ and $(\sigma_x - \sigma_y) < 0$, then $\pi < 2\alpha < (3/2)\pi$, where α is the angle, measured counterclockwise, from the x plane to the plane of maximum normal stress.

1-13 By using Mohr's circle, show that the following quantities are invariant for a two-dimensional state of stress with $\sigma_z = \tau_{xz} = \tau_{yz} = 0$;

(a) $\sigma_{x'} + \sigma_{y'}$

(b) $\sigma_{x'}\sigma_{y'} - \tau_{x'y'}^2$.

1-14 Using Mohr's circle, show that the magnitude of the stress vector (p) on any plane must satisfy the relation $\sigma_2 \le p \le \sigma_1$ thus demonstrating that the principal stresses are also the maximum and the minimum magnitudes of the stress vector on any plane in the two-dimensional case.

1-15 Given a three-dimensional state of stress with

$$\begin{aligned} \sigma_x &= +10 \text{ psi} & \tau_{xy} &= +5 \text{ psi} \\ \sigma_y &= +20 \text{ psi} & \tau_{xz} &= -10 \text{ psi} \\ \sigma_z &= -10 \text{ psi} & \tau_{yz} &= -15 \text{ psi} \end{aligned}$$

(a) Find the magnitude and direction of the stress vector p on the x' plane where the x' direction is defined by

$$a_{11} = +1/2 \qquad a_{21} = +1/\sqrt{2} \qquad a_{31} \text{ is positive.}$$

(b) Find σ and τ on this plane.
(c) Determine the angle between p and σ.
(d) Solve for $\tau_{x'y'}$ and $\tau_{x'z'}$ if $a_{12} = \frac{1}{2}$ and a_{32} is negative. Note that the resultant of $\tau_{x'y'}$ and $\tau_{x'z'}$ is τ from part (b).
(e) Evaluate all of the stress components acting on the x', y', and z' planes.

1-16 Given the normal and shearing stress components in the directions x, y, and z,

$$\sigma_x = -5C, \quad \sigma_y = C, \quad \sigma_z = C, \quad \tau_{xy} = -3C, \quad \tau_{yz} = \tau_{zx} = 0$$

where $C = 1{,}000$ psi.
(a) Set up the equation for the principal stresses.
(b) By studying the given stress components, can you pick out one of the principal stresses? Find the other two principal stresses and their directions.

1-17 If $\tau_{zx} = \tau_{yz} = 0$, then we have the two-dimensional case with a normal stress in the third direction. Making the substitution in Eq. (1.33), show that the following equation is again obtained. This will show that the three roots are real. Hint: From the given data, determine one root of the cubic equation.

$$\left.\begin{matrix}\sigma_1\\ \sigma_2\end{matrix}\right\} = \frac{\sigma_x + \sigma_y}{2} \pm \sqrt{\left(\frac{\sigma_x - \sigma_y}{2}\right)^2 + \tau_{xy}^2}$$

1-18 Given the normal and shearing stress components in the directions of x, y, and z,

$$\begin{aligned} \sigma_x &= 500 & \sigma_y &= 500 & \sigma_z &= 2000 \\ \tau_{xy} &= -1500 & \tau_{yz} &= -500\sqrt{2} & \tau_{zx} &= 500\sqrt{2} \end{aligned}$$

determine (a) the principal stresses and (b) the direction cosines of one of the principal axes.

1-19 If x, y, and z are principal directions, then the octahedral plane is defined as the plane making equal angles with x, y, and z. Find the normal and shear stress components on an octahedral plane.

1-20 Verify the fact that $\tau_{xy} = \tau_{yx}$ in the three-dimensional case.

1-21 By writing $\partial\sigma/\partial a_{11} = \partial\sigma/\partial a_{21} = 0$, where σ is given by Eq. (1.25), verify Eq. (1.30). Hint: Note that a_{11}, a_{21}, and a_{31} are not independent since

$$a_{11}^2 + a_{21}^2 + a_{31}^2 = 1$$

and a_{31} may be considered as a function of a_{11} and a_{21}. Thus

$$\left.\frac{\partial\sigma}{\partial a_{11}}\right|_{a_{21}\ \text{const.}} = \left.\frac{\partial\sigma}{\partial a_{11}}\right|_{a_{21}, a_{31}\ \text{const.}} + \frac{\partial\sigma}{\partial a_{31}}\frac{\partial a_{31}}{\partial a_{11}}$$

Also note that the sign for a_{31} is immaterial in this case.

1-22 Verify the fact that the planes of maximum shear stress in the three-dimensional case bisect the planes of maximum and minimum normal stress by writing $\partial\tau/\partial a_{11} = \partial\tau/\partial a_{21} = 0$. Use Eq. (1.26) for τ, expressing τ in terms of the principal stresses. Use the hint of Problem 1-21.

1-23 Given the following three-dimensional state of stress:

$$\sigma_x = +8 \text{ psi} \qquad \sigma_z = -6 \text{ psi}$$
$$\sigma_y = -2 \text{ psi} \qquad \tau_{xy} = \tau_{xz} = \tau_{yz} = 0$$

(a) Determine the stress components corresponding to the x', y', and z' directions where

$$a_{11} = +2/\sqrt{5} \qquad a_{12} = -1/\sqrt{5} \qquad a_{33} = +1$$

(b) Find the stress components corresponding to the x'', y'', and z'' directions where

$$a_{11} = +1 \qquad a_{22} = +1/2 \qquad a_{23} = \sqrt{3}/2$$

(c) Form Eq. (1.33) using the stresses on x', y', z'. Do the same for x'', y'', z'' as an illustration of the invariance of the quantities in parentheses in Eq. (1.33).

1-24 Is the following stress distribution possible for a body in equilibrium? A, B, and C are constants. (Body forces are zero.)

$$\sigma_x = -Axy$$
$$\tau_{xy} = (A/2)(B^2 - y^2) + Cz$$
$$\tau_{xz} = -Cy$$
$$\sigma_y = \sigma_z = \tau_{yz} = 0$$

1-25 Show that the following state of stress is in equilibrium.

$$\sigma_x = 3x^2 + 3y^2 - z \qquad \tau_{xy} = z - 6xy - \tfrac{3}{4}$$
$$\sigma_y = 3y^2 \qquad \tau_{xz} = x + y - \tfrac{3}{2}$$
$$\sigma_z = 3x + y - z + \tfrac{5}{4} \qquad \tau_{yz} = 0$$

1-26 For the state of stress in problem 25 at the specific point $x = \frac{1}{2}$, $y = 1$, $z = \frac{3}{4}$, determine the principal stresses.

1-27 Given the following stress distribution,

$$\begin{aligned}\sigma_x &= 3x^2 + 4xy - 8y^2\\ \sigma_y &= 2x^2 + xy + 3y^2\\ \tau_{xy} &= -\tfrac{1}{2}x^2 - 6xy - 2y^2\\ \sigma_z &= \tau_{zx} = \tau_{yz} = 0\end{aligned}$$

determine, in the absence of body forces, whether equilibrium exists.

2

Strain and Displacement

2.1. Introduction

In Chapter 1 it was pointed out that the equations of equilibrium, Eqs. (1.21), are not sufficient to solve for the stress components in terms of x, y, and z in a body under an external force system. This is due to the fact that there are six independent unknown quantities (σ_x, σ_y, σ_z, τ_{xy}, τ_{yz}, and τ_{zx}) and only three equations relating these unknowns. Obviously, then, the stress components depend upon other factors, and thus further information is required in order to solve a problem in elasticity. This information is supplied, for an elastic material, by generalized Hooke's law and the strain-displacement equations. In this chapter we shall study strain and in the next chapter, generalized Hooke's law.

The analyses given in this chapter apply to any continuum which satisfies the assumption of infinitesimal deformations (Section 2.2) and are not influenced by the properties of the material. We only limit our discussion to continuous materials, i.e., the material is present at each point in the medium, and to continuous displacements; thus the originally continuous material cannot contain gaps after it is displaced. Also, for any given state of deformation, the displacement functions must be single-valued, since no single material particle may occupy two positions in space. The definition of strain components and their geometrical significance will be delineated and the compatibility equations, which relate the strain components, will then be derived. The effects of strain on the geometric changes in a body's dimensions, as well as the effects of rigid body movements, will be examined. It will also be shown that we may speak of a state of strain, analogous to our discussion of a state of stress, i.e., the strain components (defined in the next section) for any directions may be found if those corresponding to two known directions (in a state of plane strain) are given.

2.2. Strain-Displacement Relations

A body is said to be strained or deformed when the *relative* positions of points in the body are changed. This is in contrast to a rigid body movement where the distance between any two points remains fixed. When forces are applied to a body, the position of any point of the body, in general, is altered. The displacement of a point is defined as the vector distance from the initial to the final location of the point. We let the x, y, and z components of displacement be denoted by u, v, and w, respectively. Therefore the point initially situated at (x, y, z) will be displaced to $(x + u, y + v, z + w)$. In general, u, v, and w are all functions of x, y, and z.

Before presenting the general strain-displacement relations, let us first consider a one-dimensional model to obtain a clearer picture of the concept of normal strain. For example, consider a bar subjected to uniaxial stress as shown in Fig. 2.1. Initially, points A and B, which are

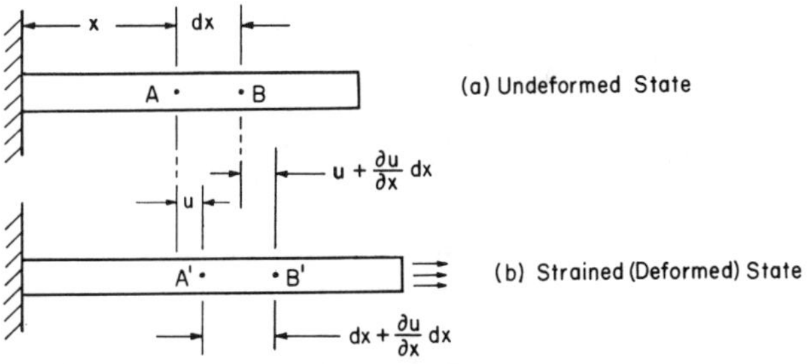

Fig. 2.1 Normal Strain in a Bar

on the center line of the bar, are situated as shown in Fig. 2.1(a). After stress is applied, A and B move to A' and B' (Fig. 2.1(b)), respectively. We see that point B is *displaced* slightly more ($(\partial u/\partial x)dx$) than A since it is farther from the fixed end. Length AB, therefore, increases by this amount. Defining normal strain (ϵ) as the unit change in length, we get

$$\epsilon_x = \frac{(\partial u/\partial x)dx}{dx} = \frac{\partial u}{\partial x}$$

We now consider a body in a state of plane strain, which is defined by

$$u = u(x, y) \qquad v = v(x, y) \qquad w = 0$$

In this case, all points originally lying in the xy plane remain in this plane after the body is strained. For example, consider the displacement of the infinitesimal element $ABCD$ shown in Fig. 2.2. This element is displaced to the final configuration $A'B'C'D'$, where the element as a whole is translated as shown by the dotted lines and also deformed. The deformation consists of two distinct types:

(a) the sides change length and

(b) the sides rotate with respect to each other.

Guided by (a) and (b), we define normal and shearing strain as follows: the normal strain ϵ in a given direction is defined as the unit change in length (change in length per unit length) of a line which was originally oriented in the given direction. The normal strain is positive if the line increases in length, negative if the line is shortened. Shear strain, like shear stress, is associated with two orthogonal directions. The shear strain (γ) is defined as the change in the original right angle between two axes, the angle being measured in radians. Shear strain is positive if the right angle between the positive directions of the two axes decreases.[1] The sign of the shear strain, like that of shear stress, depends on the coordinate system. Referring again to Fig. 2.2, we see that the strain components referred to the x and y coordinate axes are

$$\epsilon_x = \frac{A'B' - AB}{AB} = \frac{A'B' - dx}{dx}$$

$$\epsilon_y = \frac{A'D' - AD}{AD} = \frac{A'D' - dy}{dy}$$

$$\gamma_{xy} = \frac{\pi}{2} - \beta = \theta - \lambda$$

where the negative sign in front of λ accounts for the fact that counterclockwise angles of rotation are defined as positive. Strictly speaking, the shear strain defined here is engineering shear strain. The mathematical shear strain, which will be introduced in Chapter 10, is defined as $\frac{1}{2}(\theta - \lambda)$.

The displacements shown in Fig. 2.2 are greatly exaggerated in the interest of clarity. In the linear theory of elasticity, we only consider problems where the strains and the space derivatives of the displacement components are small.[2] These restrictions of infinitesimal deformations

[1] If the shear strain is positive, the right angle between the negative extensions of the coordinate axes decreases, while the angle between a positive axis and the negative extension of the other axis increases.

[2] By "small" we mean that products of these quantities constitute second-order terms.

limit the generality of our solutions. However, there is a vast number of engineering problems which are treated with sufficient accuracy by using these assumptions. In finite elasticity, where strains and displacements are relatively large, these assumptions are not valid and the strain

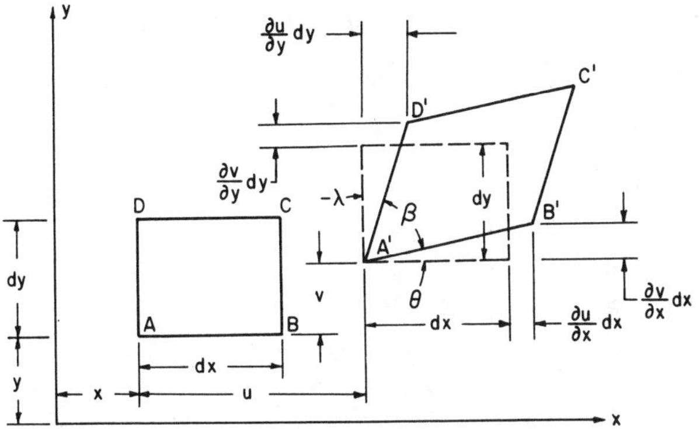

Fig. 2.2 Translation and Deformation of a Two-Dimensional Element

components must be redefined. Now if the displacement components of point A are u and v, point B will be displaced by $u + (\partial u/\partial x)dx$ and $v + (\partial v/\partial x)dx$, since y is constant along line AB. Similarly, the displacement components of point D are $u + (\partial u/\partial y)dy$ and $v + (\partial v/\partial y)dy$. Thus we may write

$$(A'B')^2 = [dx(1 + \epsilon_x)]^2 = \left(dx + \frac{\partial u}{\partial x}dx\right)^2 + \left(\frac{\partial v}{\partial x}dx\right)^2$$

so that

$$\epsilon_x{}^2 + 2\epsilon_x + 1 = 1 + 2\frac{\partial u}{\partial x} + \left(\frac{\partial u}{\partial x}\right)^2 + \left(\frac{\partial v}{\partial x}\right)^2$$

Since we are considering small strains and displacement derivatives, the squared terms are negligible with respect to the term raised to the first power, i.e., they constitute higher-order terms. Ignoring these terms, we find that

$$\epsilon_x = \frac{\partial u}{\partial x}$$

and similarly,

$$\epsilon_y = \frac{\partial v}{\partial y}$$

From Fig. 2.2, we note that

$$\theta = \frac{(\partial v/\partial x)dx}{dx + (\partial u/\partial x)dx}$$

since, conforming to our assumption of small displacement derivatives, θ is a small angle so that $\tan\theta = \theta$. Neglecting $\partial u/\partial x$ as being small compared to 1, we find that

$$\theta = \frac{\partial v}{\partial x}$$

and similarly,

$$\lambda = -\frac{\partial u}{\partial y}$$

Thus the shear strain γ_{xy} is given by

$$\gamma_{xy} = \frac{\partial u}{\partial y} + \frac{\partial v}{\partial x}$$

where the two partial derivatives are positive if AB and AD rotate inward as shown in the figure, i.e., u and v increase with increasing y and x, respectively. In the three-dimensional case, where the original element is a rectangular prism, we have the strain components

$$\begin{aligned} \epsilon_x &= \frac{\partial u}{\partial x} & \gamma_{xy} &= \frac{\partial u}{\partial y} + \frac{\partial v}{\partial x} \\ \epsilon_y &= \frac{\partial v}{\partial y} & \gamma_{yz} &= \frac{\partial v}{\partial z} + \frac{\partial w}{\partial y} \\ \epsilon_z &= \frac{\partial w}{\partial z} & \gamma_{zx} &= \frac{\partial w}{\partial x} + \frac{\partial u}{\partial z} \end{aligned} \tag{2.1}$$

We also observe that

$$\gamma_{xy} = \gamma_{yx} \qquad \gamma_{yz} = \gamma_{zy} \quad \text{and} \quad \gamma_{zx} = \gamma_{xz}$$

The expressions (2.1) are called the strain-displacement relations, since they define the strain components in terms of the displacement components. Figure 2.3 illustrates the deformation and displacement of a small rectangular prism. Referring to this figure, some of the strain components defined by the linear theory of strain are

$$\epsilon_x = \frac{E'F' - EF}{EF} = \frac{A'B' - AB}{AB} = \frac{H'G' - HG}{HG} = \frac{D'C' - DC}{DC}$$

$$\begin{aligned} \gamma_{xy} &= (\pi/2 - \angle F'E'H') = (\pi/2 - \angle F'G'H') = (\pi/2 - \angle B'A'D') \\ &= (\pi/2 - \angle B'C'D') \end{aligned}$$

It should be observed at this time, that in our development of the equilibrium equations and strain-displacement relations, there exists an inconsistency. That is, since the stress components are actually distributed over a *deformed* body, the coordinates x, y, and z in the equilibrium equations should be taken as the coordinates of particles in the

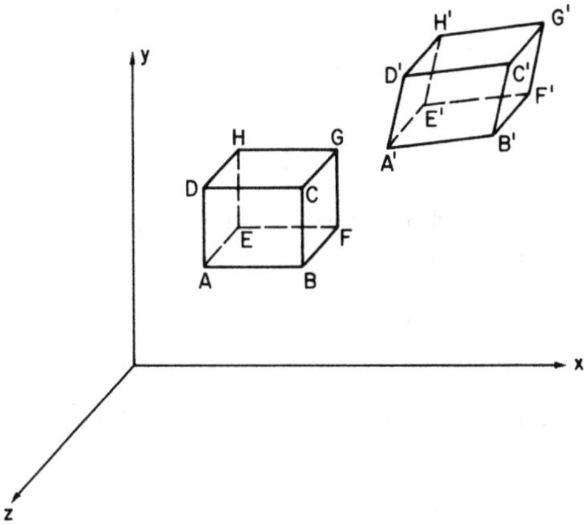

Fig. 2.3 Strain in a Three-Dimensional Element

deformed body; in the strain-displacement relations, the coordinates x, y, and z refer to position in the *undeformed* body. However, due to the assumption of infinitesimal deformations, letting x, y, and z represent the undeformed coordinates in both sets of equations leads to only minor errors. In finite elasticity, this approximation is not valid, and both sets of equations must be expressed in the same coordinates.

2.3. Compatibility Equations

We observe that Eqs. (2.1) represent six equations for the strain components as a function of only three displacement components. Thus if we specify u, v, and w as functions of x, y, and z, we may derive the strain components from Eqs. (2.1). However, we might think of the six strain components as given functions of x, y, and z, in which case we have six equations for the solution of three unknowns, u, v, and w. This system of equations does not, in general, possess a solution for u, v, and w unless the six strain components are somehow related. In other words, all six strain components cannot be arbitrarily prescribed if we are to maintain single-valued, continuous displacement functions; such displacements will only result if we have three additional equations relating the strain components. We now proceed to investigate these equations.

If we differentiate the first of Eqs. (2.1) twice with respect to y and the second twice with respect to x and add the results, we have

$$\frac{\partial^2 \epsilon_x}{\partial y^2} + \frac{\partial^2 \epsilon_y}{\partial x^2} = \frac{\partial^3 u}{\partial y^2 \partial x} + \frac{\partial^3 v}{\partial x^2 \partial y}$$

Differentiating the fourth equation with respect to x and y, we get

$$\frac{\partial^2 \gamma_{xy}}{\partial x \partial y} = \frac{\partial^2}{\partial x \partial y}\left(\frac{\partial u}{\partial y} + \frac{\partial v}{\partial x}\right)$$

and since the order of differentiation for single-valued, continuous functions is immaterial, we see that

$$\frac{\partial^2 \epsilon_x}{\partial y^2} + \frac{\partial^2 \epsilon_y}{\partial x^2} = \frac{\partial^2 \gamma_{xy}}{\partial x \partial y}$$

By similar approaches we can develop five additional equations. The six expressions

$$\begin{aligned}
\frac{\partial^2 \epsilon_x}{\partial y^2} + \frac{\partial^2 \epsilon_y}{\partial x^2} &= \frac{\partial^2 \gamma_{xy}}{\partial x \partial y} \\
\frac{\partial^2 \epsilon_y}{\partial z^2} + \frac{\partial^2 \epsilon_z}{\partial y^2} &= \frac{\partial^2 \gamma_{yz}}{\partial y \partial z} \\
\frac{\partial^2 \epsilon_z}{\partial x^2} + \frac{\partial^2 \epsilon_x}{\partial z^2} &= \frac{\partial^2 \gamma_{zx}}{\partial z \partial x} \\
2\frac{\partial^2 \epsilon_x}{\partial y \partial z} &= \frac{\partial}{\partial x}\left(-\frac{\partial \gamma_{yz}}{\partial x} + \frac{\partial \gamma_{xz}}{\partial y} + \frac{\partial \gamma_{xy}}{\partial z}\right) \\
2\frac{\partial^2 \epsilon_y}{\partial z \partial x} &= \frac{\partial}{\partial y}\left(\frac{\partial \gamma_{yz}}{\partial x} - \frac{\partial \gamma_{xz}}{\partial y} + \frac{\partial \gamma_{xy}}{\partial z}\right) \\
2\frac{\partial^2 \epsilon_z}{\partial x \partial y} &= \frac{\partial}{\partial z}\left(\frac{\partial \gamma_{yz}}{\partial x} + \frac{\partial \gamma_{xz}}{\partial y} - \frac{\partial \gamma_{xy}}{\partial z}\right)
\end{aligned} \tag{2.2}$$

are called the Saint-Venant compatibility equations, or compatibility equations in terms of strain. The strain components must satisfy these expressions in order that solutions for the displacement components exist. Although six compatibility equations are written, they are equivalent to three independent fourth-order equations. To illustrate this fact, the first equation of Eqs. 2.2 is differentiated twice with respect to z, the second equation twice with respect to x, and the third twice with respect to y. The fourth equation is then differentiated with respect to y and z, the fifth with respect to z and x, and the last one with respect to x and y. The first three fourth-order equations are seen to be equivalent to the last three. It is usually more convenient, however, to use the six second-order equations rather than the three fourth-order equations. Furthermore, since solutions of high-order partial differential equations involve more arbitrary functions than lower-order equations,[3] the three fourth-order compatibility equations are too general and additional restrictions must be imposed if they are to be used. (See Problem 2-11.)

The geometric significance of the compatibility equations is seen from the following discussion. Imagine a body divided up into infinitesimal cubes. Each cube is then subjected to arbitrary strain components (corresponding to arbitrarily prescribed strain functions). If we attempt to reassemble the body by piecing the deformed elements together, we find that these elements, in general, do not fit perfectly; they will be separated by spaces. A perfect matching is achieved only when the strain components satisfy the compatibility equations.[4] We should note, however, that if we specify *displacement* components which are single-valued, continuous functions, the strain components will automatically satisfy the compatibility equations.

[3] R. Courant and D. Hilbert, *Methods of Mathematical Physics*, Vol. 2, John Wiley & Sons, Inc., New York, 1962, p. 3.

To demonstrate the relations between a system of several partial differential equations and the corresponding system of higher-order equations, let us consider the function $u(x, y)$ governed by the equations

$$\partial u/\partial x = w(x) \qquad \partial u/\partial y = v(y)$$

where w and v are given functions of x and y, respectively. These two equations are not independent of each other, since if the first one is differentiated with respect to y and the second with respect to x, then both result in the second-order equation $\partial^2 u/\partial x \partial y = 0$. Thus these two equations may be considered as one independent equation for finding u. However, the general solution of the single second-order equations is

$$u = f(x) + g(y)$$

where f and g are arbitrary functions of x and y, respectively, whereas the general solution of the two first-order equations is

$$u = \int w\, dx + \int v\, dy + \text{const.}$$

which contains only an arbitrary constant. Therefore, additional restrictions are required if the second-order equation is to be used. That is, in order to determine the specific form of f and g, the two first-order equations must be considered.

[4] The compatibility equations must be satisfied by the strain components in any linearly deformed body. They are sufficient (sufficiency will be proved in Chapter 10) to guarantee single-valued, continuous displacement functions, however, only in a simply connected region. A simply connected region is one in which any closed curve can be continuously shrunk to a point without passing outside of the boundaries of the region.

2.4. State of Strain at a Point

It has been shown that the state of stress at a point is uniquely determined if the stress components on two planes are given (in the two-dimensional case). The same statement can be made in reference to strain, i.e., the state of strain at a point in the plane strain case is characterized by the strain components on two planes; given the strain components ϵ_x, ϵ_y, γ_{xy} at a point, it is possible to determine the strains on an element oriented in any direction at the point.

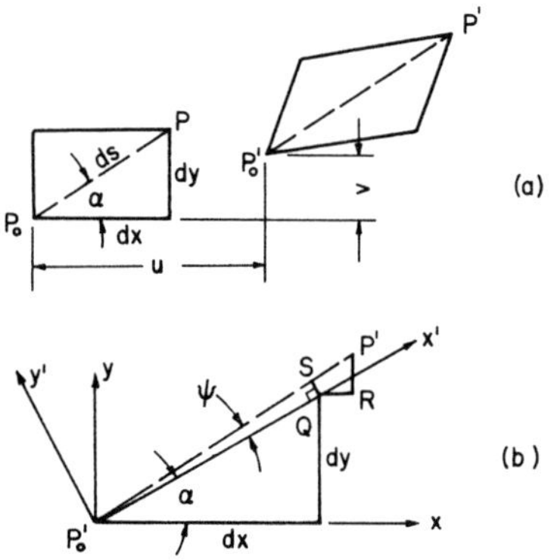

Fig. 2.4 State of Strain at a Point

Consider the displacement of two corners, P_0, and P, of the element shown in Fig. 2.4(a). If the displacement components of P_0 are u and v, those for P are

$$\left(u+\frac{\partial u}{\partial x}dx+\frac{\partial u}{\partial y}dy\right) \quad \text{and} \quad \left(v+\frac{\partial v}{\partial y}dy+\frac{\partial v}{\partial x}dx\right)$$

The relative displacement (the deformation) of P with respect to P_0 is shown in Fig. 2.4(b). Letting the $x'y'$ coordinate axes be aligned as shown, we now proceed to find the strain components referred to the $x'y'$ coordinate system, i.e., $\epsilon_{x'}$, $\epsilon_{y'}$, and $\gamma_{x'y'}$. From Fig. 2.4(b) we see that

$$QR=\frac{\partial u}{\partial x}dx+\frac{\partial u}{\partial y}dy$$

and

$$RP'=\frac{\partial v}{\partial y}dy+\frac{\partial v}{\partial x}dx$$

By projecting QR and RP' in the x' and y' directions, we find that

$$P'S = \frac{QR\cos\alpha + RP'\sin\alpha}{\cos\psi}$$
$$= QR\cos\alpha + RP'\sin\alpha$$

since $\cos\psi \cong 1$ for the small angle. The normal strain in the x' direction is, by definition,

$$\epsilon_{x'} = \frac{P'S}{ds}$$
$$= \left(\frac{\partial u}{\partial x}\frac{dx}{ds} + \frac{\partial u}{\partial y}\frac{dy}{ds}\right)\cos\alpha + \left(\frac{\partial v}{\partial y}\frac{dy}{ds} + \frac{\partial v}{\partial x}\frac{dx}{ds}\right)\sin\alpha$$

or

$$\epsilon_{x'} = \epsilon_x\cos^2\alpha + \epsilon_y\sin^2\alpha + \gamma_{xy}\sin\alpha\cos\alpha$$

since

$$\frac{dy}{ds} = \sin\alpha \qquad \text{and} \qquad \frac{dx}{ds} = \cos\alpha$$

In terms of 2α, the normal strain $\epsilon_{x'}$ becomes

$$\epsilon_{x'} = \frac{\epsilon_x + \epsilon_y}{2} + \frac{\epsilon_x - \epsilon_y}{2}\cos 2\alpha + \frac{\gamma_{xy}}{2}\sin 2\alpha \tag{2.3}$$

The normal strain in the y' direction is found by substituting $(\alpha + \pi/2)$ for α in Eq. (2.3), which gives

$$\epsilon_{y'} = \frac{\epsilon_x + \epsilon_y}{2} - \frac{\epsilon_x - \epsilon_y}{2}\cos 2\alpha - \frac{\gamma_{xy}}{2}\sin 2\alpha \tag{2.4}$$

In order to solve for the shear strain $\gamma_{x'y'}$, we first determine the angular displacement ψ of line PP_0. Referring again to Fig. 2.4(b), we observe that

$$\tan\psi \cong \psi = \frac{QS}{ds} = \frac{RP'\cos\alpha - QR\sin\alpha - (P'S)\psi}{ds}$$

However, $(P'S)\psi = \epsilon_x' \, ds\,\psi$, and since we are dealing with small strain components, this product is negligible with respect to the other terms. Therefore,

$$\psi = \left(\frac{\partial v}{\partial y}\frac{dy}{ds} + \frac{\partial v}{\partial x}\frac{dx}{ds}\right)\cos\alpha - \left(\frac{\partial u}{\partial x}\frac{dx}{ds} + \frac{\partial u}{\partial y}\frac{dy}{ds}\right)\sin\alpha$$

or

$$\psi = -(\epsilon_x - \epsilon_y)\sin\alpha\cos\alpha + \frac{\partial v}{\partial x}\cos^2\alpha - \frac{\partial u}{\partial y}\sin^2\alpha \tag{2.5}$$

This expression defines the angular displacement of x'. The angular displacement of y' may now be found, since this quantity is equal to ψ evaluated at $(\alpha + \pi/2)$, or $\psi|_{\alpha+(\pi/2)}$. Replacing α by $(\alpha + \pi/2)$ in Eq. (2.5), then, we have the result

$$\psi|_{\alpha+(\pi/2)} = -(\epsilon_y - \epsilon_x)\cos\alpha\sin\alpha + \frac{\partial v}{\partial x}\sin^2\alpha - \frac{\partial u}{\partial y}\cos^2\alpha$$

and the shear strain $\gamma_{x'y'}$ is given by

$$\gamma_{x'y'} = \psi - \psi|_{\alpha+(\pi/2)}$$

Thus we have

$$\gamma_{x'y'} = 2(\epsilon_y - \epsilon_x)\sin\alpha\cos\alpha + \left(\frac{\partial v}{\partial x} + \frac{\partial u}{\partial y}\right)(\cos^2\alpha - \sin^2\alpha)$$

In terms of 2α, and substituting $\gamma_{xy} = \partial v/\partial x + \partial u/\partial y$, we obtain

$$\gamma_{x'y'} = (\epsilon_y - \epsilon_x)\sin 2\alpha + \gamma_{xy}\cos 2\alpha \tag{2.6}$$

Equations (2.3), (2.4), and (2.6) are called the two-dimensional transformation of strain equations. Again, as in the case of transformation of stress, we have seven variables and three equations relating them. Thus if four of these quantities are given, the other three can be found from Eqs. (2.3), (2.4), and (2.6).[5] Therefore the state of strain is uniquely determined (the strain components can be found for any directions) if the strain components on two planes are given.

We also note the similarity between Eqs. (2.3), (2.4), and (2.6) and the corresponding stress relations, Eqs. (1.9) and (1.10). By simply replacing σ by ϵ and τ by $\gamma/2$, the equations for stress are converted to the strain relations. These substitutions can be made (see Problem 6) in all the analogous relations.[6] For example, the directions of principal strain (directions for which γ vanishes) are given by

$$\tan 2\alpha = \frac{\gamma_{xy}}{\epsilon_x - \epsilon_y} \tag{2.7}$$

and the magnitudes of the principal strains are

$$\left.\begin{matrix}\epsilon_1\\ \epsilon_2\end{matrix}\right\} = \frac{\epsilon_x + \epsilon_y}{2} \pm \tfrac{1}{2}\sqrt{(\epsilon_x - \epsilon_y)^2 + \gamma_{xy}^2} \tag{2.8}$$

[5] An exception, analogous to that pointed out in Problem 1-7, exists in the transformation of strain equations.

[6] These similarities do not occur simply by coincidence. As will be shown in Chapter 8, all "second-order tensors" transform according to the same law.

In fact, due to these mathematical similarities, we see that the quantities ϵ and $\gamma/2$ for any axes at a point are related by a Mohr circle.

In the three-dimensional case, the transformation of strain equations may also be deduced from the corresponding stress relations by replacing σ and τ by ϵ and $\gamma/2$, respectively. Thus, we have from Eq. (1.25), for example,

$$\epsilon_{x'} = \epsilon_x a_{11}^2 + \epsilon_y a_{21}^2 + \epsilon_z a_{31}^2 + \gamma_{xy} a_{11} a_{21} + \gamma_{yz} a_{21} a_{31} + \gamma_{zx} a_{31} a_{11}$$

2.5. General Displacements

If we specify the displacement functions u, v, and w (as functions of x, y, and z) of a body, the strain and the geometry of any infinitesimal element in its deformed state can be completely described, i.e., all of the dimensions of Fig. 2.2, including the angles θ and λ, may be found from equations developed in Section 2.2. If the distribution of the strain components is given, however, we cannot completely determine the displacement functions u, v, and w. In integrating the strain-displacement relations (2.1) to obtain the displacements, there are certain constants of integration, which will be shown to be equivalent to rigid body translations and rotations. The relation between these translations and rotations and the displacement functions u, v, and w will now be given.

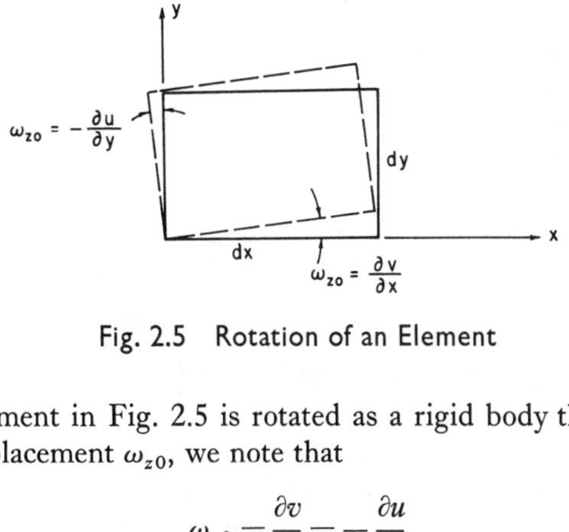

Fig. 2.5 Rotation of an Element

If the element in Fig. 2.5 is rotated as a rigid body through a *small* angular displacement ω_{z0}, we note that

$$\omega_{z0} = \frac{\partial v}{\partial x} = -\frac{\partial u}{\partial y}$$

During this rigid body movement, of course, no strain occurs. If both rigid body displacements and deformation (strain) occur, we define

$$\omega_z = \frac{1}{2}\left(\frac{\partial v}{\partial x} - \frac{\partial u}{\partial y}\right) \tag{2.9}$$

The quantity ω_z is seen to represent the average of the angular displacement of dx and the angular displacement of dy and is called the *rotation*. Referring again to Fig. 2.2, if the x component of displacement at A is u, then at C we have the displacement $u + du$, where

$$du = \frac{\partial u}{\partial x}\,dx + \frac{\partial u}{\partial y}\,dy \tag{2.10}$$

Equation (2.10) may be rewritten in the form

$$du = \frac{\partial u}{\partial x}\,dx + \frac{1}{2}\left(\frac{\partial u}{\partial y} + \frac{\partial v}{\partial x}\right) dy + \frac{1}{2}\left(\frac{\partial u}{\partial y} - \frac{\partial v}{\partial x}\right) dy$$

or

$$du = \epsilon_x\,dx + \tfrac{1}{2}\gamma_{xy}\,dy - \omega_z\,dy \tag{2.11}$$

The first two terms of Eq. (2.11) represent the x component of displacement of point C relative to A due to strain components ϵ_x and γ_{xy}, i.e., due to the state of strain (pure deformation) shown in Fig. 2.6. The last

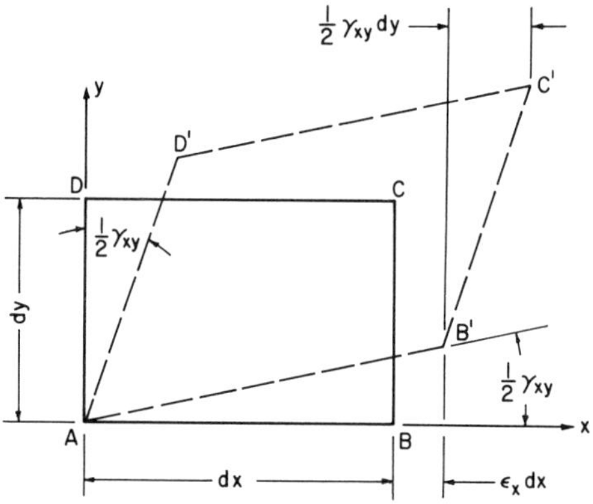

Fig. 2.6 Pure Deformation of an Element

term of Eq. (2.11) is seen to represent a displacement due to a rotation, so that if we superpose the displacement due to pure deformation and the displacement due to the rotation, the element will be oriented in its final position. It can be shown (see Problem 2–13) that ω_z represents the angular displacement of the principal axes. Similarly the y component of the displacement of point C relative to A can be written as

$$dv = \epsilon_y\,dy + \tfrac{1}{2}\gamma_{xy}\,dx + \omega_z\,dx \tag{2.12}$$

Equations (2.11) and (2.12), which give the total differentials of u and

v, may be integrated if the given strain components satisfy the compatibility equations (2.2). The exact procedure for this integration will be shown in Chapter 10. We simply point out in this section that in integrating these two equations the solution contains certain arbitrary functions of the form

$$\begin{aligned} u^* &= u_0 - \omega_{z0} y \\ v^* &= v_0 + \omega_{z0} x \end{aligned} \tag{2.13}$$

The functions u^* and v^* are arbitrary in the sense that they produce no strain, as can be verified by substituting them into Eqs. (2.1); hence they can be added to any displacement field without changing the strain distribution. These are the formulas for the displacement of a rigid body by a translation (u_0, v_0) and a small rotation ω_{z0}.

If the three strain components ϵ_x, ϵ_y, and γ_{xy} are given, the corresponding displacement (u, v) is arbitrary to the extent of an additional rigid body displacement expressed by Eq. (2.13). There are three arbitrary constants in Eqs. (2.13), u_0, v_0, and ω_{z0}, which means that if the displacement components (u_0, v_0) at one point in the body and the rotation ω_{z0} (or equivalent information) are specified, then the expression for the displacement with given strains will be unique.

One should observe the distinction between the function ω_z in Eqs. (2.11) and (2.12) and the quantity ω_{z0} in Eq. (2.13). The latter refers to the motion of a body as a whole and, as such, is a rigid body rotation. The function ω_z, however, is a space dependent rotation, i.e., it defines the rotation of an infinitesimal element in the body as a function of the position of the element.

In the three-dimensional case, we have

$$\begin{aligned} du &= \epsilon_x\,dx + \tfrac{1}{2}\gamma_{xy}\,dy + \tfrac{1}{2}\gamma_{xz}\,dz - \omega_z\,dy + \omega_y\,dz \\ dv &= \epsilon_y\,dy + \tfrac{1}{2}\gamma_{xy}\,dx + \tfrac{1}{2}\gamma_{yz}\,dz - \omega_x\,dz + \omega_z\,dx \\ dw &= \epsilon_z\,dz + \tfrac{1}{2}\gamma_{xz}\,dx + \tfrac{1}{2}\gamma_{yz}\,dy - \omega_y\,dx + \omega_x\,dy \end{aligned} \tag{2.14}$$

where

$$\begin{aligned} \omega_x &= \frac{1}{2}\left(\frac{\partial w}{\partial y} - \frac{\partial v}{\partial z}\right) \\ \omega_y &= \frac{1}{2}\left(\frac{\partial u}{\partial z} - \frac{\partial w}{\partial x}\right) \\ \omega_z &= \frac{1}{2}\left(\frac{\partial v}{\partial x} - \frac{\partial u}{\partial y}\right) \end{aligned}$$

are the small angles of rotation about axes parallel to x, y, and z and are called the *components of rotation*. Each of these is positive if counterclockwise when viewed from the positive extension of the axis of rotation toward the origin.

In integrating Eqs. (2.14), the arbitrary functions are of the form

$$\begin{aligned} u^* &= u_0 - \omega_{z0}\, y + \omega_{y0}\, z \\ v^* &= v_0 - \omega_{x0}\, z + \omega_{z0}\, x \\ w^* &= w_0 - \omega_{y0}\, x + \omega_{x0}\, y \end{aligned} \tag{2.15}$$

It should be noted that the rigid body displacement represented by (u_0, v_0, w_0) and $(\omega_{x0}, \omega_{y0}, \omega_{z0})$ does not affect the strain components, as evidenced by substituting Eqs. (2.15) into the strain-displacement relations (2.1). Therefore, in treating problems in elasticity, we very often neglect the rigid body displacement.

2.6. Principle of Superposition

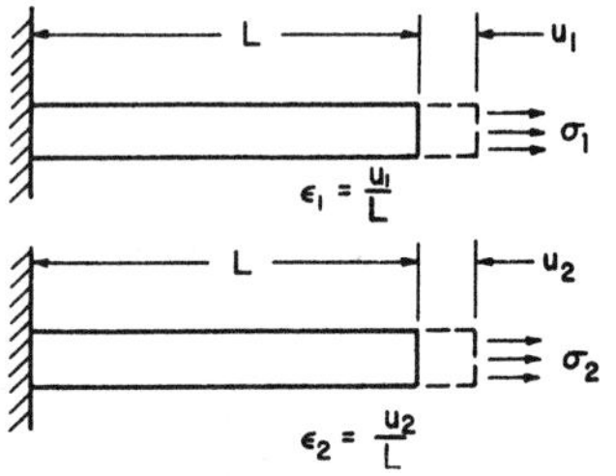

Fig. 2.7 Superposition of Displacements

The assumption of infinitesimal deformations leads to one of the basic fundamentals of the theory of linear elasticity, called the *principle of superposition*. This principle states that two strain fields may be combined by direct superposition, and the order of application of the two has no effect on the final configuration of the body in question. As an example, refer to Fig. 2.7. The bar has an elongation u_1 due to a uniform strain ϵ_1, another elongation u_2 due to ϵ_2, if each strain is applied separately. Then, if ϵ_1 is applied first, and ϵ_2 is applied in addition to ϵ_1, the final displacement of the end is given by

$$\begin{aligned} u &= u_1 + \epsilon_2(L + u_1) \\ &= u_1 + \epsilon_2(L + \epsilon_1 L) \\ &= u_1 + \epsilon_2 L + \epsilon_1 \epsilon_2 L \end{aligned}$$

or

$$u = u_1 + u_2$$

The final result is achieved since the product $\epsilon_1 \epsilon_2$ is small compared to either ϵ_1 or ϵ_2. If the order is reversed, or if ϵ_1 and ϵ_2 are applied simultaneously, the same final result will be obtained. The principle of superposition is applicable to all linear equations. A more general discussion of this principle, as applied to all of the basic elasticity equations, will be presented in Chapter 4.

2.7. Summary

Continuing to explore the solution of problems in elasticity, we have defined the components of strain and displacement. The interrelations between these quantities (2.1) provide us with six more equations to describe the behavior of a deformed body. It was also found that the strain components must satisfy the compatibility equations (2.2) in order that the displacement components exist and be single-valued, continuous functions.

Given the state of strain at a point, the strain components on arbitrary planes at the point may be determined from the transformation of strain equations. The substitutions

$$\sigma \rightarrow \epsilon \qquad \tau \rightarrow \tfrac{1}{2}\gamma$$

converts any stress transformation equation to the corresponding relation for strain. It is possible to isolate the displacement due to pure deformation and rotation at any point in the linear theory, as demonstrated in Section 2.5. The effect of rigid body displacement can also be extracted from the displacement components.

The fact that the order of application of several strain fields is immaterial in the linear theory was demonstrated in Section 2.6 for a special case. This statement is known as the principle of superposition.

PROBLEMS

2-1 Derive the last five of Eqs. (2.2).

2-2 Explain the use of Mohr's circle of strain in determining the strain components for any direction in a two-dimensional element. Determine the convention for $\gamma/2$ to determine whether a point on the Mohr circle is to be placed above or below the horizontal axis if the angle α, which defines a certain direction in space, and the double angle 2α, which locates the corresponding point on the Mohr circle, are measured in the same direction.

2-3 Derive Eqs. (2.14).

2-4 Show that the displacement components u, v, and w of Eqs. (2.15) represent rigid body components of displacement if ω_{x0}, ω_{y0}, and ω_{z0} are small angles.

2-5 Derive the equations which define the principal directions of strain and the values of the principal strains at a point (two-dimensional).

2-6 Derive the equations which define the directions and magnitude of maximum shear strain at a point (two-dimensional). Check the relations by replacing σ by ϵ and τ by $\gamma/2$ in the corresponding stress equations.

2-7 The following displacement field is applied to a certain body

$$u = k(2x + y^2) \qquad v = k(x^2 - 3y^2) \qquad w = 0$$

where $k = 10^{-4}$.

(a) Show the distorted configuration of a two-dimensional element with sides dx and dy and its lower left corner (point A) initially at the point (2, 1, 0), i.e., determine the new length and angular position of each side.

(b) Determine the coordinates of point A after the displacement field is applied.

(c) Find ω_z at this point.

2-8 Given the following system of strains,

$$\begin{aligned} \epsilon_x &= 5 + x^2 + y^2 + x^4 + y^4 \\ \epsilon_y &= 6 + 3x^2 + 3y^2 + x^4 + y^4 \\ \gamma_{xy} &= 10 + 4xy(x^2 + y^2 + 2) \\ \epsilon_z &= \gamma_{xz} = \gamma_{yz} = 0 \end{aligned}$$

determine if the system of strains is possible. If this strain distribution is possible, find the displacement components in terms of x and y, assuming that the displacement and rotation at the origin are zero.

2-9 Determine ϵ_n, ϵ_t, and γ_{tn} if $\gamma_{xy} = 0.002828$ and $\epsilon_x = \epsilon_y = 0$, for the element shown.

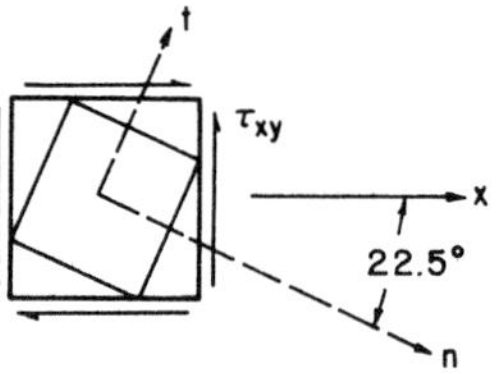

Problem 2–9

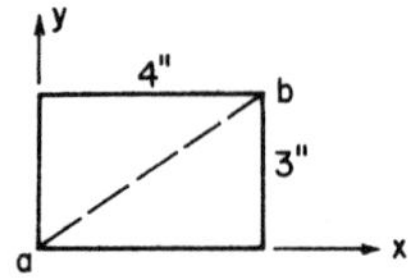

Problem 2–10

2-10 A thin rectangular plate 3″ by 4″ is acted upon by a stress distribution which results in the uniform strains

$$\epsilon_x = 0.0025, \quad \epsilon_y = 0.0050, \quad \epsilon_z = 0, \quad \gamma_{xy} = 0.001875, \quad \gamma_{xz} = \gamma_{yz} = 0$$

as shown in the figure. Determine the change in length of diagonal *ab*.

2-11 The compatibility equations in terms of strain may be given by three independent fourth-order equations as follows:

$$2\frac{\partial^4 \epsilon_x}{\partial y^2\,\partial z^2} = \frac{\partial^3}{\partial x \partial y \partial z}\left(-\frac{\partial \gamma_{yz}}{\partial x} + \frac{\partial \gamma_{xz}}{\partial y} + \frac{\partial \gamma_{xy}}{\partial z}\right)$$

$$2\frac{\partial^4 \epsilon_y}{\partial z^2\,\partial x^2} = \frac{\partial^3}{\partial x \partial y \partial z}\left(\frac{\partial \gamma_{yz}}{\partial x} - \frac{\partial \gamma_{xz}}{\partial y} + \frac{\partial \gamma_{xy}}{\partial z}\right)$$

$$2\frac{\partial^4 \epsilon_z}{\partial x^2\,\partial y^2} = \frac{\partial^3}{\partial x \partial y \partial z}\left(\frac{\partial \gamma_{yz}}{\partial x} + \frac{\partial \gamma_{xz}}{\partial y} - \frac{\partial \gamma_{xy}}{\partial z}\right)$$

What restrictions must be placed on the arbitrary functions which result from the integration of these equations in order to obtain the six second-order compatibility equations?

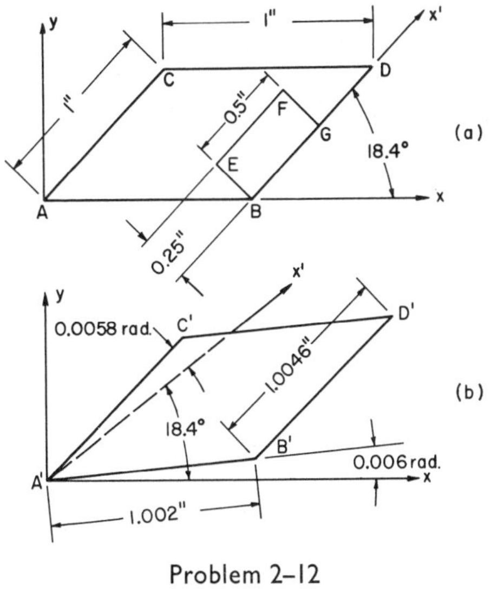

Problem 2–12

2-12 The parallelogram *ABCD* and rectangle *BEFG* are scribed on the surface of the flat plate shown in the figure. When loads are applied to the plate, the parallelogram deforms as shown in (b). If the strain condition is uniform, show the deformed configuration of rectangle *BEFG*, evaluating the length and angular position of each line. Also compute γ_{xy}. Hint: First determine ϵ_y from the geometry of $A'B'C'D'$.

2-13 By considering Eqs. (2.5) and (2.7), show that the quantity ω_z as defined by Eq. (2.9) gives the angular displacement of the principal axes of strain in the two-dimensional case.

2-14 If the strain distribution in a body is specified, the displacement components contain the arbitrary functions given by Eqs. (2.15). Demonstrate this fact by integrating Eqs. (2.1) with all strain components set equal to zero.

3

Stress-Strain Relations

3.1. Introduction

We recall that two sets of field equations were developed in the two preceding chapters, the equilibrium equations (1.21), which were based on statics and continuity of stress, and the strain-displacement relations (2.1), which were based on continuity of displacement and infinitesimal deformations. These two sets of equations were developed independently, one set involving stress components only, the other, strain and displacement. It is logical then, to ask ourselves *how* the strain components and the stress components are interrelated. This topic will be treated in this chapter.

Under the assumption of infinitesimal deformations, Eqs. (1.21) and (2.1) are valid for any solid body. The relationship between the stress components and the strain components, however, depends on the properties of the particular solid under consideration. In this chapter we shall consider a specific class of solids, those which are elastic. An elastic body is defined as one which regains its original dimensions after the forces acting on it are removed. We shall further restrict our attention to materials possessing linear stress-strain relations. The range of stress and strain for which the behavior is linearly elastic will be called the *elastic range*. In subsequent discussions, we shall use the term "elastic" to stand for "linearly elastic." Most solid materials exhibit this behavior throughout a restricted range of stress (and strain). Although some materials do not possess a well-defined elastic range, the assumption of this behavior gives results which are within engineering accuracy.

The equilibrium equations (1.21) and the strain-displacement relations, Eqs. (2.1), together constitute nine equations. We have introduced, however, 15 unknown quantities: six stress components, six strain components, and three components of displacement. The six additional equations relating the stress and strain components shall be developed in this chapter and will bring the total number of equations to 15, which along with the prescribed boundary conditions, represent the conditions which must be satisfied by an elastic body in equilibrium.

We shall also limit our discussion in this chapter to the behavior of isotropic homogeneous material. A material is isotropic if the properties relating to its behavior under stress are the same in any direction at a point. A metal that is rolled or drawn is not isotropic, since these processes produce a preferential orientation of the crystalline structure. For materials of cellular structure, such as wood, the properties in the longitudinal direction differ from those in other directions, and these materials are anisotropic (not isotropic). The condition of homogeneity is expressed by the fact that the material properties are independent of position in the medium. If a medium is homogeneous, it is necessarily continuous.

It should be pointed out that engineering materials are seldom truly isotropic or homogeneous, because the crystalline or molecular structure of material is not continuous and may be randomly oriented. However, the assumptions of isotropy and homogeneity usually lead to results consistent with experiments (subject to exceptions as noted in the previous paragraph). This is because in experimental measurements the stresses and strains are averaged over dimensions, which although small, are still much larger than the dimensions of crystals and molecules. From a macroscopic point of view, therefore, the behavior of the real material can be treated (at least to a first approximation) as being isotropic and homogeneous.

The equations relating stress, strain, stress-rate (increase of stress per unit time), and strain-rate are called the *constitutive equations*, since they depend upon the material properties of the medium under discussion. In the case of elastic solids, the constitutive equations take the form of generalized Hooke's law, which involves only stress and strain and is independent of the stress-rate or strain-rate.

3.2. Generalized Hooke's Law

As pointed out in the introduction, most engineering materials exhibit a well-defined elastic range under a condition of uniaxial normal stress. If the normal stress acts in the x direction, we have the relation known as Hooke's law:

$$\sigma_x = E\epsilon_x \tag{3.1}$$

in the elastic range. The constant E is called the *modulus of elasticity*, or

Young's modulus.[1] This constant may be determined experimentally for a given material by performing a uniaxial tension test and recording simultaneous values of stress and strain in the specimen. The modulus of elasticity is therefore equal to the slope of the straight-line portion of the curve drawn by plotting stress as the ordinate and strain as the abscissa.

It is reasonable now to ask the question: What is the stress-strain relationship under a three-dimensional state of stress? In other words, we wish to relate the stress components σ_x, σ_y, σ_z, τ_{xy}, τ_{yz}, and τ_{zx} at a point to the strain components ϵ_x, ϵ_y, ϵ_z, γ_{xy}, γ_{yz}, and γ_{zx} at the same point.

There are two approaches which can be used for the determination of the stress-strain relations, the "mathematical" approach and the "semiempirical" approach, the latter being guided by experimental evidence. In the mathematical approach, the relation between the stress and strain components is expressed by the linear equations

$$\begin{aligned}
\sigma_x &= c_{11}\epsilon_x + c_{12}\epsilon_y + c_{13}\epsilon_z + c_{14}\gamma_{xy} + c_{15}\gamma_{yz} + c_{16}\gamma_{zx} \\
\sigma_y &= c_{21}\epsilon_x + c_{22}\epsilon_y + c_{23}\epsilon_z + c_{24}\gamma_{xy} + c_{25}\gamma_{yz} + c_{26}\gamma_{zx} \\
\sigma_z &= c_{31}\epsilon_x + c_{32}\epsilon_y + c_{33}\epsilon_z + c_{34}\gamma_{xy} + c_{35}\gamma_{yz} + c_{36}\gamma_{zx} \\
\tau_{xy} &= c_{41}\epsilon_x + c_{42}\epsilon_y + c_{43}\epsilon_z + c_{44}\gamma_{xy} + c_{45}\gamma_{yz} + c_{46}\gamma_{zx} \\
\tau_{yz} &= c_{51}\epsilon_x + c_{52}\epsilon_y + c_{53}\epsilon_z + c_{54}\gamma_{xy} + c_{55}\gamma_{yz} + c_{56}\gamma_{zx} \\
\tau_{zx} &= c_{61}\epsilon_x + c_{62}\epsilon_y + c_{63}\epsilon_z + c_{64}\gamma_{xy} + c_{65}\gamma_{yz} + c_{66}\gamma_{zx}
\end{aligned} \tag{3.2}$$

Each strain component may also be expressed as a linear combination of the stress components, in which case another set of 36 constants will be present in the equations. The set of equations (3.2) is a logical generalization of Eq. (3.1); that is, we assume that each stress component is a linear function of all the strain components. The coefficients $c_{11} \ldots c_{66}$ represent material properties, functions only of the material for which the equations are written. If we assume an isotropic material, these constants must be the same for an orthogonal coordinate system of any orientation at the point in question. For example, isotropy demands that c_{11}, which measures the stress component σ_x caused by ϵ_x, must also relate $\sigma_{x'}$ and $\epsilon_{x'}$.

It should be noted that the assumption of a linear relationship in Eqs. (3.2) is strongly influenced by the fact that we do not wish to introduce

[1] The modulus of elasticity of mild steel is approximately 30,000,000 psi, and its proportional limit (highest stress in the elastic range) is approximately 30,000 psi. Thus, the maximum *elastic* strain in mild steel is about 0.001, under a condition of uniaxial stress. This gives one an idea as to the magnitude of the strains with which we will be dealing.

undue mathematical complexity in our solutions. Fortunately, this assumption usually predicts the behavior of real elastic materials with good accuracy. It will be shown in Chapter 10 that, for an isotropic material, many of the 36 coefficients are zero and that the remaining ones may be represented by two independent coefficients. For a homogeneous material, each of these coefficients has the same value at all points in the body.

In this chapter, we shall use the "semiempirical" approach to derive the stress-strain relations. In this approach, we shall make certain assumptions which are based on experimental evidence for most engineering materials under small strains. These assumptions are that a normal stress σ_x does not produce shear strain on the x, y, or z planes; a shear stress τ_{xy} does not cause normal strain on the x, y, or z planes; and furthermore, a shear stress component τ_{xy} causes only one shear strain component, γ_{xy}. If the strain components are small quantities (assumption of infinitesimal deformations), the principle of superposition may be applied to determine the strain components produced by more than one stress component.

Consider an elemental parallelepiped under a uniaxial stress σ_x shown in Fig. 3.1. The strain component ϵ_x, from Eq. (3.1), has the value

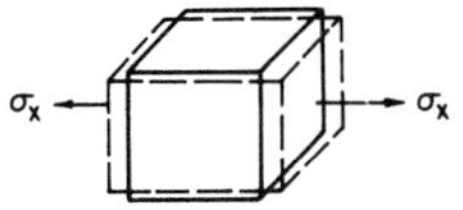

Fig. 3.1 Element Under Uniaxial Stress σ_x

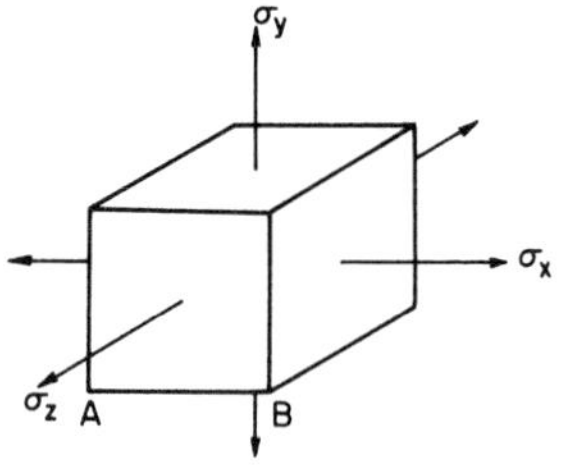

Fig. 3.2 Element Under Triaxial State of Stress

$\epsilon_x = \sigma_x/E$. Accompanying this elongation in the x direction, there will be contractions in the y and z directions. These are given by

$$\epsilon_y = \epsilon_z = -\nu(\sigma_x/E)$$

For most materials, ν is a constant within the elastic range and is called *Poisson's ratio*. We now consider an element subjected to the triaxial state of stress shown in Fig. 3.2, where the initial length of AB is taken as unity. The strain component ϵ_x is determined by assuming σ_x is applied first, thus changing the length of AB by the amount $(1/E)\sigma_x$, then σ_y is applied, which effects an additional change in length AB equal to

$$-(\nu/E)\sigma_y(1 + \sigma_x/E)$$

But since $(1/E)\sigma_x$ is an elastic strain, it is negligible with respect to unity, and is therefore dropped. When σ_z is applied then, again ignoring small quantities of higher order, length AB is changed by

$$-(\nu/E)\sigma_z$$

The total strain in the x direction therefore, is given by

$$\epsilon_x = \frac{1}{E}[\sigma_x - \nu(\sigma_y + \sigma_z)] \tag{3.3}$$

Similarly

$$\epsilon_y = \frac{1}{E}[\sigma_y - \nu(\sigma_x + \sigma_z)]$$

and

$$\epsilon_z = \frac{1}{E}[\sigma_z - \nu(\sigma_x + \sigma_y)]$$

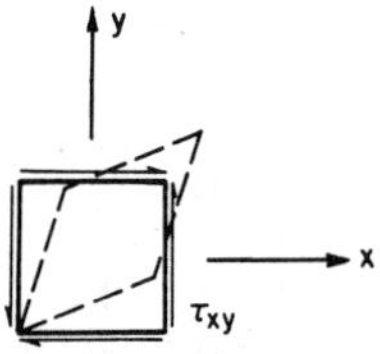

Fig. 3.3 Element Under Pure Shear

The elastic stress-strain relation under a two-dimensional state of pure shear (Fig. 3.3) is found experimentally to take the form

$$\gamma_{xy} = \frac{1}{G}\tau_{xy} \tag{3.4}$$

Similarly,

$$\gamma_{yz} = \frac{1}{G}\tau_{yz}$$

$$\gamma_{xz} = \frac{1}{G}\tau_{xz}$$

where the elastic constant G is called the modulus of elasticity in shear, or the modulus of rigidity. According to the assumptions made in this semiempirical approach, Eqs. (3.3) and (3.4) are valid for any three-dimensional state of stress. Thus we have introduced three elastic constants E, ν, and G, which govern the stress-strain relations for an isotropic, elastic material. We shall now proceed to demonstrate that only two of these are independent.

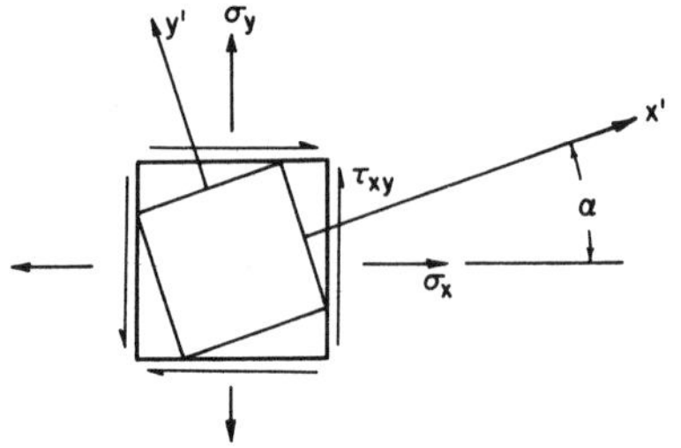

Fig. 3.4 State of Plane Stress on an Element

Consider a state of plane stress (Fig. 3.4) defined by

$$\sigma_z = \tau_{xz} = \tau_{yz} = 0$$

The strain components on the x and y planes are given by

$$\epsilon_x = \frac{1}{E}[\sigma_x - \nu\sigma_y] \tag{3.5}$$

$$\epsilon_y = \frac{1}{E}[\sigma_y - \nu\sigma_x] \tag{3.6}$$

$$\gamma_{xy} = \frac{1}{G}\tau_{xy} \tag{3.7}$$

The stress components referred to the $x'y'$ coordinate system, from Eqs. (1.9), are given by

$$\sigma_{x'} = \frac{\sigma_x + \sigma_y}{2} + \frac{\sigma_x - \sigma_y}{2}\cos 2\alpha + \tau_{xy}\sin 2\alpha \tag{3.8}$$

$$\sigma_{y'} = \frac{\sigma_x + \sigma_y}{2} - \frac{\sigma_x - \sigma_y}{2}\cos 2\alpha - \tau_{xy}\sin 2\alpha \tag{3.9}$$

and the transformation of strain equations (2.3) and (2.4) are

$$\epsilon_{x'} = \frac{\epsilon_x + \epsilon_y}{2} + \frac{\epsilon_x - \epsilon_y}{2}\cos 2\alpha + \frac{\gamma_{xy}}{2}\sin 2\alpha \tag{3.10}$$

$$\epsilon_{y'} = \frac{\epsilon_x + \epsilon_y}{2} - \frac{\epsilon_x - \epsilon_y}{2}\cos 2\alpha - \frac{\gamma_{xy}}{2}\sin 2\alpha \tag{3.11}$$

Now, due to isotropy, the elastic constants referred to $x'y'$ must be the same as those referred to xy. Thus the stress-strain relations for the $x'y'$ axes are given by

$$\epsilon_{x'} = \frac{1}{E}[\sigma_{x'} - \nu\sigma_{y'}] \tag{3.12}$$

$$\epsilon_{y'} = \frac{1}{E}[\sigma_{y'} - \nu\sigma_{x'}] \tag{3.13}$$

Substituting Eqs. (3.8), (3.9), and (3.10) into Eq. (3.12) and subtracting the resulting equation from the equation obtained by substituting Eqs. (3.8), (3.9), and (3.11) into Eq. (3.13), we have

$$(\epsilon_x - \epsilon_y)\cos 2\alpha + \gamma_{xy}\sin 2\alpha = \frac{1}{E}[(\sigma_x - \sigma_y)(1+\nu)\cos 2\alpha + 2\tau_{xy}(1+\nu)\sin 2\alpha] \tag{3.14}$$

Combining Eqs. (3.5) and (3.6) yields

$$(\epsilon_x - \epsilon_y)\cos 2\alpha = \frac{1}{E}[(\sigma_x - \sigma_y)(1+\nu)\cos 2\alpha] \tag{3.15}$$

Equation (3.15) is then subtracted from Eq. (3.14), which gives

$$\gamma_{xy} = \frac{2(1+\nu)}{E}\tau_{xy}$$

Comparing this equation with Eq. (3.7), we see that

$$G = \frac{E}{2(1+\nu)} \tag{3.16}$$

Thus, there are only two independent elastic constants for an isotropic material. For a general state of stress, then, the stress-strain relations, which are known as generalized Hooke's law, or simply as Hooke's law, consist of the equations

$$\begin{aligned}
\epsilon_x &= \frac{1}{E}[\sigma_x - \nu(\sigma_y + \sigma_z)] \\
\epsilon_y &= \frac{1}{E}[\sigma_y - \nu(\sigma_z + \sigma_x)] \\
\epsilon_z &= \frac{1}{E}[\sigma_z - \nu(\sigma_x + \sigma_y)] \\
\gamma_{xy} &= \frac{1}{G}\tau_{xy} \\
\gamma_{yz} &= \frac{1}{G}\tau_{yz} \\
\gamma_{zx} &= \frac{1}{G}\tau_{zx}
\end{aligned} \tag{3.17}$$

where $G = E/[2(1+\nu)]$. As defined in Chapter 2, the directions along which the shear strains vanish are called *principal directions of strain.* The normal strains along these directions are called the principal strains. If x, y, and z are chosen in the directions of principal stress, we have

$$\tau_{xy} = \tau_{yz} = \tau_{zx} = 0$$

so that, from Eq. (3.17),

$$\gamma_{xy} = \gamma_{yz} = \gamma_{zx} = 0$$

which indicates that x, y, and z are also the axes of principal strain. For isotropic, elastic materials, then, the principal axes of stress and principal axes of strain coincide.

Equations (3.17) may be solved for the stress components in terms of the strain components, with the result

$$\begin{aligned} \sigma_x &= 2G\epsilon_x + \lambda(\epsilon_x + \epsilon_y + \epsilon_z) \\ \sigma_y &= 2G\epsilon_y + \lambda(\epsilon_x + \epsilon_y + \epsilon_z) \\ \sigma_z &= 2G\epsilon_z + \lambda(\epsilon_x + \epsilon_y + \epsilon_z) \\ \tau_{xy} &= G\gamma_{xy} \\ \tau_{yz} &= G\gamma_{yz} \\ \tau_{zx} &= G\gamma_{zx} \end{aligned} \tag{3.18}$$

where the constants G (modulus of rigidity) and λ are called Lamé's constants and

$$\lambda = \frac{\nu E}{(1+\nu)(1-2\nu)} \tag{3.19}$$

3.3. Bulk Modulus of Elasticity

Another important elastic constant is called the bulk modulus of elasticity. The physical significance of this constant is seen by referring to a state of hydrostatic pressure, as shown in Fig. 3.5. In this case we have

$$\sigma_x = \sigma_y = \sigma_z = -p \qquad (p > 0)$$

$$\tau_{xy} = \tau_{yz} = \tau_{zx} = 0$$

and from Eqs. (3.17) we obtain the strain components

$$\epsilon_x = \epsilon_y = \epsilon_z = -\left(\frac{1-2\nu}{E}\right)p$$

$$\gamma_{xy} = \gamma_{yz} = \gamma_{zx} = 0$$

Now the dilatation (volumetric strain) ϵ, defined as the unit change in volume (change in volume divided by the original volume) is given by

$$\epsilon = \epsilon_x + \epsilon_y + \epsilon_z \tag{3.20}$$

which can be shown by finding the change in volume of a prism with initial dimensions dx, dy, and dz. After straining, the volume becomes

$$[dx(1 + \epsilon_x)][dy(1 + \epsilon_y)][dz(1 + \epsilon_z)] = (1 + \epsilon_x + \epsilon_y + \epsilon_z)dx\,dy\,dz$$

neglecting higher-order terms. Thus the dilatation is given by Eq. (3.20).

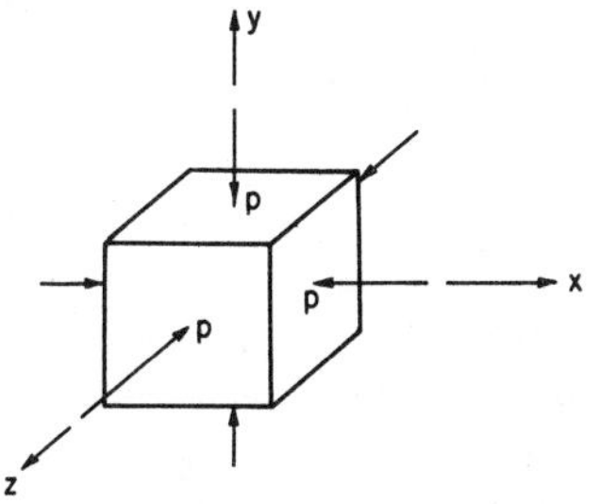

Fig. 3.5 Hydrostatic State of Stress

In the case of hydrostatic pressure, then, we see that

$$\begin{aligned}\epsilon &= -\frac{3}{E}(1 - 2\nu)p \\ &= -\frac{1}{K}p\end{aligned} \tag{3.21}$$

where $K = E/[3(1 - 2\nu)]$ is the bulk modulus of elasticity. We see that this quantity represents the negative of the ratio of the hydrostatic pressure to the resulting dilatation.

The quantity ϵ, as well as σ_m, which is defined by the relation

$$\sigma_m = \tfrac{1}{3}(\sigma_x + \sigma_y + \sigma_z) \tag{3.22}$$

is of particular interest in the study of plasticity. The fact that the dilatation under any state of stress is given by Eq. (3.20) is apparent, since shear strains cause no volume change. Therefore, adding the first three of Eqs. (3.17), and using Eq. (3.22), we have

$$\epsilon = \epsilon_x + \epsilon_y + \epsilon_z = \frac{1 - 2\nu}{E}(3\sigma_m)$$

so that

$$\epsilon = \frac{1}{K}\sigma_m \tag{3.23}$$

applies for any state of stress. The quantity σ_m is called the *hydrostatic component of stress*, or *spherical component of stress*. We observe that both ϵ and σ_m are invariant quantities with respect to any orthogonal transformation of axes.

We have thus defined five elastic constants in our discussion, but they are interrelated so that only two of them are independent. The constants E and G can be easily determined experimentally for a given material, and ν, K, and λ can be found from relations developed in this chapter, e.g., Eqs. (3.16) and (3.19).

3.4. Summary

In summary, the stress-strain relations (3.17), or equivalently, (3.18), complete the system of field equations necessary to formulate a problem in elasticity. We have developed 15 equations, (1.21), (2.1), and (3.17), which must be satisfied by the 15 stress, strain, and displacement variables. The remaining conditions which must be satisfied are expressed in terms of the external surface forces and/or the boundary displacements. These will be discussed in the next chapter.

The stress-strain relations (Hooke's law) for an isotropic, elastic medium can be written in terms of five material constants, E, ν, G, K, and λ, two of which are independent. As an example of this, refer to Problem 3-11.

PROBLEMS

3-1 One of the applications of generalized Hooke's law is found in the use of rosette gages in the field of experimental stress analysis. The state of strain at a point is determined experimentally by determining the rosette gage readings, which give the normal strain in three directions in a plane. Since rosette gages are applied to a free surface, the stress components σ_z, τ_{zx}, and τ_{zy} are zero, where the z axis is normal to the free surface. By using generalized Hooke's law, then, one can define the state of stress.

By using the rosette gage shown in the figure in one experiment, the following strains are recorded on the surface of a steel bar: $\epsilon_{0-1} = +10^{-4}$, $\epsilon_{0-2} = +4 \times 10^{-4}$, $\epsilon_{0-3} = +6 \times 10^{-4}$. Given the material properties of this steel, $E = 30 \times 10^6$ psi, $\nu = 0.25$, determine the principal stresses and maximum shear stress at point O. Give the directions of these stresses.

3-2 If a medium is initially unstrained and is then subjected to a constant positive temperature change, the normal strains are expressed by

$$\epsilon_x = \frac{1}{E}[\sigma_x - \nu(\sigma_y + \sigma_z)] + \alpha T$$
$$\cdots$$
$$\cdots$$

where α is the coefficient of linear expansion and T is the temperature rise. The temperature change does not affect the shear strain components.

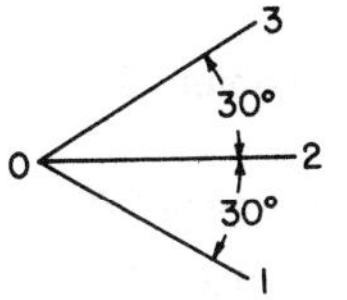

Problem 3–1

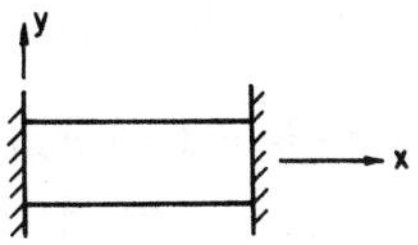

Problem 3–2

A bar restrained in the x direction only, and free to expand in the y and z directions as shown, is subjected to a uniform temperature rise T. Show that the only nonvanishing stress (the bar is in a state of uniform stress) and strain components are

$$\sigma_x = -E\alpha T$$
$$\epsilon_y = \epsilon_z = \alpha T(1 + \nu).$$

3-3 Determine the stress and strain components if the bar in the preceding problem is restrained in the x and y directions but is free to expand in the z direction.

3-4 Repeat Problem 3-3 for a bar restrained in all directions.

3-5 Verify Eqs. (3.18).

3-6 Show that

$$\sigma_x = \frac{E}{(1+\nu)(1-2\nu)}[(1-\nu)\epsilon_x + \nu(\epsilon_y + \epsilon_z)] - \frac{E}{1-2\nu}\alpha T, \text{ etc.}$$

$$\tau_{xy} = G\gamma_{xy}, \text{ etc.}$$

for the state of stress at a point with temperature increase T. Refer to the equations given in the first paragraph of Problem 3-2.

3-7 A square Duralumin plate is loaded as shown, where

$$\sigma_x = \sigma_y = \tau_{xy} = 15{,}000 \text{ psi}.$$

If $E = 10^7$ psi and $\nu = 0.3$, determine the change in length of the diagonal ab. Solve by two methods: 1) by using Mohr's circle of stress, find the stresses on planes normal and parallel to ab, and then apply generalized Hooke's law, and 2) by finding the strain components on the given planes and using the transformation of strain equations or Mohr's circle of strain.

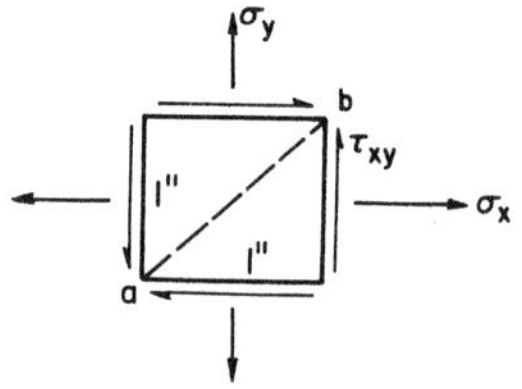

Problem 3–7

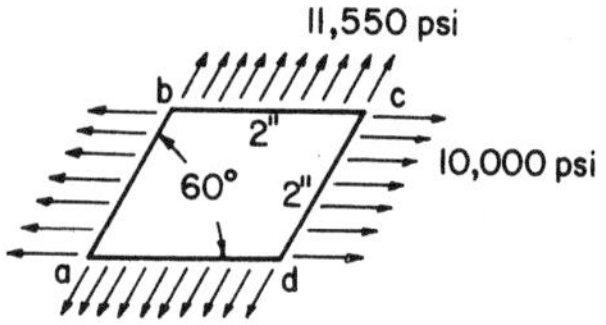

Problem 3–8

3-8 A thin plate is under a uniform state of stress as shown (see Problem 1-4).

$$E = 3 \times 10^7 \text{ psi}, \qquad \nu = 0.3.$$

Find: (a) the change in length of *dc*.
(b) the maximum shear strain between two orthogonal directions in the plane of the plate.
(c) the greatest of all shear strains (three-dimensional).

3-9 A thin plate is under a uniform state of stress, with $p = 14{,}140$ psi, $E = 30 \times 10^6$ psi, and $\nu = 0.3$, as shown.

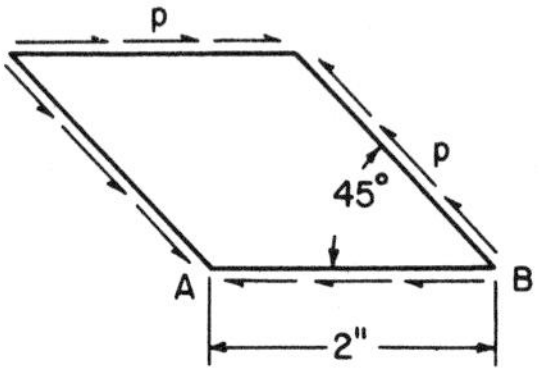

Problem 3–9

(a) Find the change in length of *AB*.
(b) Find the principal strains and their directions.

3-10 Determine the slope of the σ_x vs. ϵ_x curve in the elastic range if a material is tested under the following state of stress:

$$\sigma_x = 2\sigma_y = 3\sigma_z.$$

3-11 Prove the following relations among various elastic constants:

$$\nu = \frac{3K - E}{6K} \qquad \lambda = \frac{3K - 2G}{3} \qquad E = \frac{\lambda(1+\nu)(1-2\nu)}{\nu}$$

4

Formulation of Problems in Elasticity

4.1. Introduction

As we have pointed out previously, the solution of a given problem in elasticity consists of the determination of the stress, strain, and displacement components as functions of the space coordinates of the elastic body. The necessary equations which the stress, strain, and displacement components must satisfy have also been developed in Chapters 1, 2, and 3. These will be restated here for clarity, i.e., we have the equilibrium equations (1.21),

$$\frac{\partial \sigma_x}{\partial x} + \frac{\partial \tau_{xy}}{\partial y} + \frac{\partial \tau_{zx}}{\partial z} + F_x = 0 \qquad (x, y, z) \tag{4.1}$$

where the expression (x, y, z) at the end of the equation indicates that there are two more equations which are obtainable by cyclic permutation of x, y, and z. We also have the strain displacement relations (2.1),

$$\left.\begin{aligned} \epsilon_x &= \frac{\partial u}{\partial x} \\ \gamma_{xy} &= \frac{\partial u}{\partial y} + \frac{\partial v}{\partial x} \end{aligned}\right\} (x, y, z; u, v, w) \tag{4.2}$$

and the stress-strain relations (generalized Hooke's law), Eqs. (3.18),

$$\left.\begin{aligned} \sigma_x &= 2G\epsilon_x + \lambda\epsilon \\ \tau_{xy} &= G\gamma_{xy} \end{aligned}\right\} (x, y, z) \tag{4.3}$$

where $\epsilon = \epsilon_x + \epsilon_y + \epsilon_z$. These 15 equations, (4.1), (4.2), and (4.3), are the governing equations for the 15 variables (stress, strain, and displacement). They must be satisfied at all points inside an elastic body in equilibrium and are sometimes referred to as *field* equations.

The compatibility equations (2.2) are derived from the strain displacement equations (4.2) and, therefore, cannot be counted as governing equations. The compatibility equations will be satisfied automatically if the 15 governing equations are satisfied.

In this chapter, we shall enumerate various combinations of the governing equations which can be used to solve various types of boundary-value problems in elasticity. For example, it will be shown that the 15 equations can be reduced to six equations in terms of stress, or three expressions in terms of displacement. The two-dimensional state of plane strain will be considered first.

Any set of stress, strain, and displacement functions satisfying the governing equations will represent the solution to some problem in elasticity. Suppose, however, that we are to solve a specific problem, for example, a plate subjected to prescribed surface forces. In this case, the stresses must not only satisfy the field equations, but the stresses evaluated at any point on the boundary surface must be in equilibrium with the prescribed surface force at the same point. Thus, the solution of a specific problem in elasticity consists of the determination of the stress, strain, and displacement functions satisfying the field equations and the conditions prescribed on the boundary (surface forces or displacements). It will also be shown that the solution satisfying all of these conditions for a given problem is unique, i.e., it represents the *only* solution to this problem.

In this chapter we shall concentrate on the mathematical formulation of elasticity problems without detailed discussion of their solutions. Some of the methods of solution will be presented in Chapters 5 and 6.

4.2. Boundary Conditions

Consider the body shown in Fig. 4.1. The surface force[1] (stress) *distribution* is specified by the components T_x^μ, T_y^μ, and T_z^μ, where $\boldsymbol{\mu}$ is a unit vector normal to the surface and pointing outward. The coordinates of points on the boundary surface are denoted by x_0, y_0, z_0; these are related by the equation of the boundary surface. By isolating the infinitesimal tetrahedron $OABC$ shown in Fig. 4.2, where the inclined face is on the boundary, and expressing the equilibrium of forces, we

[1] When speaking of surface forces applied on a boundary, we shall adopt the common practice of interpreting the term "surface force" as meaning surface force per unit area, instead of force per se. It is also common to call the surface forces per unit area the surface tractions.

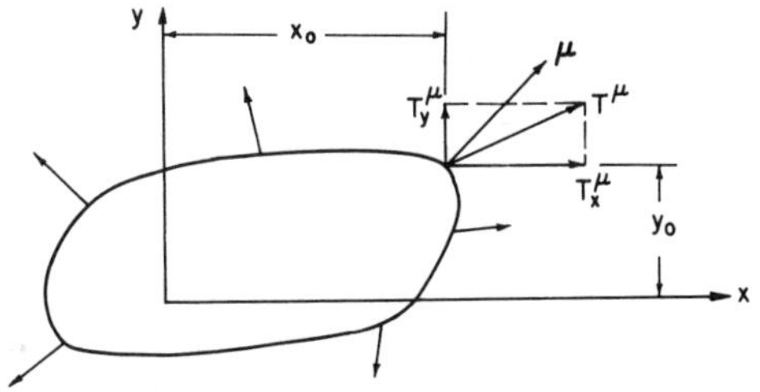

Fig. 4.1 Surface Forces

have the relations

$$T_x^{\mu} = \sigma_{x0}\mu_x + \tau_{xy0}\mu_y + \tau_{xz0}\mu_z$$

$$T_y^{\mu} = \tau_{xy0}\mu_x + \sigma_{y0}\mu_y + \tau_{yz0}\mu_z \tag{4.4}$$

$$T_z^{\mu} = \tau_{xz0}\mu_x + \tau_{yz0}\mu_y + \sigma_{z0}\mu_z$$

where σ_{x0}, τ_{xy0}, etc. are stress components evaluated at the boundary (x_0, y_0, z_0), and μ_x, μ_y, μ_z are the direction cosines of the unit outward normal vector $\boldsymbol{\mu}$ with respect to x, y, and z, respectively (or simply the

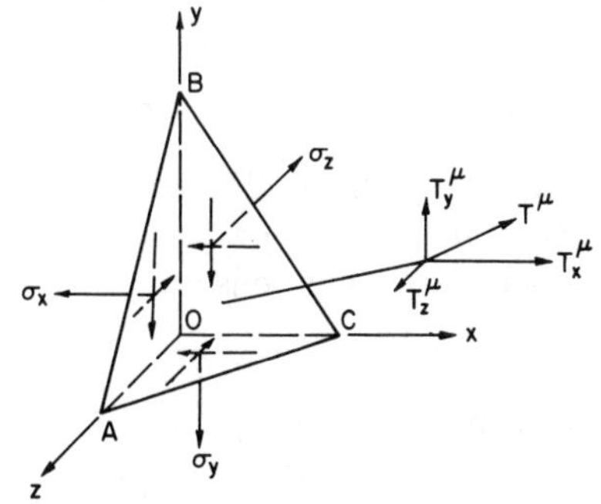

Fig. 4.2 Equilibrium on the Boundary

components of $\boldsymbol{\mu}$ along the three coordinate axes). On those parts of the boundary surface where the applied stress $(T_x^{\mu}, T_y^{\mu}, T_z^{\mu})$ is prescribed, we see that the stress components must satisfy these relations. Notice that Eqs. (4.4) are the same as Eqs. (1.24) except that, in this case, the inclined face is on the boundary.

On the other hand, the displacement components may be prescribed over the boundary surface or a portion of the boundary surface. In this case, we have the relations

$$
\begin{aligned}
u(x_0, y_0, z_0) &= u_b \\
v(x_0, y_0, z_0) &= v_b \\
w(x_0, y_0, z_0) &= w_b
\end{aligned}
\tag{4.5}
$$

where u_b, v_b, and w_b are the prescribed displacement components on the boundary, and $u(x_0, y_0, z_0) \ldots$ are the displacement functions in the body, evaluated at the boundary. The loading conditions of a problem in elasticity are thus expressed by prescribing either the stress components or the displacement components (or a combination of these) on each surface of the body. The body forces are assumed to be given in all cases.

We shall call T_x^μ, T_y^μ, T_z^μ the prescribed surface forces and Eqs. (4.4) the stress boundary conditions. Similarly, we refer to u_b, v_b, w_b as prescribed displacements and to Eqs. (4.5) as the displacement boundary conditions. It will be understood that the terms $\sigma_{x_0} \ldots \tau_{xz_0}$ are functions determined from the field equations and evaluated on the boundary (x is replaced by x_0, etc.).

A problem in which the stress is prescribed over the entire boundary is known as *a first boundary-value problem* in elasticity. An example of this is a rectangular prism with edges along the x, y, and z axes under uniaxial stress p in the x direction (longitudinal). On the ends of the bar the applied (or prescribed) surface forces are

$$T_x^\mu = \pm p \qquad T_y^\mu = T_z^\mu = 0$$

where the positive sign is for the positive x plane and the negative sign, the negative x plane. Since on the end surfaces of the bar, $\mu_x = \pm 1$, $\mu_y = 0$, $\mu_z = 0$, the boundary conditions are

$$\sigma_{x_0} = p \qquad \tau_{xy_0} = 0 \qquad \tau_{xz_0} = 0$$

that is, if we denote the positive x plane by $x = \ell$, we have $\sigma_x(\ell, y, z) = p$, $\tau_{xy}(\ell, y, z) = \tau_{xz}(\ell, y, z) = 0$ on this plane. On the lateral surfaces, the prescribed surface forces are

$$T_x^\mu = T_y^\mu = T_z^\mu = 0$$

and the direction cosines of the lateral surfaces with outward normals in the $\pm y$ directions are $\mu_x = 0$, $\mu_y = \pm 1$, $\mu_z = 0$. From Eqs. (4.4) we see that the boundary conditions on these planes are

$$\sigma_{y_0} = \tau_{xy_0} = \tau_{yz_0} = 0$$

Similarly, it can be shown that the boundary conditions on the z surfaces are $\sigma_{z_0} = \tau_{yz_0} = \tau_{xz_0} = 0$.

In the case in which the displacement components are prescribed over the entire boundary surface, the problem is classified as *a second boundary-value problem* in elasticity. An example of this type of problem is a block which is cemented inside a rigid box so that its surfaces cannot move and is then subjected to a given temperature distribution. If the edges of the block are parallel to the x, y, and z axes, the prescribed boundary displacements are

$$u_b = v_b = w_b = 0$$

on all surfaces.

A mixed boundary-value problem is one in which the displacement components are prescribed over part of the boundary and the stress components over the rest of the boundary. An example of this type of problem is illustrated in Problem 4–10.

It might now be advisable to again point out the essential difference between the handling of problems in elasticity as opposed to the methods of strength of materials. In elasticity solutions, we treat all problems essentially in the same manner. We seek a solution of the governing (field) equations which satisfies the given loading conditions (boundary conditions). In strength of materials, however, each problem is usually treated separately, and the solution is based on assumed stress or strain variations, the assumptions being exact only in special cases of loading and shape of the body under consideration.

4.3. Governing Equations in Plane Strain Problems

A state of plane strain is said to exist in a body where the displacement components in the body take the form

$$\begin{aligned} u &= u(x, y) \\ v &= v(x, y) \\ w &= 0 \end{aligned} \tag{4.6}$$

that is, the only nonvanishing components, u and v, are functions of x and y only. In the only problems of practical importance satisfying Eqs. (4.6), the body is cylindrical (prismatic) with its longitudinal axis in the z direction.

We shall now proceed to develop the equations governing the elastic behavior of bodies under plane strain, as well as the restrictions which must be imposed on an actual body under load, in order that plane strain can exist. First let us look at the governing equations. By substituting Eqs. (4.6) into Eqs. (4.2), we find that the strain components are given by

$$\begin{aligned} \epsilon_x &= \frac{\partial u}{\partial x} \\ \epsilon_y &= \frac{\partial v}{\partial y} \\ \gamma_{xy} &= \frac{\partial u}{\partial y} + \frac{\partial v}{\partial x} \\ \epsilon_z &= \gamma_{yz} = \gamma_{xz} = 0 \end{aligned} \tag{4.7}$$

where ϵ_x, ϵ_y, and γ_{xy} are functions of x and y only. From generalized Hooke's law, Eqs. (4.3), we find the stress components

$$\begin{aligned} \sigma_x &= 2G\epsilon_x + \lambda(\epsilon_x + \epsilon_y) \\ \sigma_y &= 2G\epsilon_y + \lambda(\epsilon_x + \epsilon_y) \\ \tau_{xy} &= G\gamma_{xy} \\ \tau_{xz} &= \tau_{yz} = 0 \end{aligned} \tag{4.8}$$

These stresses are accompanied by a stress component σ_z given by

$$\begin{aligned} \sigma_z &= \lambda(\epsilon_x + \epsilon_y) \\ &= \nu(\sigma_x + \sigma_y) \end{aligned} \tag{4.9}$$

However, σ_z will not appear in any of the other governing equations so that it is not necessary to consider it as an unknown quantity, even though it is not zero. For plane strain problems, ϵ_z vanishes, but σ_z does not.

In the case of plane strain, the equations of equilibrium (4.1) reduce to

$$\begin{aligned} \frac{\partial \sigma_x}{\partial x} + \frac{\partial \tau_{xy}}{\partial y} + F_x = 0 \\ \frac{\partial \sigma_y}{\partial y} + \frac{\partial \tau_{xy}}{\partial x} + F_y = 0 \end{aligned} \tag{4.10}$$

the third of which becomes

$$F_z = 0$$

since σ_z is only a function of x and y, and $\tau_{xz} = \tau_{yz} = 0$. Thus, in order for a state of plane strain to exist, we can have no body force component in the z direction.

In addition to the restriction of $F_z = 0$, in order for plane strain to exist, the body must be cylindrical and the prescribed surface forces must be independent of z and have no z component. The cylindrical (prismatic) bodies may be of either infinite length, or finite length with fixed ends, as shown in Fig. 4.3. (The term "fixed" here implies a smooth

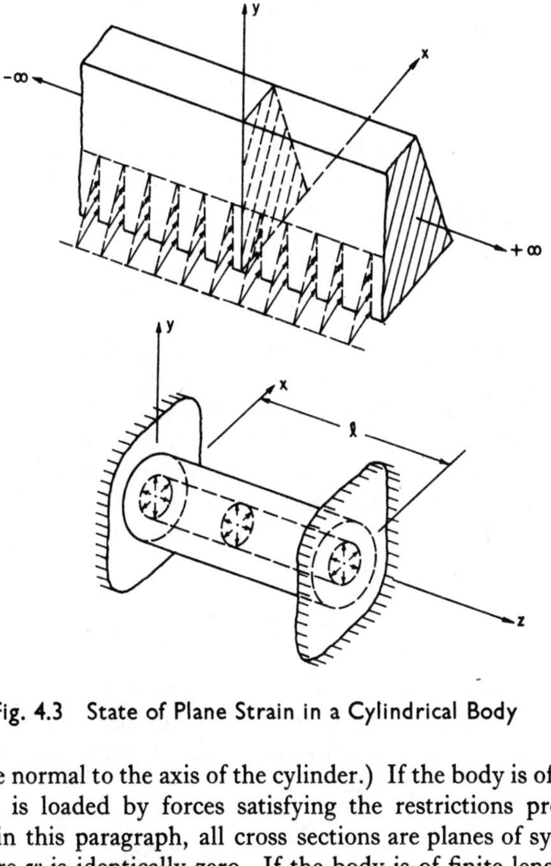

Fig. 4.3 State of Plane Strain in a Cylindrical Body

rigid surface normal to the axis of the cylinder.) If the body is of infinite length, and is loaded by forces satisfying the restrictions previously mentioned in this paragraph, all cross sections are planes of symmetry and therefore w is identically zero. If the body is of finite length with fixed ends, then the boundary conditions on the ends are

$$w_b = w(x, y, 0) = w(x, y, \ell) = \tau_{xz_0} = \tau_{yz_0} = 0$$

which are identically satisfied by the state of plane strain. Therefore, the only boundary conditions required for plane strain problems are those specified on the lateral surface. The prescribed surface forces on the lateral surface, which are functions of x and y only, may be expressed as

$$T_x^\mu = T_x^\mu(x_0, y_0)$$
$$T_y^\mu = T_y^\mu(x_0, y_0)$$
$$T_z^\mu = 0$$

But on the lateral surface, $\mu_z = 0$; therefore the boundary conditions, Eqs. (4.4), become

$$\begin{aligned} T_x^\mu(x_0, y_0) &= \sigma_{x0}\mu_x + \tau_{xy0}\mu_y \\ T_y^\mu(x_0, y_0) &= \sigma_{y0}\mu_y + \tau_{xy0}\mu_x \end{aligned} \tag{4.11}$$

Notice that the third of Eqs. (4.4) is identically satisfied. If displacements, instead of forces, are specified on the lateral surface, we have the displacement boundary conditions

$$\begin{aligned} u(x_0, y_0) &= u_b \\ v(x_0, y_0) &= v_b \end{aligned} \tag{4.12}$$

The solution of a particular plane strain problem then consists of the determination of the eight variables σ_x, σ_y, τ_{xy}, ϵ_x, ϵ_y, γ_{xy}, u, and v which satisfy Eqs. (4.7), (4.8), and (4.10), and the given boundary conditions expressed either by Eqs. (4.11) (prescribed T_x^μ and T_y^μ), or by Eqs. (4.12). Since σ_z is, in general, different from zero, a restraining force is required on the end planes of the finite-length body. If the restraining force is desired, after all the other stresses are calculated, it may be computed from the relation

$$P_z = \iint \sigma_z \, dx \, dy$$

We shall now reduce the eight governing equations of plane strain problems to two equations in terms of the displacement components u and v. This is accomplished by substituting the strain components given by Eqs. (4.7) into Eqs. (4.8), which yields

$$\begin{aligned} \sigma_x &= \lambda\left(\frac{\partial u}{\partial x} + \frac{\partial v}{\partial y}\right) + 2G\frac{\partial u}{\partial x} \\ \sigma_y &= \lambda\left(\frac{\partial u}{\partial x} + \frac{\partial v}{\partial y}\right) + 2G\frac{\partial v}{\partial y} \\ \tau_{xy} &= G\left(\frac{\partial v}{\partial x} + \frac{\partial u}{\partial y}\right) \end{aligned} \tag{4.13}$$

Combining these equations with Eqs. (4.10) and eliminating the stress variables, we have

$$G\nabla^2 u + (\lambda + G)\frac{\partial}{\partial x}\left(\frac{\partial u}{\partial x} + \frac{\partial v}{\partial y}\right) + F_x = 0$$
$$G\nabla^2 v + (\lambda + G)\frac{\partial}{\partial y}\left(\frac{\partial u}{\partial x} + \frac{\partial v}{\partial y}\right) + F_y = 0 \qquad (4.14)$$

where

$$\nabla^2 = \frac{\partial^2}{\partial x^2} + \frac{\partial^2}{\partial y^2}$$

These are the most convenient forms of governing equations for second boundary-value problems in elasticity for plane strain.

The system of eight equations can also be reduced to three equations in terms of the stress components, σ_x, σ_y, and τ_{xy}. By differentiating the first of Eqs. (4.7) twice with respect to y, the second twice with respect to x, and the third with respect to x and y, and adding, we obtain the compatibility equation for plane strain

$$\frac{\partial^2 \epsilon_x}{\partial y^2} + \frac{\partial^2 \epsilon_y}{\partial x^2} = \frac{\partial^2 \gamma_{xy}}{\partial x\, \partial y} \qquad (4.15$$

This is actually the same as the first of Eqs. (2.2), the remainder of these being identically satisfied for plane strain problems. By solving Eqs. (4.8) for the strain components in terms of stress (or by using the alternate form of generalized Hooke's law), substituting the results in Eq. (4.15), and combining with Eqs. (4.10), we have the compatibility equation in terms of stress, i.e.,

$$\nabla^2(\sigma_x + \sigma_y) = -\frac{1}{1-\nu}\left(\frac{\partial F_x}{\partial x} + \frac{\partial F_y}{\partial y}\right) \qquad (4.16)$$

Eqs. (4.10) and (4.16) give a complete system of three equations in terms of σ_x, σ_y, and τ_{xy}. This system of equations, which are most useful in solving first boundary value problems, may be reduced further to one equation and one dependent variable, as will be shown in Chapter 5.

The stress components in any body in a state of plane strain must satisfy Eqs. (4.10) and (4.16) and the prescribed boundary conditions. If the solution for the stress components is obtained for a given problem, we may determine the strain components from generalized Hooke's law and the displacement components from the strain-displacement relations.[2] If, on the other hand, we determine the displacement

[2] The displacement components obtained will be defined except for constants of integration which represent rigid body displacements. If we specify the displacement components of one point and either u or v for another point, where the two points are not on a line parallel to the z axis, or the displacement and rotation of a single point, the constants of integration can be evaluated.

Table 4.1

DEVELOPMENT OF PLANE STRAIN EQUATIONS

$$u = u(x, y), \quad v = v(x, y), \quad w = 0$$

---

2 Equilibrium (4.10) $\quad \dfrac{\partial \sigma_x}{\partial x} + \dfrac{\partial \tau_{xy}}{\partial y} + F_x = 0 \quad (x, y)$

3 Hooke's law (4.8) $\quad \sigma_x = 2G\epsilon_x + \lambda(\epsilon_x + \epsilon_y) \quad (x, y)$

$\tau_{xy} = G\gamma_{xy}$

3 Strain-displacement (4.7) $\quad \epsilon_x = \dfrac{\partial u}{\partial x}; \quad \epsilon_y = \dfrac{\partial v}{\partial y}; \quad \gamma_{xy} = \dfrac{\partial u}{\partial y} + \dfrac{\partial v}{\partial x}$

—

8

$\sigma_x, \sigma_y, \tau_{xy}, \epsilon_x, \epsilon_y, \gamma_{xy}, u, v$

---

2 Equilibrium (4.10)

3 Hooke's law (4.8)

1 Compatibility (strain) (4.15)

—

6

$$\frac{\partial^2 \epsilon_x}{\partial y^2} + \frac{\partial^2 \epsilon_y}{\partial x^2} = \frac{\partial^2 \gamma_{xy}}{\partial x \partial y}$$

$\sigma_x, \sigma_y, \tau_{xy}, \epsilon_x, \epsilon_y, \gamma_{xy}$

---

2 Equilibrium (4.10)

3 Stress-displacement (4.13)

—

5

$$\sigma_x = \lambda\left(\frac{\partial u}{\partial x} + \frac{\partial v}{\partial y}\right) + 2G\frac{\partial u}{\partial x} \quad (x, y; u, v)$$

$$\tau_{xy} = G\left(\frac{\partial v}{\partial x} + \frac{\partial u}{\partial y}\right)$$

$\sigma_x, \sigma_y, \tau_{xy}, u, v$

---

2 Equilibrium (4.10)

1 Compatibility (stress) (4.16)

—

3

$$\nabla^2(\sigma_x + \sigma_y) = -\frac{1}{1-\nu}\left(\frac{\partial F_x}{\partial x} + \frac{\partial F_y}{\partial y}\right)$$

$\sigma_x, \sigma_y, \tau_{xy}$

---

2 Equilibrium equations in terms of displacements (4.14)

$$G\nabla^2 u + (\lambda + G)\frac{\partial}{\partial x}\left(\frac{\partial u}{\partial x} + \frac{\partial v}{\partial y}\right) + + F_x = 0 \quad (x, y; u, v)$$

u, v

$$\epsilon_z = \gamma_{xz} = \gamma_{yz} = \tau_{xz} = \tau_{yz} = 0$$

$$\sigma_z = \nu(\sigma_x + \sigma_y)$$

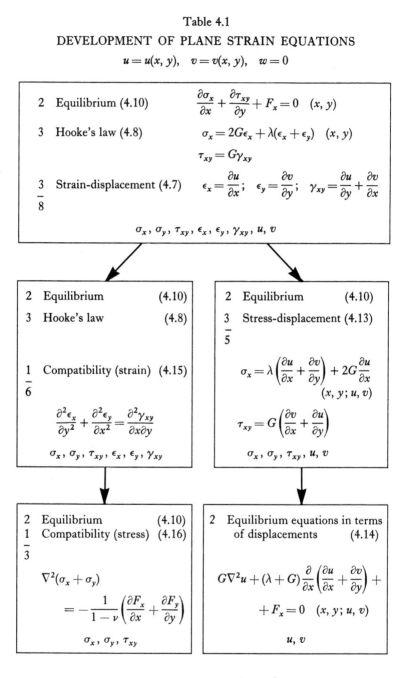

components from Eqs. (4.14), the strain components may be determined from the strain-displacement relations, and the stress components from generalized Hooke's law.

A summary of the development of equations is listed in Table 4.1. The box on top gives the system of eight governing field equations and eight dependent variables. The two boxes on the left hand side give the stress formulation, those on the right hand side give the displacement formulation. The equations in each of these boxes must be satisfied by the variables in the same box. Furthermore, each set of variables are required to satisfy only those equations within the same box. For instance, if we are working with σ_x, σ_y, τ_{xy}, u and v as our variables, they are required to satisfy only Eqs. (4.10) and (4.13); if σ_x, σ_y, and τ_{xy} are the variables, they are required to satisfy Eqs. (4.10) and (4.16).

Further discussion of plane strain and on the comparison between plane stress and plane strain problems will be given in Chapter 5.

4.4. Governing Equations in Three-Dimensional Problems

In three-dimensional problems, the governing equations are Eqs. (4.1), (4.2), and (4.3), i.e., we have 15 equations and 15 unknown quantities. One method of solution, therefore, is to seek expressions for the stress, strain, and displacement components which satisfy these equations and the prescribed boundary conditions. This system of equations, however, is not convenient to apply, so we shall reduce it to systems of equations which will be more convenient in the various types of boundary value problems. This reduction procedure is similar to the one discussed for plane strain problems in the previous section.

We shall begin with the displacement formulation. By substituting Eqs. (4.2), into Eqs. (4.3), we obtain six stress-displacement relations of the form

$$\begin{aligned}
\sigma_x &= \lambda\epsilon + 2G\frac{\partial u}{\partial x} \\
\sigma_y &= \lambda\epsilon + 2G\frac{\partial v}{\partial y} \\
\sigma_z &= \lambda\epsilon + 2G\frac{\partial w}{\partial z} \\
\tau_{xy} &= G\left(\frac{\partial u}{\partial y} + \frac{\partial v}{\partial x}\right) \\
\tau_{yz} &= G\left(\frac{\partial w}{\partial y} + \frac{\partial v}{\partial z}\right) \\
\tau_{zx} &= G\left(\frac{\partial w}{\partial x} + \frac{\partial u}{\partial z}\right)
\end{aligned} \tag{4.17}$$

where $\epsilon = \epsilon_x + \epsilon_y + \epsilon_z = \partial u/\partial x + \partial v/\partial y + \partial w/\partial z$. Together with the three equations of equilibrium (4.1) then, we have a system of nine equations and nine unknowns (six stress components and three displacement components). After these quantities are determined, the strain components may be found from the strain-displacement relations or generalized Hooke's law.

These nine equations may be further reduced to three equations in terms of three displacement components. Eliminating the stress variables by substituting Eqs. (4.17) into Eqs. (4.1), we have

$$
\begin{aligned}
(\lambda + G)\frac{\partial \epsilon}{\partial x} + G\nabla^2 u + F_x = 0 \\
(\lambda + G)\frac{\partial \epsilon}{\partial y} + G\nabla^2 v + F_y = 0 \\
(\lambda + G)\frac{\partial \epsilon}{\partial z} + G\nabla^2 w + F_z = 0
\end{aligned}
\tag{4.18}
$$

where

$$\nabla^2 = \frac{\partial^2}{\partial x^2} + \frac{\partial^2}{\partial y^2} + \frac{\partial^2}{\partial z^2}$$

These are called the *equations of equilibrium in terms of displacement*, or *Navier's equations*. If this system of equations is utilized in the solution of a problem in which the surface forces are prescribed over a portion of the boundary, it will be convenient to express the right-hand sides of the boundary condition equations (4.4) in terms of displacement derivatives, or (see Problem 4-2)

$$
\begin{aligned}
T_x^{\mu} &= \lambda\epsilon\mu_x + G\left(\frac{\partial u}{\partial x}\mu_x + \frac{\partial u}{\partial y}\mu_y + \frac{\partial u}{\partial z}\mu_z\right) + G\left(\frac{\partial u}{\partial x}\mu_x + \frac{\partial v}{\partial x}\mu_y + \frac{\partial w}{\partial x}\mu_z\right) \\
T_y^{\mu} &= \lambda\epsilon\mu_y + G\left(\frac{\partial v}{\partial y}\mu_y + \frac{\partial v}{\partial z}\mu_z + \frac{\partial v}{\partial x}\mu_x\right) + G\left(\frac{\partial v}{\partial y}\mu_y + \frac{\partial w}{\partial y}\mu_z + \frac{\partial u}{\partial y}\mu_x\right) \\
T_z^{\mu} &= \lambda\epsilon\mu_z + G\left(\frac{\partial w}{\partial z}\mu_z + \frac{\partial w}{\partial x}\mu_x + \frac{\partial w}{\partial y}\mu_y\right) + G\left(\frac{\partial w}{\partial z}\mu_z + \frac{\partial u}{\partial z}\mu_x + \frac{\partial v}{\partial z}\mu_y\right)
\end{aligned}
\tag{4.19}
$$

Once we obtain the solution for u, v, and w, the strain components can be determined from Eqs. (4.2) and the stress components determined from Eqs. (4.3) or (4.17).

Another approach in reducing the number of equations and variables is the stress formulation, i.e., to eliminate all variables except the stress components. To accomplish this, just as in the plane strain case, we eliminate the variables u, v, and w from the strain-displacement equations (4.2) and obtain the compatibility equations (2.2). As explained in Section 2.3, these six compatibility equations are equivalent to three independent fourth-order equations. Together with Eqs. (4.1) and (4.3), these constitute a set of equations which can be solved for the twelve variables (six stress components and six strain components).

The next step is to eliminate the strain components from this set of equations, i.e., (4.1), (4.3), and (2.2). To do this we write the last two of Eqs. (4.1) in the form

$$\frac{\partial \tau_{yz}}{\partial z} = -\left(\frac{\partial \sigma_y}{\partial y} + \frac{\partial \tau_{xy}}{\partial x} + F_y\right)$$

and

$$\frac{\partial \tau_{yz}}{\partial y} = -\left(\frac{\partial \sigma_z}{\partial z} + \frac{\partial \tau_{zx}}{\partial x} + F_z\right)$$

By differentiating the first of these with respect to y and the second with respect to z, and adding, we obtain

$$-2\frac{\partial^2 \tau_{yz}}{\partial y\,\partial z} = \frac{\partial^2 \sigma_z}{\partial z^2} + \frac{\partial^2 \sigma_y}{\partial y^2} + \frac{\partial}{\partial x}\left(\frac{\partial \tau_{xz}}{\partial z} + \frac{\partial \tau_{xy}}{\partial y}\right) + \frac{\partial F_z}{\partial z} + \frac{\partial F_y}{\partial y}$$

which becomes, using the first of Eqs. (4.1),

$$-2\frac{\partial^2 \tau_{yz}}{\partial y\,\partial z} = -\frac{\partial^2 \sigma_x}{\partial x^2} + \frac{\partial^2 \sigma_y}{\partial y^2} + \frac{\partial^2 \sigma_z}{\partial z^2} - \frac{\partial F_x}{\partial x} + \frac{\partial F_y}{\partial y} + \frac{\partial F_z}{\partial z} \qquad (4.20)$$

Now the second of Eqs. (2.2), with the aid of generalized Hooke's law (Eqs. (4.3)), may be written in the form

$$\frac{\partial^2}{\partial z^2}[(1+\nu)\sigma_y - \nu\Theta] + \frac{\partial^2}{\partial y^2}[(1+\nu)\sigma_z - \nu\Theta] = \frac{\partial^2}{\partial y\,\partial z}[2(1+\nu)\tau_{yz}] \qquad (4.21)$$

where

$$\Theta = \sigma_x + \sigma_y + \sigma_z$$

Eliminating the terms containing τ_{yz} from Eqs. (4.20) and (4.21), we have

$$(1+\nu)\left(\nabla^2\Theta - \nabla^2\sigma_x - \frac{\partial^2\Theta}{\partial x^2}\right) - \nu\left(\nabla^2\Theta - \frac{\partial^2\Theta}{\partial x^2}\right) = (1+\nu)\left(\frac{\partial F_x}{\partial x} - \frac{\partial F_y}{\partial y} - \frac{\partial F_z}{\partial z}\right) \qquad (4.22)$$

In the same manner, we can develop two other relations similar to Eq. (4.22). Thus, the three expressions are

$$
\begin{aligned}
&(1+\nu)\left(\nabla^2\Theta-\nabla^2\sigma_x-\frac{\partial^2\Theta}{\partial x^2}\right)-\nu\left(\nabla^2\Theta-\frac{\partial^2\Theta}{\partial x^2}\right)\\
&\qquad\qquad=(1+\nu)\left(\frac{\partial F_x}{\partial x}-\frac{\partial F_y}{\partial y}-\frac{\partial F_z}{\partial z}\right)\\
&(1+\nu)\left(\nabla^2\Theta-\nabla^2\sigma_y-\frac{\partial^2\Theta}{\partial y^2}\right)-\nu\left(\nabla^2\Theta-\frac{\partial^2\Theta}{\partial y^2}\right)\\
&\qquad\qquad=(1+\nu)\left(\frac{\partial F_y}{\partial y}-\frac{\partial F_x}{\partial x}-\frac{\partial F_z}{\partial z}\right)\\
&(1+\nu)\left(\nabla^2\Theta-\nabla^2\sigma_z-\frac{\partial^2\Theta}{\partial z^2}\right)-\nu\left(\nabla^2\Theta-\frac{\partial^2\Theta}{\partial z^2}\right)\\
&\qquad\qquad=(1+\nu)\left(\frac{\partial F_z}{\partial z}-\frac{\partial F_y}{\partial y}-\frac{\partial F_x}{\partial x}\right)
\end{aligned}
\tag{4.23}
$$

Now adding Eqs. (4.23), we have the result

$$
\nabla^2\Theta=-\frac{(1+\nu)}{(1-\nu)}\left(\frac{\partial F_x}{\partial x}+\frac{\partial F_y}{\partial y}+\frac{\partial F_z}{\partial z}\right)
\tag{4.24}
$$

and substituting this expression for $\nabla^2\Theta$ into Eq. (4.22), we obtain the first of the following equations. The next two are found in an analogous manner.

$$
\begin{aligned}
\nabla^2\sigma_x+\frac{1}{1+\nu}\frac{\partial^2\Theta}{\partial x^2}&=-\frac{\nu}{1-\nu}\left[\frac{\partial F_x}{\partial x}+\frac{\partial F_y}{\partial y}+\frac{\partial F_z}{\partial z}\right]-2\frac{\partial F_x}{\partial x}\\
\nabla^2\sigma_y+\frac{1}{1+\nu}\frac{\partial^2\Theta}{\partial y^2}&=-\frac{\nu}{1-\nu}\left[\frac{\partial F_x}{\partial x}+\frac{\partial F_y}{\partial y}+\frac{\partial F_z}{\partial z}\right]-2\frac{\partial F_y}{\partial y}\\
\nabla^2\sigma_z+\frac{1}{1+\nu}\frac{\partial^2\Theta}{\partial z^2}&=-\frac{\nu}{1-\nu}\left[\frac{\partial F_x}{\partial x}+\frac{\partial F_y}{\partial y}+\frac{\partial F_z}{\partial z}\right]-2\frac{\partial F_z}{\partial z}
\end{aligned}
\tag{4.25a}
$$

Similarly, we may obtain three equations in terms of the shear stress components, i.e.,

$$
\begin{aligned}
\nabla^2\tau_{yz}+\frac{1}{1+\nu}\frac{\partial^2\Theta}{\partial y\,\partial z}&=-\left(\frac{\partial F_z}{\partial y}+\frac{\partial F_y}{\partial z}\right)\\
\nabla^2\tau_{zx}+\frac{1}{1+\nu}\frac{\partial^2\Theta}{\partial z\,\partial x}&=-\left(\frac{\partial F_x}{\partial z}+\frac{\partial F_z}{\partial x}\right)\\
\nabla^2\tau_{xy}+\frac{1}{1+\nu}\frac{\partial^2\Theta}{\partial x\,\partial y}&=-\left(\frac{\partial F_y}{\partial x}+\frac{\partial F_x}{\partial y}\right)
\end{aligned}
\tag{4.25b}
$$

The six relations, Eqs. (4.25), which are equivalent to three independent fourth-order equations, are called the *compatibility equations in terms of stress*, or the *Beltrami-Michell compatibility equations*. These equations and the equilibrium equations give six independent relations which can be solved for the six stress variables. Once the stresses are known, the strains may be determined from Eqs. (4.3), and the displacement components (which contain the arbitrary rigid body components) may be obtained by integrating Eqs. (4.2). If the body under consideration is simply connected, then the displacements thus obtained will be single valued, since the compatibility equations are satisfied. If the body is multiply connected, however, we cannot be sure that the displacements are single valued, even though the compatibility equations are satisfied. Further discussion on multiply connected bodies will be given in Chapter 10.

The development of the three-dimensional equations in elasticity, as discussed in this section, is summarized in Table 4.2. In this table, again, the left-hand side gives the stress formulation, and the right-hand side gives the displacement formulation. In the stress formulation, we can add one more box with a system of equations in terms of three stress-functions. These stress-functions satisfy the equilibrium equations identically, and reduce the number of dependent variables from six (stress) to three (stress-functions). In the displacement formulation, the governing equations can be expressed in terms of a vector function and a scalar potential. These topics will be covered in Chapter 13.

4.5. Principle of Superposition

In Chapter 2, we discussed the principle of superposition of strain and demonstrated for a special case, that the displacement fields due to two separate strain distributions may be added (superposed) to give the displacement field due to the combined strain distribution. This is true only when the displacements and strains are small such that the strain-displacement equations are linear. Actually, the principle is applicable to all linear systems. In this section we shall discuss this principle for the governing equations in linear elasticity.

Under the assumptions of infinitesimal deformations and the linear stress-strain relations, all the governing equations in elasticity as shown in Table (4.2), and the boundary condition equations (4.4) and (4.5), are linear equations. (An equation is linear if it contains only terms of up to the first degree in the dependent variables and their derivatives.) Furthermore, our equations are either homogeneous in the dependent variables, i.e., every term of the equation contains one dependent variable or its derivatives to the first degree, such as Eqs. (4.2) and (4.3), or nonhomogeneous with only the external loads as the nonhomogeneous terms, such as Eqs. (4.1), (4.25), and (4.4). Due to this nature of our equations, the dependent variables vary linearly with the external loads and therefore may be superposed. The principle of superposition

Table 4.2
DEVELOPMENT OF THREE-DIMENSIONAL EQUATIONS

3	Equilibrium equations	(4.1)
6	Strain-displacement equations	(4.2)
6	Hooke's law equations	(4.3)
15		
	$\sigma_x, \sigma_y, \sigma_z, \tau_{xy}, \tau_{yz}, \tau_{zx}$	
	$\epsilon_x, \epsilon_y, \epsilon_z, \gamma_{xy}, \gamma_{yz}, \gamma_{zx}$	
	u, v, w	

↙ ↘

3	Equilibrium equations	(4.1)
6	(3) Compatibility equations	(2.2)
6	Hooke's law equations	(4.3)
15 (12)		
	Six stress variables	
	Six strain variables	

3	Equilibrium equations	(4.1)
6	Stress-displacement equations	(4.17)
9		
	Six stress variables	
	Three displacement variables	

↓

3	Equilibrium equations	(4.1)
6	(3) Compatibility equations in terms of stress	(4.25)
9 (6)		
	Six stress variables	

↓

3	Equilibrium equations in terms of displacement	(4.18)
	u, v, w	

may be stated as: The dependent variables (stress, strain and displacement) determined due to each set of external loads (body force, surface force, and prescribed surface displacement) acting separately, may be superposed to give the total values due to the combined external loads.

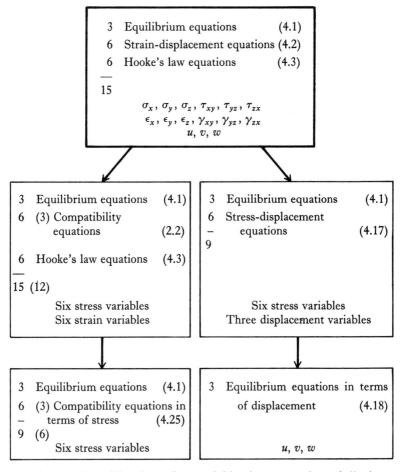

To demonstrate this, let us consider the stress formulation, where the governing equations are equilibrium equations (4.1) and the compatibility equations in terms of stress (4.25). Let $\sigma_x \ldots \tau_{xz}$ denote the stress components satisfying the governing equations and prescribed boundary conditions for a certain elastic body under body forces $F_x \ldots$ and surface forces $T_x^{\mu} \ldots$. Further, let the stress components in the same body under $F'_x \ldots, T_x^{\mu'} \ldots$, be denoted by $\sigma'_x \ldots \tau'_{xz}$. Thus the primed and unprimed stress components satisfy the following relations: the equilibrium equations,

$$\frac{\partial \sigma_x}{\partial x} + \frac{\partial \tau_{xy}}{\partial y} + \frac{\partial \tau_{xz}}{\partial z} + F_x = 0 \qquad (x, y, z)$$

$$\frac{\partial \sigma'_x}{\partial x} + \frac{\partial \tau'_{xy}}{\partial y} + \frac{\partial \tau'_{xz}}{\partial z} + F'_x = 0 \qquad (x, y, z)$$

the compatibility equations,

$$\left.\begin{aligned} \nabla^2 \sigma_x + \frac{1}{1+\nu}\frac{\partial^2 \Theta}{\partial x^2} &= -\frac{\nu}{1-\nu}\left(\frac{\partial F_x}{\partial x} + \frac{\partial F_y}{\partial y} + \frac{\partial F_z}{\partial z}\right) - 2\frac{\partial F_x}{\partial x} \\ \nabla^2 \tau_{yz} + \frac{1}{1+\nu}\frac{\partial^2 \Theta}{\partial y\,\partial z} &= -\left(\frac{\partial F_z}{\partial y} + \frac{\partial F_y}{\partial z}\right) \end{aligned}\right\} (x, y, z)$$

$$\left.\begin{aligned} \nabla^2 \sigma'_x + \frac{1}{1+\nu}\frac{\partial^2 \Theta'}{\partial x^2} &= -\frac{\nu}{1-\nu}\left(\frac{\partial F'_x}{\partial x} + \frac{\partial F'_y}{\partial y} + \frac{\partial F'_z}{\partial z}\right) - 2\frac{\partial F'_x}{\partial x} \\ \nabla^2 \tau'_{yz} + \frac{1}{1+\nu}\frac{\partial^2 \Theta'}{\partial y\,\partial z} &= -\left(\frac{\partial F'_z}{\partial y} + \frac{\partial F'_y}{\partial z}\right) \end{aligned}\right\} (x, y, z)$$

and the boundary condition equations

$$\sigma_{xo}\mu_x + \tau_{xyo}\mu_y + \tau_{xzo}\mu_z = T_x^{\mu} \qquad (x, y, z)$$

$$\sigma'_{xo}\mu_x + \tau'_{xyo}\mu_y + \tau'_{xzo}\mu_z = T_x^{\mu'} \qquad (x, y, z)$$

and by addition, we see that

$$\frac{\partial(\sigma_x + \sigma'_x)}{\partial x} + \frac{\partial(\tau_{xy} + \tau'_{xy})}{\partial y} + \frac{\partial(\tau_{xz} + \tau'_{xz})}{\partial z} + (F_x + F'_x) = 0 \qquad (x, y, z)$$

$$\left.\begin{aligned} \nabla^2(\sigma_x + \sigma'_x) + \cdots &= -\frac{\nu}{1-\nu}\left[\frac{\partial(F_x + F'_x)}{\partial x} + \cdots\right]\cdots \\ \nabla^2(\tau_{yz} + \tau'_{yz}) + \cdots &= -\left[\frac{\partial(F_z + F'_z)}{\partial y} + \cdots\right] \end{aligned}\right\} (x, y, z)$$

and

$$(\sigma_{xo} + \sigma'_{xo})\mu_x + (\tau_{xyo} + \tau'_{xyo})\mu_y + (\tau_{xzo} + \tau'_{xzo})\mu_z = (T_x{}^{\mu} + T_x{}^{\mu'}) \qquad (x, y, z)$$

The results show that the stress components $(\sigma_x + \sigma'_x) \ldots (\tau_{xz} + \tau'_{xz})$ are the solution for the given elastic body under body forces $(F_x + F'_x) \ldots$ and surface forces $(T_x{}^{\mu} + T_x^{\mu'}) \ldots$. In other words, two (or more) stress fields may be superposed to yield the results for combined loads. Since the equations governing the strain and displacement components are also linear, it follows that the principle of superposition also applies for these quantities.

4.6. Uniqueness of Elasticity Solutions

We shall now demonstrate that elasticity solutions are unique, i.e., for a given surface force and body force distribution, there is only one solution for the stress components consistent with equilibrium and compatibility.

Let us consider a body subjected to a specific *given* surface force distribution with the components $T_x{}^{\mu}$, $T_y{}^{\mu}$, $T_z{}^{\mu}$ and a *given* body force field with components F_x, F_y, F_z. Let us assume that there are two sets of stress components, each of which satisfy the governing equations and boundary conditions. We denote these two solutions by

$$\sigma'_x \ldots \tau'_{xz}$$

and

$$\sigma''_x \ldots \tau''_{xz}$$

Now the first solution must satisfy the equilibrium equations

$$\frac{\partial \sigma'_x}{\partial x} + \frac{\partial \tau'_{xy}}{\partial y} + \frac{\partial \tau'_{xz}}{\partial z} + F_x = 0 \qquad (x, y, z) \qquad (4.26a)$$

the compatibility equations (4.25),

$$\nabla^2 \sigma'_x + \cdots = \cdots - 2\frac{\partial F_x}{\partial x} \qquad (x, y, z)$$

$$\nabla^2 \tau'_{yz} + \cdots = -\left(\frac{\partial F_z}{\partial y} + \frac{\partial F_y}{\partial z}\right) \qquad (x, y, z) \qquad (4.26b)$$

and the boundary conditions

$$T_x{}^{\mu} = \sigma'_{xo}\mu_x + \tau'_{xyo}\mu_y + \tau'_{xzo}\mu_z \qquad (x, y, z) \qquad (4.26c)$$

Similarly, the second solution must satisfy the relations

$$\frac{\partial \sigma_x''}{\partial x} + \frac{\partial \tau_{xy}''}{\partial y} + \frac{\partial \tau_{xz}''}{\partial z} + F_x = 0 \qquad (x, y, z) \qquad (4.27a)$$

$$\nabla^2 \sigma_x'' + \cdots = \cdots - 2\frac{\partial F_x}{\partial x} \qquad (x, y, z)$$

$$\nabla^2 \tau_{yz}'' + \cdots = -\left(\frac{\partial F_z}{\partial y} + \frac{\partial F_y}{\partial z}\right) \qquad (x, y, z) \qquad (4.27b)$$

and

$$T_x^{\mu} = \sigma_{x0}'' \mu_x + \tau_{xy0}'' \mu_y + \tau_{xz0}'' \mu_z \qquad (x, y, z) \qquad (4.27c)$$

Introducing the notations $\sigma_x = \sigma'_x - \sigma''_x$, $\sigma_{x0} = \sigma'_{x0} - \sigma''_{x0}$, etc., and subtracting the set of equations with double primes, Eqs. (4.27), from those with single primes, Eqs. (4.26), we have

$$\frac{\partial \sigma_x}{\partial x} + \frac{\partial \tau_{xy}}{\partial y} + \frac{\partial \tau_{xz}}{\partial z} = 0 \qquad (x, y, z)$$

$$\nabla^2 \sigma_x + \frac{1}{1+\nu}\frac{\partial^2 \Theta}{\partial x^2} = 0 \qquad (x, y, z)$$

$$\nabla^2 \tau_{yz} + \frac{1}{1+\nu}\frac{\partial^2 \Theta}{\partial y\, \partial z} = 0 \qquad (x, y, z) \qquad (4.28)$$

$$\sigma_{x0}\mu_x + \tau_{xy0}\mu_y + \tau_{xz0}\mu_z = 0 \qquad (x, y, z)$$

Notice that due to the linear character of the equations and due to the fact that the same body forces and surface forces appear in both sets of equations, (4.26) and (4.27), the body forces and surface forces terms do not appear in Eqs. (4.28). Therefore, the stresses $\sigma_x \ldots \tau_{zx}$ represent the state of stress in the body with no body forces and no surface forces. It is evident that a body with zero body forces and zero surface forces is in an unstressed state and the stresses at all points throughout the volume of the body are zero.[3] Or, we have

$$\sigma_x = \sigma'_x - \sigma''_x = 0 \qquad (x, y, z)$$

$$\tau_{xz} = \tau'_{xz} - \tau''_{xz} = 0 \qquad (x, y, z)$$

[3] If the surface and body forces are zero, there is no work done on the body, so that the strain energy (defined in Chapter 7) must vanish. However, as shown by Eq. (7.6), the strain energy is a positive, definite, quadratic form in terms of the strain components. Thus the strain energy can vanish only if each strain component, and consequently each stress component, is identically zero. In certain cases, stresses may exist in a body while external forces are absent. These stresses could be due to (a) shrink and force fits of cylinders or (b) nonuniform plastic forming processes such as rolling or

Thus, the two states of stress are identical so that the solution is unique. The proof given here is limited to the first boundary-value problem, where the stresses are prescribed on the entire boundary. The uniqueness theorem can similarly be proved for the second boundary-value problem or mixed boundary-value problem. See Problem 4-4.

4.7. Saint-Venant's Principle

The principle of Saint-Venant extends considerably the practical use of elasticity solutions. This principle states that the stresses due to two statically equivalent loadings applied over a small area are significantly different only in the vicinity of the area on which the loadings are applied; and at distances which are large in comparison with the linear dimensions of the area on which the loadings are applied, the effects due to these two loadings are the same.[4] This may be illustrated by the following example.

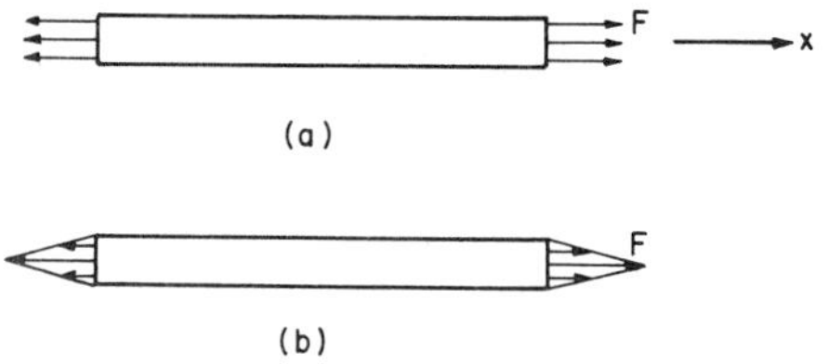

Fig. 4.4 Bar Under Statically Equivalent Loadings

Consider the thin bar under the uniform tensile stress p which produces a resultant force F as shown in Fig. 4.4(a). Since the stress components

$$\begin{gathered}\sigma_x = p \\ \tau_{xy} = \tau_{yz} = \tau_{zx} = \sigma_y = \sigma_z = 0\end{gathered} \tag{4.29}$$

satisfy the governing equations (4.1) and (4.25) as well as the boundary conditions (refer to the example of Section 4.2), we have the exact

---

forging. Note, however, that for problems such as (a) we must write the governing equations for *each* body, since there may be a discontinuity in stress at an interface between two bodies, and we are not discussing problems such as (b), where plastic strains occur. Also, as long as the governing equations are linear such that the principle of superposition is applicable, the additional stresses and strains produced by external forces are not effected by these stresses.

[4] This principle is verified for some special cases in S. Timoshenko and J. N. Goodier, *Theory of Elasticity*, Second Edition, McGraw-Hill Book Company, Inc., New York, 1951, p. 52, and I. S. Sokolnikoff, *Mathematical Theory of Elasticity*, Second Edition, McGraw-Hill Book Company, Inc., New York, 1956, p. 190.

solution to the problem. Furthermore, the uniqueness theorem asserts that it is the *only* solution for the stresses in this bar.

Suppose that the applied force F is distributed as shown in Fig. 4.4(b). The stress components given by Eqs. (4.29) no longer represent the solution, since the boundary conditions on the end planes are not satisfied. Saint-Venant's principle states, however, that these stresses closely approximate the actual stress distribution except near the ends of the bar. This principle is used in some problems to express the boundary conditions in terms of resultant force over a particular region, rather than specifying the exact distribution of stress.

4.8. Summary

We have found that the solution of a given problem in elasticity consists of the determination of stress, strain, and displacement components which satisfy the 15 governing field equations, and the prescribed stress or displacement boundary conditions. Satisfaction of the boundary conditions demands that the stress components on the boundary must be in equilibrium with the applied surface forces in regions where the surface forces are prescribed, and the displacement components on the boundary are equal to the prescribed displacements in regions where displacement is prescribed.

In the solution of a particular problem, there are five equivalent systems of equations (the five boxes in Table 4.2) to utilize, the choice depending upon the given data in the problem.

The equations of elasticity are greatly simplified in certain classes of physical problems, which possess specialized stresses or displacements. One such class is the plane strain problem. Other classes will be presented in subsequent chapters. We have seen that the solution of a given problem in elasticity is unique, i.e., it is the *only* solution. Finally, Saint-Venant's principle, which is of the utmost importance in practical problems, was stated.

PROBLEMS

4-1 Derive Eqs. (4.14) and (4.16).

4-2 Derive the stress boundary condition in terms of displacement equations (4.19), by substituting Eqs. (4.17) into Eqs. (4.4).

4-3 Fill in the missing steps in the derivation of Eqs. (4.25).

4-4 Prove the uniqueness theorem for the second boundary value problem by using the same procedure as given in Section 4.6.

4-5 Describe a physical situation where the boundary conditions on a certain region of an elastic body are such that two components of surface force and one component of displacement are prescribed.

4-6 An irregularly shaped body is subjected to a constant pressure p at all points on its surface. Determine the stress, strain, and displacement components in the body. Assume that the displacement and rotation components at the origin are zero. Hint: Make a reasonable assumption regarding the stress components, and show that the assumed stresses satisfy the governing equations and the boundary conditions.

4-7 Verify the following relations for plane strain problems with F_x and F_y constant by considering Eqs. (4.14)

$$\frac{\partial}{\partial y}\nabla^2 u = \frac{\partial}{\partial x}\nabla^2 v$$

$$\frac{\partial}{\partial x}\nabla^2 u = -\frac{\partial}{\partial y}\nabla^2 v$$

$$\nabla^2 \epsilon = \nabla^2 \omega_z = 0$$

$$\nabla^4 u = \nabla^4 v = 0$$

where

$$\nabla^4 = \nabla^2\nabla^2 = \left(\frac{\partial^2}{\partial x^2} + \frac{\partial^2}{\partial y^2}\right)\left(\frac{\partial^2}{\partial x^2} + \frac{\partial^2}{\partial y^2}\right)$$

It is interesting to note that, from the complex variable theory, the functions $\nabla^2 u$ and $\nabla^2 v$ satisfy the Cauchy-Riemann conditions so that the function $\nabla^2 u + i\nabla^2 v$ is analytic.

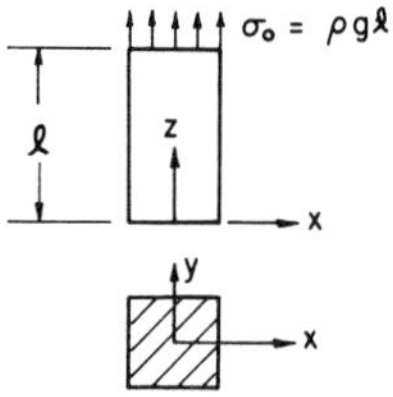

Problem 4–8

4-8 A bar of constant mass density ρ hangs under its own weight and is supported by the uniform stress σ_0 as shown in the figure. Assume that the stresses σ_x, σ_y, τ_{xy}, τ_{xz}, and τ_{yz} vanish identically.

(a) Based on the above assumption, reduce the 15 governing equations to seven equations in terms of σ_z, ϵ_x, ϵ_y, ϵ_z, u, v, and w.

(b) Integrate the equilibrium equation to show that

$$\sigma_z = \rho g z$$

where g is the acceleration due to gravity. Also show that the precribed boundary conditions are satisfied by this solution.

(c) Find ϵ_x, ϵ_y, and ϵ_z from Hooke's law.

(d) If the displacement and rotation components are zero at the point $(0, 0, \ell)$, determine the displacement components u and v.

(e) Show that

$$w = \frac{\rho g}{2E}(z^2 + \nu x^2 + \nu y^2 - \ell^2)$$

4-9 Show that the stresses

$$\begin{aligned}
\sigma_x &= kxy \\
\sigma_y &= kx^2 \\
\sigma_z &= \nu kx(x+y) \\
\tau_{xy} &= -\tfrac{1}{2}ky^2 \\
\tau_{xz} &= \tau_{yz} = 0
\end{aligned}$$

where k is a constant, represent the solution to a plane strain problem with no body forces if the displacement and rotation components at the origin are zero. Determine the displacement components and the restraining force in the z direction. What surface force distribution must be applied to the prism shown in the figure so that the given stress components give the solution for this body?

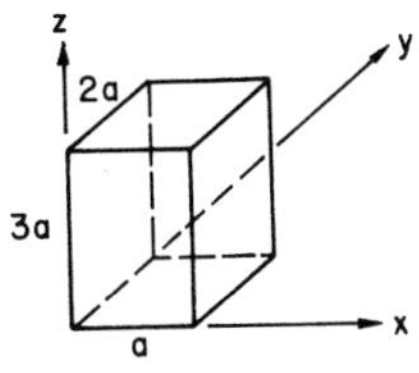

Problem 4–9

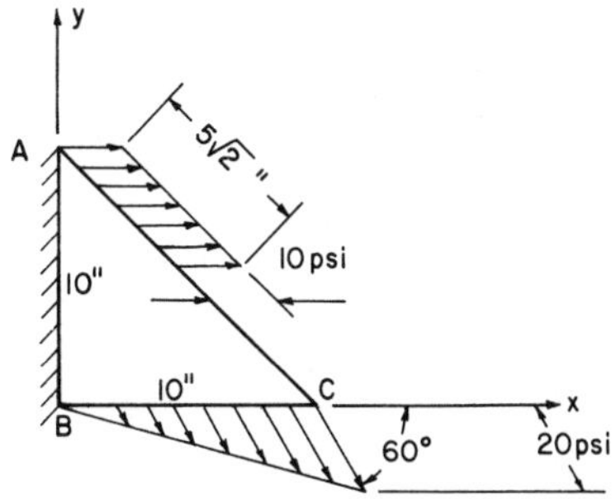

Problem 4–10

4-10 Express the boundary conditions for the plate shown. The surface forces are functions of x and y only. The displacement on surface AB is zero (plane strain problem).

4-11 Show that the stresses

$$\sigma_x = ky$$

$$\sigma_y = \sigma_z = \tau_{xy} = \tau_{xz} = \tau_{yz} = 0$$

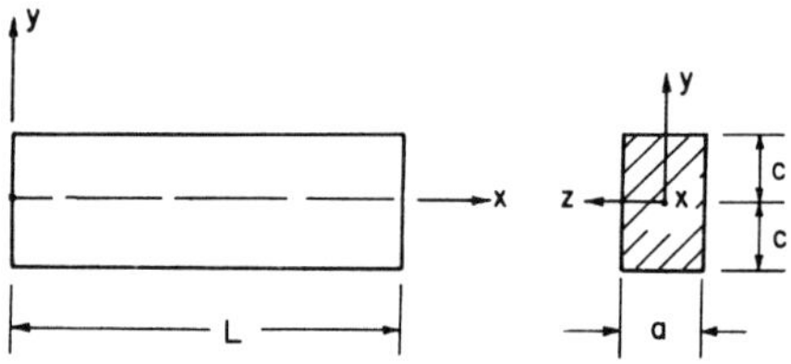

Problem 4–11

represent the solution to a problem in elasticity. Determine the loading which must be applied to the bar in the figure so that the given stress components result. Express the maximum stress in terms of the applied load (a couple) and the dimensions of the cross section.

5

Two-Dimensional Problems

5.1. Introduction

We have already discussed one case in which the nonvanishing stress, strain, and displacement components are functions of x and y only, i.e., in plane strain problems. In this chapter we shall consider another condition, called plane stress, in which the stress distribution is essentially two-dimensional. In this case, as in plane strain problems, the governing equations of elasticity are considerably simplified. We shall also show that the governing equations for two-dimensional problems can be reduced to one equation with one dependent variable. This governing equation is applicable to both plane strain and plane stress problems. Thus, solutions of a plane strain problem are also solutions of the corresponding plane stress problem if certain constants in the equations are given different values. The governing equations of elasticity will also be developed in terms of polar coordinates.

5.2. Plane Stress Problems

In this section, we shall define the state of plane stress and present the governing equations for plane stress problems. Similar to the procedures used in Chapter 4 for plane strain and three-dimensional problems, the system of basic equations will be reduced to systems with less dependent variables by two methods, the stress formulation and the displacement formulation. It will be pointed out that the plane stress equations are approximate. The detailed nature of the approximation involved and the conditions under which the elastic behavior of an actual body may be approximated by the plane stress equations will be discussed in Section 5.3.

A state of plane stress is said to exist when the stress components are of the form

$$\begin{aligned} \sigma_x &= \sigma_x(x, y) \\ \sigma_y &= \sigma_y(x, y) \\ \tau_{xy} &= \tau_{xy}(x, y) \\ \tau_{xz} &= \tau_{yz} = \sigma_z = 0 \end{aligned} \tag{5.1}$$

that is, the three nonvanishing stress components are functions of x and y alone. According to these equations, the governing equations in elasticity simplify as follows: The equilibrium equations (4.1) become

$$\begin{aligned} \frac{\partial \sigma_x}{\partial x} + \frac{\partial \tau_{xy}}{\partial y} + F_x &= 0 \\ \frac{\partial \tau_{xy}}{\partial x} + \frac{\partial \sigma_y}{\partial y} + F_y &= 0 \end{aligned} \tag{5.2}$$

where the body forces F_x and F_y must be functions of x and y only, and F_z must be zero in order for plane stress to exist. The strain-stress relations (Eqs. (4.3)) take the form

$$\begin{aligned} \epsilon_x &= \epsilon_x(x, y) = \frac{1}{E}[\sigma_x - \nu\sigma_y] \\ \epsilon_y &= \epsilon_y(x, y) = \frac{1}{E}[\sigma_y - \nu\sigma_x] \\ \gamma_{xy} &= \gamma_{xy}(x, y) = \frac{1}{G}\tau_{xy} \end{aligned} \tag{5.3}$$

The two shear strains γ_{xz} and γ_{yz} vanish identically. These strains are accompanied by the component ϵ_z, given by

$$\epsilon_z = \epsilon_z(x, y) = -\frac{\nu}{E}(\sigma_x + \sigma_y) \tag{5.4}$$

However, ϵ_z does not appear in any of the other governing equations, thus it will not be considered as an unknown. Finally, the strain-displacement relations (Eqs. (4.2)) are simplified to[1]

$$\begin{aligned} \epsilon_x &= \frac{\partial u}{\partial x} \qquad \epsilon_y = \frac{\partial v}{\partial y} \\ \gamma_{xy} &= \frac{\partial u}{\partial y} + \frac{\partial v}{\partial x} \end{aligned} \tag{5.5}$$

[1] The omission of some of the strain-displacement equations leads to approximations which will be discussed in the next section. In plane stress problems, the displacement w, although not zero, is not considered as one of the independent variables.

Thus, we have formulated eight equations, i.e., Eqs. (5.2), (5.3), and (5.5), in terms of the eight unknown quantities, σ_x, σ_y, τ_{xy}, ϵ_x, ϵ_y, γ_{xy}, u, and v. Notice that the equilibrium equations (5.2) and strain-displacement equations (5.5) are identical with those of plane strain; however, the Hooke's law equations (5.3) are different. Also note that in the case of plane stress, $\sigma_z = 0$ and $\epsilon_z \neq 0$, whereas in plane strain, $\sigma_z \neq 0$ and $\epsilon_z = 0$.

Following the procedure of stress formulation, we eliminate the displacements u and v from Eqs. (5.5) and obtain the compatibility equation

$$\frac{\partial^2 \epsilon_x}{\partial y^2} + \frac{\partial^2 \epsilon_y}{\partial x^2} = \frac{\partial^2 \gamma_{xy}}{\partial x \partial y} \tag{5.6}$$

This is identical to the corresponding equation in the plane strain case, Eq. (4.15). Substituting the Hooke's law equations (5.3) into Eq. (5.6) and combining with Eqs. (5.2), we have the compatibility equation in terms of stress:

$$\nabla^2(\sigma_x + \sigma_y) = -(\nu + 1)\left(\frac{\partial F_x}{\partial x} + \frac{\partial F_y}{\partial y}\right) \tag{5.7}$$

Equations (5.7) and (5.2) are the governing equations for the three stress variables σ_x, σ_y and τ_{xy}. For problems with no body forces, $F_x = F_y = 0$, Eq. (5.7) is identical with the corresponding plane strain equation (4.16).

For those problems where the body force field is conservative, there exists a potential function V such that

$$F_x = \frac{\partial V}{\partial x} \qquad F_y = \frac{\partial V}{\partial y} \tag{5.8}$$

and Eqs. (5.2) and (5.7) can be further reduced to one equation with one dependent variable. Let us, then, introduce a new variable $\phi = \phi(x, y)$, called a *stress function* (also known as *Airy's stress function*), defined by

$$\begin{aligned} \sigma_x + V &= \frac{\partial^2 \phi}{\partial y^2} \\ \sigma_y + V &= \frac{\partial^2 \phi}{\partial x^2} \\ \tau_{xy} &= -\frac{\partial^2 \phi}{\partial x \partial y} \end{aligned} \tag{5.9}$$

The introduction of ϕ implies that the equilibrium equations are identically satisfied, as is shown by substituting Eqs. (5.9) into Eqs. (5.2). It can be shown that a stress function defined by Eqs. (5.9) can be derived for every stress field which satisfies Eqs. (5.2) provided the body force components satisfy Eqs. (5.8) (see Prob. 5-14). In terms of this stress function, Eq. (5.7) becomes

$$\nabla^4\phi = (1-\nu)\nabla^2 V \tag{5.10}$$

where

$$\nabla^4 = \nabla^2(\nabla^2) = \frac{\partial^4}{\partial x^4} + 2\frac{\partial^4}{\partial x^2 \partial y^2} + \frac{\partial^4}{\partial y^4}$$

is called the *biharmonic operator*. Equation (5.10) is the governing field equation for plane stress problems in which the body forces are conservative. For a given plane stress problem, if a function $\phi(x, y)$ is found, such that it satisfies Eq. (5.10) and the proper prescribed boundary conditions, it represents the solution of the problem. The corresponding stresses and strains may be determined from Eqs. (5.9) and (5.3), respectively. If the body forces are constant, or if V is a harmonic[2] function, Eq. (5.10) reduces to

$$\nabla^4\phi = 0 \tag{5.11}$$

which is a particular type of partial differential equation called a *biharmonic equation*. In these cases (constant body forces or harmonic potential function), Eq. (5.7) is identical with Eq. (4.16), and it can be shown that Eq. (5.11) is true for both plane stress and plane strain.

Now let us follow the displacement formulation, starting again with the eight equations (5.2), (5.3), and (5.5). Eliminating the strain components from Eqs. (5.3) and (5.5), we have

$$\begin{aligned} \frac{\partial u}{\partial x} &= \frac{1}{E}(\sigma_x - \nu\sigma_y) \\ \frac{\partial v}{\partial y} &= \frac{1}{E}(\sigma_y - \nu\sigma_x) \\ \frac{\partial u}{\partial y} + \frac{\partial v}{\partial x} &= \frac{1}{G}\tau_{xy} \end{aligned} \tag{5.12}$$

or, after algebraic manipulation,

$$\begin{aligned} \sigma_x &= \frac{E}{1-\nu^2}(\epsilon_x + \nu\epsilon_y) = \frac{E}{1-\nu^2}\left(\frac{\partial u}{\partial x} + \nu\frac{\partial v}{\partial y}\right) \\ \sigma_y &= \frac{E}{1-\nu^2}(\epsilon_y + \nu\epsilon_x) = \frac{E}{1-\nu^2}\left(\frac{\partial v}{\partial y} + \nu\frac{\partial u}{\partial x}\right) \\ \tau_{xy} &= G\gamma_{xy} = G\left(\frac{\partial u}{\partial y} + \frac{\partial v}{\partial x}\right) \end{aligned} \tag{5.13}$$

[2] A harmonic function V is defined as a function which satisfies Laplace's equation, i.e., $\nabla^2 V = 0$.

Substitution Eqs. of (5.13) into Eq. (5.2) yields the equilibrium equations in terms of displacement, or Navier's equations for plane stress,

$$G\nabla^2 u + \frac{E}{2(1-\nu)}\frac{\partial}{\partial x}\left(\frac{\partial u}{\partial x}+\frac{\partial v}{\partial y}\right)+F_x=0$$
$$G\nabla^2 v + \frac{E}{2(1-\nu)}\frac{\partial}{\partial y}\left(\frac{\partial u}{\partial x}+\frac{\partial v}{\partial y}\right)+F_y=0 \tag{5.14}$$

The development of the plane stress equations is summarized in Table 5.1, where the left-hand column gives the stress formulation and the right-hand column gives the displacement formulation.

All of the plane stress equations in Table 5.1 may be changed into the corresponding equations for plane strain in Table 4.1 if every E and ν in Table 5.1 are replaced by E_1 and ν_1, respectively, where

$$E_1=\frac{E}{1-\nu^2} \qquad \nu_1=\frac{\nu}{1-\nu} \tag{5.15}$$

Conversely, all the plane strain equations in Table 4.1 may be changed into those for plane stress if E and ν in Table 4.1 are replaced by E_2 and ν_2, respectively, where

$$E_2=\frac{E(1+2\nu)}{(1+\nu)^2} \qquad \nu_2=\frac{\nu}{1+\nu} \tag{5.16}$$

As a result, the solution for a plane stress problem can be determined from the solution for the corresponding plane strain problem, and vice versa.

5.3. Approximate Character of Plane Stress Equations

In deriving the plane stress equations in the preceding section, we made the assumptions that the stresses on all z planes vanish and the other three stresses, σ_x, σ_y, and τ_{xy}, are functions of x and y only, Eqs. (5.1). Incorporating these assumptions with the general three-dimensional governing equations, we arrived at the eight plane stress governing equations. However, we did not satisfy all the 15 basic equations; some of the strain-displacement equations were not used, thus they cannot normally be satisfied by the plane stress assumptions (5.1). Consequently, the plane stress equations are only approximately true. We shall investigate the nature of the approximation in this section, where, for simplicity, we shall discuss the equations with zero body forces. To avoid confusion, we shall use the notations

$$\nabla^2=\frac{\partial^2}{\partial x^2}+\frac{\partial^2}{\partial y^2}+\frac{\partial^2}{\partial z^2}$$

$$\nabla_1^{\,2} = \frac{\partial^2}{\partial x^2} + \frac{\partial^2}{\partial y^2}$$

in this section.

Table 5.1

DEVELOPMENT OF PLANE STRESS EQUATIONS

σ_x, σ_y, τ_{xy} are functions of x, y only.

$\tau_{xz} = \tau_{yz} = \sigma_z = 0$

2	Equilibrium equations	(5.2)
3	Strain-displacement equations	(5.5)
3	Hooke's law	(5.3)
8		

σ_x, σ_y, τ_{xy}, ϵ_x, ϵ_y, γ_{xy}, u, and v.

↓ (left branch)

2	Equilibrium equations	(5.2)
1	Compatibility	(5.6)
3	Hooke's law	(5.3)
6		

σ_x, σ_y, τ_{xy}, ϵ_x, ϵ_y, and γ_{xy}

↓

2	Equilibrium equations	(5.2)
1	Compatibility in terms of stress	(5.7)
3		

σ_x, σ_y, τ_{xy}

↓ $\left(\text{If } F_x = \dfrac{\partial V}{\partial x},\ F_y = \dfrac{\partial V}{\partial y}\right)$

$$\nabla^4\phi = (1-\nu)\nabla^2 V \qquad (5.10)$$

where

$$\frac{\partial^2\phi}{\partial x^2} = \sigma_y + V, \quad \frac{\partial^2\phi}{\partial y^2} = \sigma_x + V, \quad \frac{\partial^2\phi}{\partial x \partial y} = -\tau_{xy}$$

↓ (right branch)

2	Equilibrium equations	(5.2)
3	Stress-displacement equations	(5.13)
5		

σ_x, σ_y, τ_{xy}, u, and v

↓

2	Equilibrium equations in terms of displacement	(5.14)

u, v

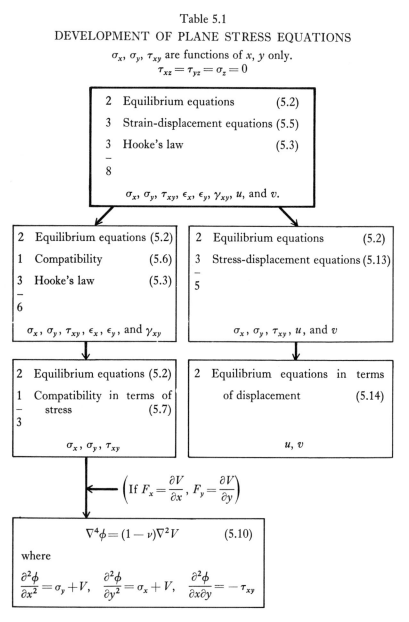

Since the assumptions of Eqs. (5.1) are written in terms of stress, it is more convenient to use the governing equations (4.1) and (4.25). (See Table 4.2.) Without body forces, Eqs. (4.25) are

$$
\begin{aligned}
\nabla^2\sigma_x + \frac{1}{1+\nu}\frac{\partial^2\Theta}{\partial x^2} &= 0 \\
\nabla^2\sigma_y + \frac{1}{1+\nu}\frac{\partial^2\Theta}{\partial y^2} &= 0 \\
\nabla^2\sigma_z + \frac{1}{1+\nu}\frac{\partial^2\Theta}{\partial z^2} &= 0 \\
\nabla^2\tau_{yz} + \frac{1}{1+\nu}\frac{\partial^2\Theta}{\partial y\partial z} &= 0 \\
\nabla^2\tau_{xz} + \frac{1}{1+\nu}\frac{\partial^2\Theta}{\partial x\partial z} &= 0 \\
\nabla^2\tau_{xy} + \frac{1}{1+\nu}\frac{\partial^2\Theta}{\partial x\partial y} &= 0
\end{aligned}
\tag{5.17}
$$

and the equilibrium equations (4.1) are

$$
\begin{aligned}
\frac{\partial\sigma_x}{\partial x} + \frac{\partial\tau_{xy}}{\partial y} + \frac{\partial\tau_{xz}}{\partial z} &= 0 \\
\frac{\partial\tau_{yx}}{\partial x} + \frac{\partial\sigma_y}{\partial y} + \frac{\partial\tau_{yz}}{\partial z} &= 0 \\
\frac{\partial\tau_{zx}}{\partial x} + \frac{\partial\tau_{zy}}{\partial y} + \frac{\partial\sigma_z}{\partial z} &= 0
\end{aligned}
\tag{5.18}
$$

Introducing the assumptions of Eqs. (5.1), we see that the third, fourth, and fifth of Eqs. (5.17) and the third of Eqs. (5.18) are identically satisfied. The introduction of the stress-function ϕ, defined by

$$
\frac{\partial^2\phi}{\partial x^2} = \sigma_y, \qquad \frac{\partial^2\phi}{\partial y^2} = \sigma_x, \qquad \frac{\partial^2\phi}{\partial x\partial y} = -\tau_{xy} \tag{5.19}
$$

satisfies the other two of Eqs. (5.18). Adding the first two of Eqs. (5.17) yields the governing biharmonic equation (5.11). The sixth of Eqs. (5.17) (also the first and second individually), in general, cannot be satisfied by the stresses assumed in Eqs. (5.1).

We say "in general" because there is a certain type of stress distribution satisfying the assumptions of Eqs. (5.1) which also satisfies all of Eqs. (5.17) as shown below.

Noting that, with $\sigma_z = 0$,

$$\Theta = \sigma_x + \sigma_y = \nabla_1^2\phi \tag{5.20}$$

we may write a number of combinations of the first, second, and sixth of Eqs. (5.17) in the form

$$\frac{\partial^2}{\partial x^2}(\nabla_1^2\phi) = 0$$

$$\frac{\partial^2}{\partial y^2}(\nabla_1^2\phi) = 0$$

$$\frac{\partial^2}{\partial x \partial y}(\nabla_1^2\phi) = 0$$

These equations indicate that

$$\nabla_1^2\phi = Ax + By + C \tag{5.21}$$

where A, B, and C are arbitrary constants. Since in deriving Eq. (5.21) all the governing equations (three equilibrium and six compatibility) are satisfied, the stresses resulting from it are exact solutions under the assumption of Eqs. (5.1). However, the stresses derived from Eq. (5.21) represent a limited special class of plane stress problems and are of little practical importance.

Since the assumptions of Eqs. (5.1) are only approximate in nature, it is desirable to investigate the significance of the approximation involved. Let us relax the assumptions made in Eqs. (5.1) and study the two-dimensional stress distribution with

$$\sigma_z = \tau_{xz} = \tau_{yz} = 0 \tag{5.22}$$

and with no restrictions placed upon the other stress components, i.e., they may be functions of x, y, and z. If the external forces are distributed symmetrically with respect to the middle plane $z = 0$ of the plate (Fig. 5.1), it can be shown that the exact solution which satisfies all the equilibrium and compatibility equations is[3]

$$\phi = \phi_0 - \frac{1}{2}\frac{\nu}{1+\nu}(\nabla_1^2\phi_0)z^2 \tag{5.23}$$

where

$$\phi_0 = \phi_0(x, y)$$

satisfies

$$\nabla_1^4\phi_0 = 0 \tag{5.24}$$

[3] See S. Timoshenko and J. N. Goodier, *Theory of Elasticity*, McGraw-Hill Book Company, Inc., New York, 1951, pp. 241–244.

and ϕ is defined by Eqs. (5.19). The second term on the right-hand side of Eq. (5.23), however, is proportional to z^2 and may be neglected for thin plates, in which case, $\phi \cong \phi_0$ is independent of z and

$$\nabla_1{}^4\phi \cong \nabla_1{}^4\phi_0 = 0$$

That is, for thin plates, solutions of Eq. (5.11) very closely approximate the true stress distributions.

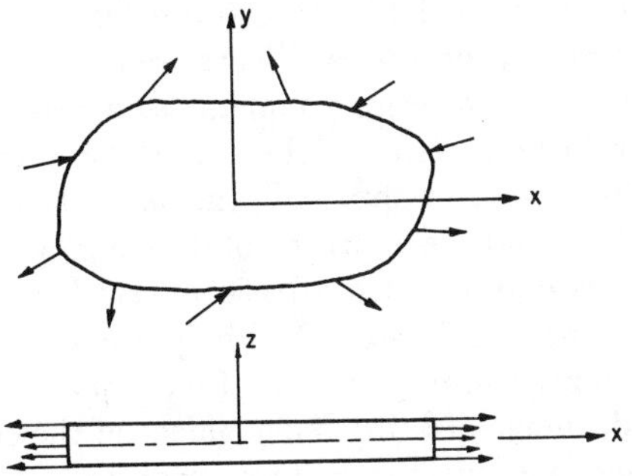

Fig. 5.1 A Thin Plate Under a State of Plane Stress

For an actual elastic body to be in a state of plane stress, the restrictions are as follows: the body must be a thin plate; the two z surfaces of the plate must be free from load; the external forces (body forces and surface forces) can have no z component; through the thickness of the plate, the external forces may be either uniformly distributed (independent of z), or distributed symmetrically with respect to the middle plane. If the external forces are not uniform, but symmetrical with respect to z, the surface forces and stress components in the governing equations are taken as the values of the corresponding quantities averaged through the thickness of the plate. For thin plates, the error involved by this averaging process is negligible.

The boundary conditions on the lateral surface for plane stress problems are

$$\begin{aligned} T_x{}^\mu &= \sigma_{xo}\mu_x + \tau_{xyo}\mu_y \\ T_y{}^\mu &= \tau_{xyo}\mu_x + \sigma_{yo}\mu_y \end{aligned} \tag{5.25}$$

if the surface forces are prescribed, or

$$\begin{aligned} u_b &= u(x_0, y_0) \\ v_b &= v(x_0, y_0) \end{aligned} \tag{5.26}$$

if u_b and v_b are prescribed.

5.4. Polar Coordinates in Two-Dimensional Problems

The solutions of many elasticity problems, especially for bodies of circular cylindrical form, are conveniently formulated in terms of cylindrical coordinates r, θ, z. In the case of plane problems (plane stress or plane strain), the shear stresses on any z plane vanish and the other dependent variables are independent of z, thus the cylindrical coordinates may be reduced to polar coordinates r, θ.

Polar coordinates or cylindrical coordinates are special cases of curvilinear orthogonal coordinates. There are four different approaches available to obtain the governing equations in elasticity for different curvilinear coordinate systems. These are the vector (and dyadic) approach, the general tensor approach, the use of transformations from Cartesian equations, and the approach of deriving every equation in the particular coordinate system from basic physical laws. The last two approaches, for polar coordinates, shall be presented in this chapter.

In the vector approach, we write all of our equations in vector form, which has the advantage of being independent of the coordinate system selected. For any particular curvilinear coordinate system, it is only necessary to find the expressions for different vector operators in that system. In the general tensor approach, all equations are derived in tensor (general tensor, not Cartesian tensor) form, which is applicable to any (orthogonal or nonorthogonal) coordinate system. The vector approach to develop the equations of elasticity will be discussed in Chapters 11 and 12.

In the approach of transformation from Cartesian equations, we first find the functional relationship between the particular curvilinear coordinates and the Cartesian coordinates x, y, z, and then transform all our equations in Cartesian coordinates accordingly. This approach is frequently very tedious and lengthy. For the polar coordinates, as shown in Fig. 5.2, we have the functional relations

$$\begin{cases} x = r\cos\theta \\ y = r\sin\theta \end{cases} \qquad \begin{cases} \theta = \tan^{-1}\dfrac{y}{x} \\ r^2 = x^2 + y^2 \end{cases} \tag{5.27}$$

and

$$\begin{aligned} &\frac{\partial r}{\partial x} = \frac{x}{r} = \cos\theta \qquad && \frac{\partial r}{\partial y} = \frac{y}{r} = \sin\theta \\ &\frac{\partial \theta}{\partial x} = -\frac{y}{r^2} = -\frac{\sin\theta}{r} \qquad && \frac{\partial \theta}{\partial y} = \frac{x}{r^2} = \frac{\cos\theta}{r} \end{aligned} \tag{5.28}$$

Any derivatives with respect to x and y in the Cartesian equations may be transformed into derivatives with respect to r and θ by

$$\begin{aligned}\frac{\partial}{\partial x} &= \frac{\partial r}{\partial x}\frac{\partial}{\partial r} + \frac{\partial \theta}{\partial x}\frac{\partial}{\partial \theta} = \cos\theta\,\frac{\partial}{\partial r} - \frac{\sin\theta}{r}\frac{\partial}{\partial \theta}\\ \frac{\partial}{\partial y} &= \frac{\partial r}{\partial y}\frac{\partial}{\partial r} + \frac{\partial \theta}{\partial y}\frac{\partial}{\partial \theta} = \sin\theta\,\frac{\partial}{\partial r} + \frac{\cos\theta}{r}\frac{\partial}{\partial \theta}\end{aligned} \tag{5.29}$$

Relations governing properties at a point and which do not contain any space derivatives, such as the transformation of stress equations and the

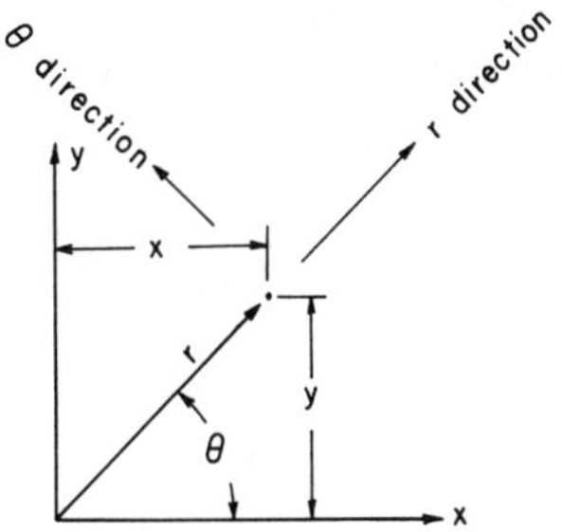

Fig. 5.2 Polar Coordinates

Hooke's law equations, are not affected by the curvilinear nature of the coordinates. These equations in orthogonal curvilinear coordinates may be obtained directly from the corresponding equations in Cartesian coordinates by changing x, y into, say, r, θ. Letting σ_r, σ_θ, and $\tau_{r\theta}$ be the stress components in the radial and tangential directions, Eqs. (1.5), (1.6), and (1.7) become

$$\begin{aligned}\sigma_x &= \sigma_r \cos^2\theta + \sigma_\theta \sin^2\theta - \tau_{r\theta}\sin 2\theta\\ \sigma_y &= \sigma_r \sin^2\theta + \sigma_\theta \cos^2\theta + \tau_{r\theta}\sin 2\theta\\ \tau_{xy} &= (\sigma_r - \sigma_\theta)\sin\theta\cos\theta + \tau_{r\theta}(\cos^2\theta - \sin^2\theta)\end{aligned} \tag{5.30}$$

Similar transformation equations may also be written for the strains ϵ_r, ϵ_θ and $\gamma_{r\theta}$. Substituting Eqs. (5.30) into the first of equilibrium Eqs. (5.2), assuming zero body forces, and transforming the derivatives according to Eqs. (5.29), we obtain

$$\begin{aligned}&\left[\frac{\partial \sigma_r}{\partial r} + \frac{1}{r}\frac{\partial \tau_{r\theta}}{\partial \theta} + \frac{\sigma_r - \sigma_\theta}{r}\right]\cos\theta\\ &\qquad - \left[\frac{1}{r}\frac{\partial \sigma_\theta}{\partial \theta} + \frac{\partial \tau_{r\theta}}{\partial r} + 2\frac{\tau_{r\theta}}{r}\right]\sin\theta = 0\end{aligned} \tag{5.31}$$

Since this equation must be satisfied for all values of θ, we must have

$$\frac{\partial\sigma_r}{\partial r}+\frac{1}{r}\frac{\partial\tau_{r\theta}}{\partial\theta}+\frac{\sigma_r-\sigma_\theta}{r}=0 \qquad \text{at } \theta=0$$
$$\frac{1}{r}\frac{\partial\sigma_\theta}{\partial\theta}+\frac{\partial\tau_{r\theta}}{\partial r}+\frac{2\tau_{r\theta}}{r}=0 \qquad \text{at } \theta=\pi/2 \tag{5.32}$$

But the choice of the x direction is arbitrary, so that Eqs. (5.32) must be valid for all values of θ. The first of Eqs. (5.32) is the equilibrium equation in the radial direction, and the second, that in the tangential direction. By a similar technique, the strain-displacement equations (5.5) may be transformed into (see Problem 5-8)

$$\epsilon_r=\frac{\partial u_r}{\partial r}$$
$$\epsilon_\theta=\frac{u_r}{r}+\frac{1}{r}\frac{\partial u_\theta}{\partial\theta} \tag{5.33}$$
$$\gamma_{r\theta}=\frac{1}{r}\frac{\partial u_r}{\partial\theta}+\frac{\partial u_\theta}{\partial r}-\frac{u_\theta}{r}$$

where u_r and u_θ are the displacement components in the radial and tangential directions, respectively. Eliminating u_r and u_θ from Eqs. (5.33), we obtain the compatibility equation

$$\frac{\partial^2\epsilon_\theta}{\partial r^2}+\frac{1}{r^2}\frac{\partial^2\epsilon_r}{\partial\theta^2}+\frac{2}{r}\frac{\partial\epsilon_\theta}{\partial r}-\frac{1}{r}\frac{\partial\epsilon_r}{\partial r}=\frac{1}{r}\frac{\partial^2\gamma_{r\theta}}{\partial r\,\partial\theta}+\frac{1}{r^2}\frac{\partial\gamma_{r\theta}}{\partial\theta} \tag{5.34}$$

Since the polar coordinate system is orthogonal, the Hooke's law equations in this system for plane problems in isotropic media may be written by simply replacing x by r and y by θ in the corresponding equations expressed in Cartesian coordinates. Thus for plane stress problems, we have

$$\epsilon_r=\frac{1}{E}[\sigma_r-\nu\sigma_\theta]$$
$$\epsilon_\theta=\frac{1}{E}[\sigma_\theta-\nu\sigma_r]$$
$$\gamma_{r\theta}=\frac{1}{G}\tau_{r\theta}$$

or (5.35a)

$$\sigma_r=\frac{E}{1-\nu^2}(\epsilon_r+\nu\epsilon_\theta)$$
$$\sigma_\theta=\frac{E}{1-\nu^2}(\epsilon_\theta+\nu\epsilon_r)$$
$$\tau_{r\theta}=G\gamma_{r\theta}$$

and for plane strain problems,

$$\epsilon_r = \frac{1-\nu^2}{E}\left[\sigma_r - \left(\frac{\nu}{1-\nu}\right)\sigma_\theta\right]$$

$$\epsilon_\theta = \frac{1-\nu^2}{E}\left[\sigma_\theta - \left(\frac{\nu}{1-\nu}\right)\sigma_r\right]$$

$$\gamma_{r\theta} = \frac{1}{G}\tau_{r\theta}$$

or (5.35b)

$$\sigma_r = 2G\epsilon_r + \lambda(\epsilon_r + \epsilon_\theta)$$

$$\sigma_\theta = 2G\epsilon_\theta + \lambda(\epsilon_r + \epsilon_\theta)$$

$$\tau_{r\theta} = G\gamma_{r\theta}$$

The plane dilatation ϵ in plane strain problems is given by

$$\epsilon = \epsilon_r + \epsilon_\theta = \frac{1}{r}\frac{\partial(ru_r)}{\partial r} + \frac{1}{r}\frac{\partial u_\theta}{\partial\theta} \tag{5.36}$$

and the plane strain equilibrium equations in terms of displacement are

$$\begin{aligned}(\lambda+2G)\frac{\partial\epsilon}{\partial r} - \frac{2G}{r}\frac{\partial\omega}{\partial\theta} + F_r = 0\\ (\lambda+2G)\frac{1}{r}\frac{\partial\epsilon}{\partial\theta} + 2G\frac{\partial\omega}{\partial r} + F_\theta = 0\end{aligned} \tag{5.37}$$

where

$$\omega = \frac{1}{2r}\left[\frac{\partial(ru_\theta)}{\partial r} - \frac{\partial u_r}{\partial\theta}\right] \tag{5.38}$$

is the rotation about the z axis.

The compatibility equation in terms of the stress function (the biharmonic equation), with no body forces, is

$$\nabla^4\phi = \nabla^2\nabla^2\phi = \left(\frac{\partial^2}{\partial r^2} + \frac{1}{r}\frac{\partial}{\partial r} + \frac{1}{r^2}\frac{\partial^2}{\partial\theta^2}\right)\left(\frac{\partial^2}{\partial r^2} + \frac{1}{r}\frac{\partial}{\partial r} + \frac{1}{r^2}\frac{\partial^2}{\partial\theta^2}\right)\phi = 0 \tag{5.39}$$

and the stress components are given by

$$\begin{aligned}\sigma_r &= \frac{1}{r}\frac{\partial\phi}{\partial r} + \frac{1}{r^2}\frac{\partial^2\phi}{\partial\theta^2}\\ \sigma_\theta &= \frac{\partial^2\phi}{\partial r^2}\\ \tau_{r\theta} &= -\frac{\partial}{\partial r}\left(\frac{1}{r}\frac{\partial\phi}{\partial\theta}\right)\end{aligned} \tag{5.40}$$

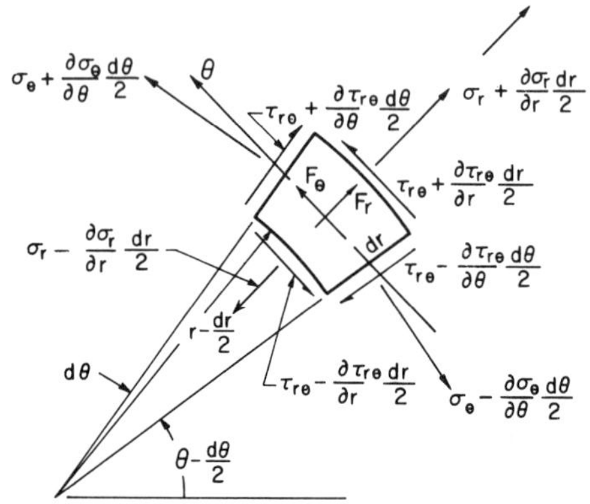

Fig. 5.3 Stresses on an Element in Polar Coordinates

Equations (5.38), (5.39), and (5.40) are valid for plane strain *and* plane stress problems.

We shall now demonstrate the approach to obtain the governing equations in polar coordinates by determining the stress, strain, and displacement relations for an infinitesimal element in the polar coordinate system. The equilibrium equations for plane stress or plane strain problems will be developed by referring to Fig. 5.3. Assuming that the dimension of the element in the z direction is unity, we have, by summing forces in the r direction,

$$\left(\sigma_r + \frac{\partial \sigma_r}{\partial r}\frac{dr}{2}\right)\left(r + \frac{dr}{2}\right) d\theta - \left(\sigma_r - \frac{\partial \sigma_r}{\partial r}\frac{dr}{2}\right)\left(r - \frac{dr}{2}\right) d\theta$$
$$+ \left(\tau_{r\theta} + \frac{\partial \tau_{r\theta}}{\partial \theta}\frac{d\theta}{2}\right) dr \cos\frac{d\theta}{2} - \left(\tau_{r\theta} - \frac{\partial \tau_{r\theta}}{\partial \theta}\frac{d\theta}{2}\right) dr \cos\frac{d\theta}{2}$$
$$- \left(\sigma_\theta + \frac{\partial \sigma_\theta}{\partial \theta}\frac{d\theta}{2}\right) dr \sin\frac{d\theta}{2} - \left(\sigma_\theta - \frac{\partial \sigma_\theta}{\partial \theta}\frac{d\theta}{2}\right) dr \sin\frac{d\theta}{2}$$
$$+ F_r r\, dr\, d\theta = 0$$

Making the small angle approximations and neglecting higher-order terms, we have

$$\frac{\partial \sigma_r}{\partial r} + \frac{1}{r}\frac{\partial \tau_{r\theta}}{\partial \theta} + \frac{\sigma_r - \sigma_\theta}{r} + F_r = 0 \tag{5.41}$$

Similarly, summing forces in the θ direction, we find that

$$\frac{1}{r}\frac{\partial \sigma_\theta}{\partial \theta} + \frac{\partial \tau_{r\theta}}{\partial r} + \frac{2\tau_{r\theta}}{r} + F_\theta = 0 \tag{5.42}$$

These equations, of course, reduce to Eqs. (5.32) when the body forces vanish.

In order to express the strain-displacement relations, let us first consider the normal strain ϵ_r and refer to Fig. 5.4(a). An element AD

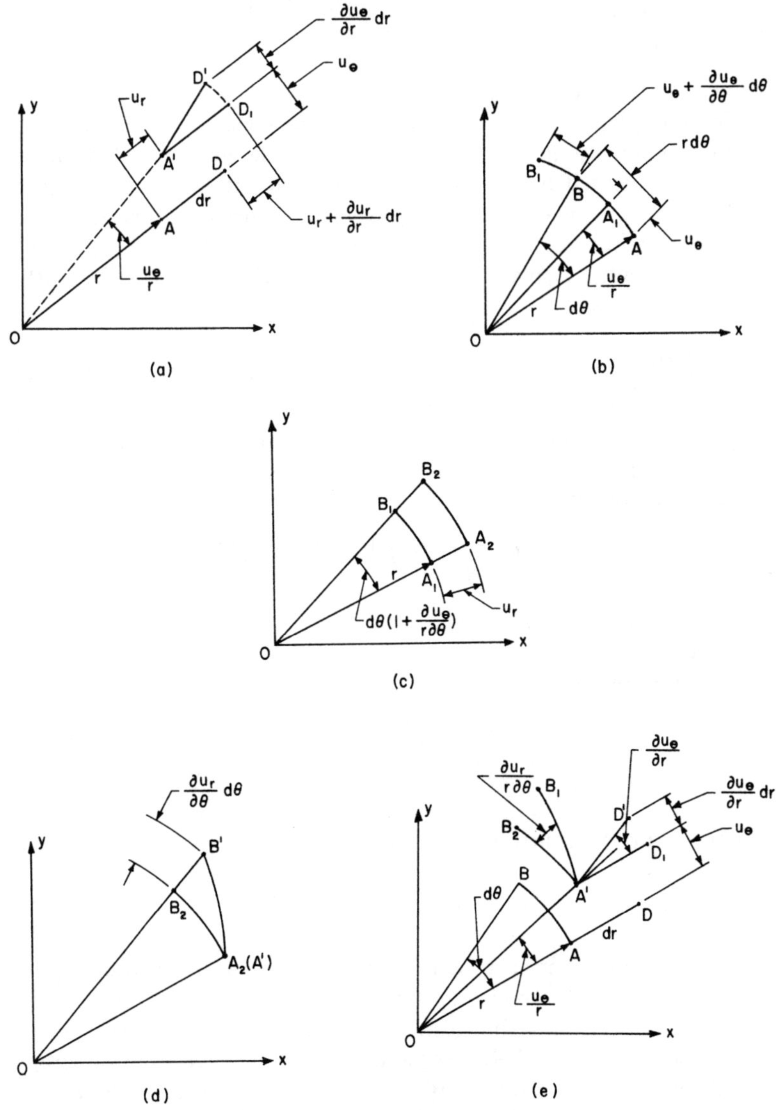

Fig. 5.4 Deformation and Displacement of an Element in Polar Coordinates

in the radial direction before deformation moves to $A'D'$ afterwards. According to the definition of normal strain in the linear theory, we have

$$\epsilon_r = \frac{A'D' - AD}{AD} = \frac{\left[\left(dr + \dfrac{\partial u_r}{\partial r}\,dr\right)^2 + \left(\dfrac{\partial u_\theta}{\partial r}\,dr\right)^2\right]^{1/2} - dr}{dr} = \frac{\partial u_r}{\partial r} \tag{5.43}$$

dropping higher-order terms.

For the relation between the strain component ϵ_θ and the displacement components u_r, u_θ, let us consider a line element AB which is originally oriented in the tangential direction and which is moved to $A'B'$ after straining occurs. It is convenient to visualize this movement as occurring in three steps, as shown in Fig. 5.4(b), (c), and (d). Due to a pure tangential displacement, AB to A_1B_1, the strain is (Fig. 5.4(b))

$$\epsilon_{\theta_1} = \frac{A_1B_1 - AB}{AB} = \frac{\left(u_\theta + \dfrac{\partial u_\theta}{\partial \theta}\,d\theta + r\,d\theta - u_\theta\right) - (r\,d\theta)}{r\,d\theta} = \frac{\partial u_\theta}{r\,\partial \theta} \tag{5.44}$$

A pure radial displacement also contributes to the strain ϵ_θ, as shown in Fig. 5.4(c). Neglecting the term $(1/r)(\partial u_\theta/\partial \theta)$ as being small compared with unity, we get

$$\epsilon_{\theta_2} = \frac{A_2B_2 - A_1B_1}{A_1B_1} = \frac{(r + u_r)\,d\theta - r\,d\theta}{r\,d\theta} = \frac{u_r}{r} \tag{5.45}$$

The contribution to the normal strain by $\partial u_r/\partial \theta$ (Fig. 5.4(d)) is a small quantity of higher order and can be neglected. This is given by (dropping higher-order terms)

$$\epsilon_{\theta_3} = \frac{\left[(r\,d\theta)^2 + \left(\dfrac{\partial u_r}{\partial \theta}\,d\theta\right)^2\right]^{1/2} - r\,d\theta}{r\,d\theta} \cong 0$$

Summing up the contributions from these three steps, we have

$$\epsilon_\theta = \epsilon_{\theta_1} + \epsilon_{\theta_2} + \epsilon_{\theta_3} = \frac{1}{r}\frac{\partial u_\theta}{\partial \theta} + \frac{u_r}{r} \tag{5.46}$$

Figure 5.4(e) illustrates the relation between the shear strain $\gamma_{r\theta}$ and the displacement components. By definition, $\gamma_{r\theta}$ is the decrease of the right angle BAD. As shown in the figure, angle $D'A'D_1$ is given by $\partial u_\theta/\partial r$, where $A'D_1$ and AD are parallel. As in the preceding paragraph, the displacement of AB to $A'B'$ is assumed to take place in three steps, AB to A_1B_1, A_1B_1 to A_2B_2, and then A_2B_2 to $A'B'$. When AB moves to A_1B_1, an angular displacement (counterclockwise) of the magnitude

u_θ/r occurs. Together with the clockwise angular displacement $\partial u_r/r\partial\theta$ from A_2B_2 to $A'B'$, the total angular displacement of AB is

$$\frac{\partial u_r}{r\,\partial\theta} - \frac{u_\theta}{r}$$

and the shear strain is

$$\gamma_{r\theta} = \frac{\partial u_\theta}{\partial r} + \frac{\partial u_r}{r\,\partial\theta} - \frac{u_\theta}{r} \tag{5.47}$$

We have thus checked Eqs. (5.33).

The Hooke's law equations in polar coordinates can be derived from a small element following the same procedure as that in the Cartesian coordinates. Thus we have the eight governing equations in polar coordinates.

5.5. Axisymmetric Plane Problems

The use of polar coordinates is particularly convenient in the solution of axisymmetric problems, i.e., the body is symmetrical about the z axis, as are the applied forces and/or displacements. We shall restrict our attention in this section to problems in which the stresses and the displacements are independent of θ, so that their derivatives with respect to θ vanish.[4] In particular, if the stress function ϕ is also independent of θ, Eq. (5.39), which governs the solution of plane stress or plane strain problems, simplifies to an ordinary differential equation,

$$\left(\frac{d^4}{dr^4} + \frac{2}{r}\frac{d^3}{dr^3} - \frac{1}{r^2}\frac{d^2}{dr^2} + \frac{1}{r^3}\frac{d}{dr}\right)\phi = 0 \tag{5.48}$$

This can be reduced to a differential equation with constant coefficients by the introduction of a new independent variable ξ, such that

$$r = e^\xi$$

and Eq. (5.48) becomes

$$\left(\frac{d^4}{d\xi^4} - 4\frac{d^3}{d\xi^3} + 4\frac{d^2}{d\xi^2}\right)\phi = 0$$

for which the general solution is

$$\phi = A\xi e^{2\xi} + Be^{2\xi} + C\xi + D$$

or

$$\phi = Ar^2\log r + Br^2 + C\log r + D \tag{5.49}$$

[4] In some problems of practical importance, the stresses may be independent of θ, while the displacements are functions of r and θ.

where A, B, C, and D are arbitrary constants and "log" denotes the natural logarithm. The stress components then, can be determined from Eqs. (5.40), which in this case reduce to

$$\sigma_r = \frac{1}{r}\frac{d\phi}{dr}$$
$$\sigma_\theta = \frac{d^2\phi}{dr^2} \tag{5.50}$$
$$\tau_{r\theta} = 0$$

Substituting Eq. (5.49) into Eq. (5.50), we see that

$$\sigma_r = 2A \log r + \frac{C}{r^2} + A + 2B$$
$$\sigma_\theta = 2A \log r - \frac{C}{r^2} + 3A + 2B \tag{5.51}$$
$$\tau_{r\theta} = 0$$

If the region is simply connected (solid circular cylinder), we must take

$$A = C = 0$$

in order that the stresses remain finite at the origin. Thus the stresses in a solid circular cylinder under axisymmetric loading are

$$\sigma_r = \sigma_\theta = 2B$$
$$\tau_{r\theta} = 0$$

which obviously solve the problem of a solid cylinder under a normal stress $2B$.

If the cylinder is multiply connected (a circular cylinder with a concentric circular hole), as stated in Chapter 2, the compatibility equations are not sufficient to guarantee single-valued displacements. Thus in this case, the displacement must be considered. Referring to Eqs. (5.33), we note that the normal strain components are given by

$$\epsilon_r = \frac{du_r}{dr}$$
$$\epsilon_\theta = \frac{u_r}{r} \tag{5.52}$$

and from Eqs. (5.52) and (5.35), we have

$$\frac{du_r}{dr} = K_1(\sigma_r - K_2\,\sigma_\theta)$$
$$\frac{u_r}{r} = K_1(\sigma_\theta - K_2\,\sigma_r) \tag{5.53}$$

where

$$K_1 = \frac{1}{E} \qquad K_2 = \nu$$

for plane stress, and

$$K_1 = \frac{1-\nu^2}{E} \qquad K_2 = \frac{\nu}{1-\nu}$$

for plane strain. Substituting Eqs. (5.51) into (5.53) then, we have

$$\frac{du_r}{dr} = K_1\left[2A \log r + \frac{C}{r^2} + A + 2B - K_2(2A \log r - \frac{C}{r^2} + 3A + 2B)\right]$$

$$\frac{u_r}{r} = K_1\left[2A \log r - \frac{C}{r^2} + 3A + 2B - K_2\left(2A \log r + \frac{C}{r^2} + A + 2B\right)\right] \tag{5.54}$$

Integration of the first of Eqs. (5.54) gives

$$u_r = K_1\left[2Ar \log r - Ar + 2Br - \frac{C}{r} - K_2\left(2Ar \log r + Ar + 2\,Br + \frac{C}{r}\right) + F\right] \tag{5.55}$$

where F is a constant. Equating the expressions for u_r given by Eq. (5.55) and the second of Eqs. (5.54) yields the result

$$4Ar - F = 0$$

and since the equation must be satisfied for all values of r in the region, we have

$$A = F = 0$$

The remaining constants, B and C, are to be determined from the boundary conditions on the inner and outer boundary surfaces.

In the displacement formulation, the same problem with axisymmetric stress and strain may be solved by taking

$$u_\theta = \frac{\partial}{\partial\theta} = \omega = 0$$

With no body forces, the second of Eqs. (5.37) is identically satisfied, and the first of Eqs. (5.37) reduces to

$$\frac{d^2u_r}{dr^2} + \frac{1}{r}\frac{du_r}{dr} - \frac{u_r}{r^2} = 0$$

The general solution of this equation is given by

$$u_r(r) = c_1 r + c_2 \frac{1}{r} \tag{5.56}$$

where c_1 and c_2 are arbitrary constants.

Let us consider for example, the hollow cylinder shown in Fig. 5.5, which is subjected to the prescribed radial displacements

$$u_r|_{r=a} = u_0$$
$$u_r|_{r=b} = 0$$

Substituting the boundary conditions

$$u_r(a) = u_0$$
$$u_r(b) = 0$$

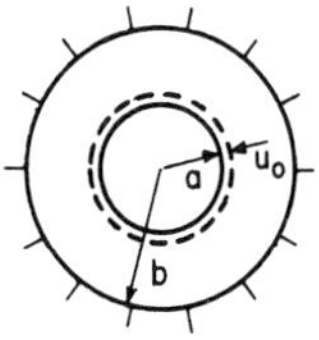

Fig. 5.5 Hollow Cylinder

into Eq. (5.56), we find that

$$c_1 = \frac{a}{a^2 - b^2} u_0$$

and

$$c_2 = \frac{-ab^2}{a^2 - b^2} u_0$$

Thus the solution is given by

$$u_r(r) = \frac{au_0}{a^2 - b^2}\left[r - \frac{b^2}{r}\right]$$

The stresses may now be computed, using Eqs. (5.33) and Hooke's law, Eqs. (5.35).

Another problem with axisymmetric stress and strain may be obtained by taking

$$\phi = \tau_0 \theta$$

where τ_0 is a constant. In this case, the only nonvanishing stress and strain components are

$$\tau_{r\theta} = \frac{\tau_0}{r^2} \qquad \gamma_{r\theta} = \frac{\tau_0}{Gr^2}$$

respectively. This represents the case of a hollow cylinder under uniform shear stresses applied on the inner and outer boundaries.

5.6. The Semi-Inverse Method

In general, closed form solutions of the governing partial differential equations with prescribed boundary conditions which occur in elasticity problems are very difficult to obtain directly, and they are frequently impossible to achieve. Because of this fact, the inverse and semi-inverse methods may be attempted in the solutions of certain problems. In the inverse method we select a specific solution which satisfies the governing equations, and then search for the boundary conditions which can be satisfied by this solution, i.e., we have the solution first and then ask what problem it can solve.

In the semi-inverse method, we assume a partial solution to a given problem. A partial solution consists of an assumed form for each dependent variable in terms of known and unknown functions. The assumed partial solution is then substituted into the original set of governing equations. As a result, these equations will be reduced to a set of simplified differential equations which govern the remaining unknown functions. This simplified set of equations, together with proper boundary conditions, is then solved by direct methods. An example of the semi-inverse method will be presented in this section and another will be discussed in the next chapter in the solution of the torsion problem. The form of the partial solution in the semi-inverse method may be determined from physical reasoning or from the known solution of a similar problem. The uniqueness theorem, of course, authenticates the solution which satisfies the governing equations and prescribed boundary conditions for a given problem.

As an example of the application of the semi-inverse method, we consider the plane strain problem in polar coordinates. The governing equation in this case is the biharmonic equation (5.11). Let us assume a partial solution for ϕ of the form

$$\phi = rF(\theta) \tag{5.57}$$

where F is an unknown function of θ. Substitution of this ϕ into Eq. (5.11) yields the following simplified equation,

$$\frac{d^4F(\theta)}{d\theta^4} + 2\frac{d^2F(\theta)}{d\theta^2} + F(\theta) = 0 \tag{5.58}$$

This is an ordinary differential equation for which the general solution may be found easily. With proper prescribed boundary conditions, the specific solutions may be obtained for certain problems. This amounts to the determination of arbitrary constants.

Practical problems that can be solved by Eq. (5.57) include a semi-infinite wedge under loads applied at the apex. Figure 5.6(a) shows a wedge under a horizontal line load P (force per unit length in the z direction). To avoid the infinite stress at the origin, we shall exclude a small wedge near the apex and consider the truncated semi-infinite wedge to the right as shown in Fig. 5.6(b). As long as the resultant of

σ_r and $\tau_{r\theta}$ over the circular arc boundary on the left side of the semi-infinite wedge is equal to a horizontal force P (per unit length), by Saint-Venant's principle, the stress distribution a small distance away from this boundary is the same as that in the original wedge.

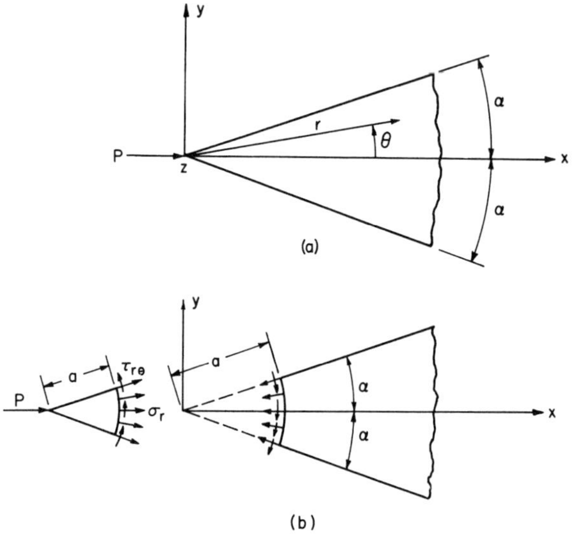

Fig. 5.6 A Semi-Infinite Wedge Loaded by a Force P

The boundary conditions for this problem then, are at $r = a$,

$$\int_{-\alpha}^{\alpha} (\sigma_r \cos\theta - \tau_{r\theta} \sin\theta) r\, d\theta + P = 0 \tag{5.59a}$$

$$\int_{-\alpha}^{\alpha} (\sigma_r \sin\theta + \tau_{r\theta} \cos\theta) r\, d\theta = 0 \tag{5.59b}$$

$$\int_{-\alpha}^{\alpha} \tau_{r\theta} r^2\, d\theta = 0 \tag{5.59c}$$

at $\theta = \pm\alpha$,
$$\sigma_\theta = \tau_{r\theta} = 0 \tag{5.60}$$

at $r = \infty$,
$$\sigma_r = \sigma_\theta = \tau_{r\theta} = 0 \tag{5.61}$$

These boundary conditions can be satisfied by the general solution of Eq. (5.58), which is

$$F(\theta) = A \sin\theta + B \cos\theta + C\theta \sin\theta + D\theta \cos\theta \tag{5.62}$$

where A, B, C, and D are arbitrary constants. The corresponding stress components are, from Eqs. (5.40),

$$\left.\begin{aligned} \sigma_\theta &= \tau_{r\theta} = 0 \\ \sigma_r &= (2/r)(C\cos\theta - D\sin\theta) \end{aligned}\right\} \tag{5.63}$$

The boundary conditions of Eqs. (5.60) and (5.61) are satisfied by Eqs. (5.63). From Eqs. (5.59), the constants C and D are shown to be

$$C = \frac{-P}{2\alpha + \sin 2\alpha}$$

$$D = 0$$

and the solution for the stress components in this problem becomes

$$\left.\begin{aligned} \sigma_r &= \frac{-2P}{2\alpha + \sin 2\alpha}\,\frac{\cos\theta}{r} \\ \sigma_\theta &= \tau_{r\theta} = 0 \end{aligned}\right\} \tag{5.64}$$

The form of the partial solution (5.57) is motivated by the following consideration. Given that $\tau_{r\theta} = 0$ in this problem, it is evident from Eq. (5.59a) that σ_r can be of the form $\sigma_r = (1/r)g(\theta)$ since Eq. (5.59a) is actually valid for any r. This is due to the fact that the force (per unit length) on any section passing entirely through the wedge is P. Problems 5-12 and 5-13 can also be solved by the stress function of Eq. (5.57).

PROBLEMS

5-1 Verify the fact that, with no body forces, the governing equation in plane strain problems is the biharmonic equation, $\nabla^4\phi = 0$.

5-2 (a) Show that the following stress distribution represents a plane stress elasticity solution for a thin plate

$$\sigma_x = pyx^3 - 2axy + by$$

$$\sigma_y = pxy^3 - 2px^3y$$

$$\tau_{xy} = -\frac{3}{2}px^2y^2 + ay^2 + \frac{p}{2}x^4 + c$$

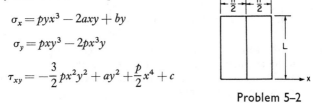

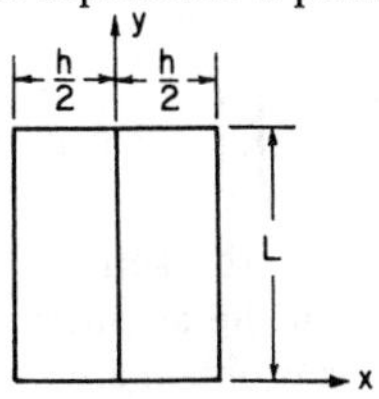

Problem 5–2

where p, a, b, and c are constants.

(b) Find the stress function ϕ corresponding to this stress distribution.

(c) If the given stresses act in the thin plate shown, find the constants a, b, and c such that there is no shear stress on the edges $x = \pm h/2$ and no normal stress on the edge $x = -h/2$.

(d) Determine the surface forces on each boundary of the plate.

5-3 The approximate solution for the stress distribution in the plate shown in the figure as given in strength of materials is $\sigma_x = \sigma_0 a/x$ and $\sigma_y = \sigma_z = \tau_{xy} = \tau_{yz} = \tau_{zx} = 0$.

(a) Show that these stresses do not satisfy the governing equations.

(b) Show that the boundary conditions are not satisfied by these stresses.

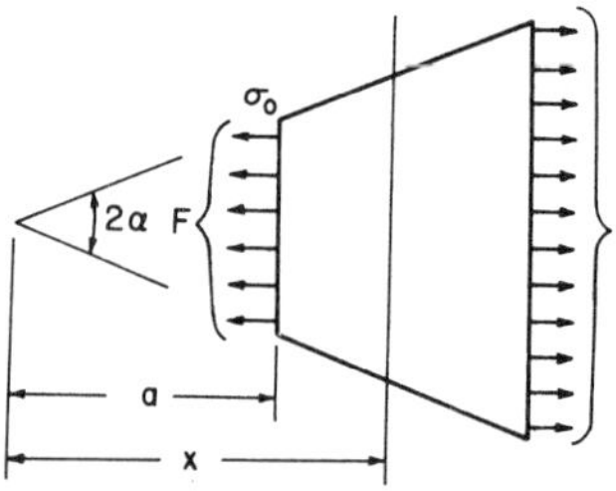

Problem 5–3

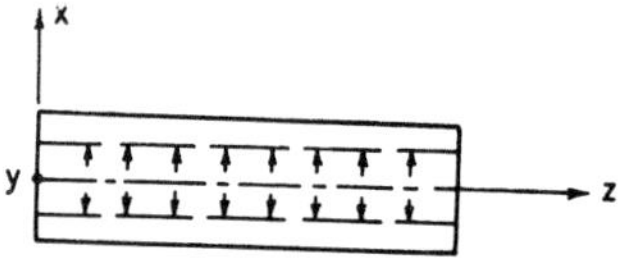

Problem 5–4

5-4 A long cylinder under surface forces which are independent of z and which have no z components constitutes a plane strain problem if both ends are fixed, i.e., $w = 0$. If the ends ($z = 0$ and $z = \ell$) are free to expand ($w \neq 0$), the desired solution may be obtained by superposing the solution of an auxiliary problem and the solution for the fixed-end problem. The auxiliary problem is for the same cylinder under a surface stress σ_z which is equal in magnitude but with opposite sign to the axial stress in the problem of the fixed-end cylinder. The combined problem is thus stress-free on the ends. According to Saint-Venant's principle, however, the auxiliary problem may be defined by the fact that the force and moment resultants on the end planes are equal and opposite to those in the fixed-end problem. Since σ_z in the auxiliary problem is a linear function of x and y (see Problem 4-11), show that the distribution of σ_z in the auxiliary problem can be determined from the equations

$$\iint \sigma_z{}^* \, dx\, dy = \iint \sigma_z{}^* x \, dx\, dy = \iint \sigma_z{}^* y \, dx\, dy = 0$$

where $\sigma_z{}^*$ is the sum of the axial stress in the fixed-end problem and the stress σ'_z in the auxiliary problem. Also show that the stresses σ_x, σ_y, and τ_{xy} for this problem are the same as the respective stresses given by the fixed-end solution (except near the ends).

5-5 Verify the statements preceding Eqs. (5.15) and (5.16).

5-6 Show that the expression for ϕ as given by Eq. (5.23) satisfies all of the three-dimensional governing equations.

5-7 Derive the strain-displacement relation equations (5.33) in polar coordinates by transformation from the corresponding equations (5.5) in Cartesian coordinates. Utilize the relations (see figure)

$$u = u_r \cos\theta - u_\theta \sin\theta$$
$$v = u_r \sin\theta + u_\theta \cos\theta$$

together with the transformation of strain equations

$$\epsilon_r = \epsilon_x \cos^2\theta + \epsilon_y \sin^2\theta + \gamma_{xy} \sin\theta\cos\theta$$
$$\epsilon_\theta = \qquad \cdots$$
$$\gamma_{r\theta} = \qquad \cdots$$

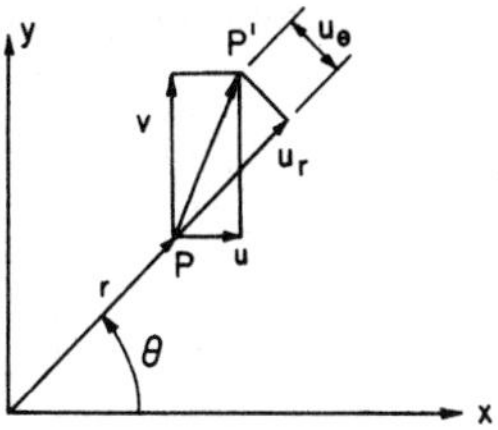

Problem 5–7

and the functional relations of Eqs. (5.28).

5-8 Fill in the missing steps in the derivation of Eqs. (5.31), (5.33), (5.36), (5.37), (5.38), (5.39), and (5.40).

5-9 Construct a table of the formulation of plane strain and plane stress problems in polar coordinates similar to Tables 4.1 and 5.1. Can we change all plane stress equations in polar coordinates into corresponding plane strain equations by replacing E and ν in the plane-stress equations with E_1 and ν_1 (Eqs. (5.15)), as is done in Cartesian coordinates?

5-10 Integrate the strain-displacement relations, Eqs. (5.33), with $\epsilon_r = \epsilon_\theta = \gamma_{r\theta} = 0$ to show that the rigid body displacement components in polar coordinates are given by

$$u_r^* = a \sin\theta + b \cos\theta$$
$$u_\theta^* = a \cos\theta - b \sin\theta + cr$$

where a, b, and c are constants. Provide a physical interpretation of the constants a, b, and c.

5-11 Find the stress and displacement distribution for the infinitely long hollow cylinder shown under the following boundary conditions:

(a) at $r = a$, $\sigma_r = \sigma_a$
at $r = b$, $\sigma_r = \sigma_b$
(b) at $r = a$, $\sigma_r = \sigma_a$
at $r = b$, $u_r = u_b$
(c) at $r = a$, $u_r = u_a$
at $r = b$, $u_r = u_b$

Repeat the problem for the case where the cylinder is very thin (plane stress).

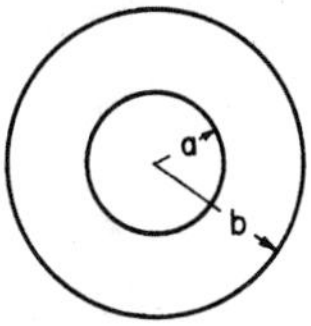

Problem 5–11

5-12 Use the analysis of Section 5.6 to determine the stress components in the semi-infinite wedge shown.

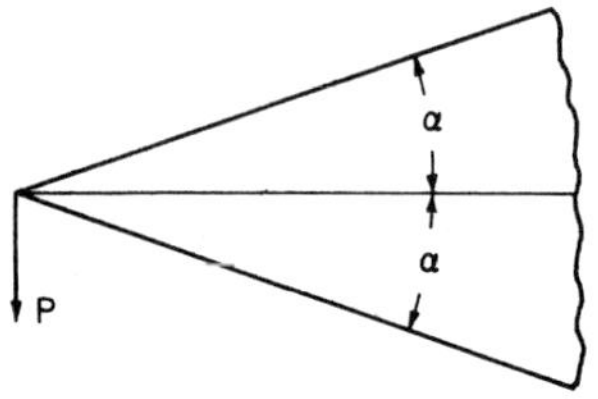

Problem 5–12

5-13 Use the results of Problem 5-12 and Eqs. (5.64) to write down the stress components for each of the problems of the figure.

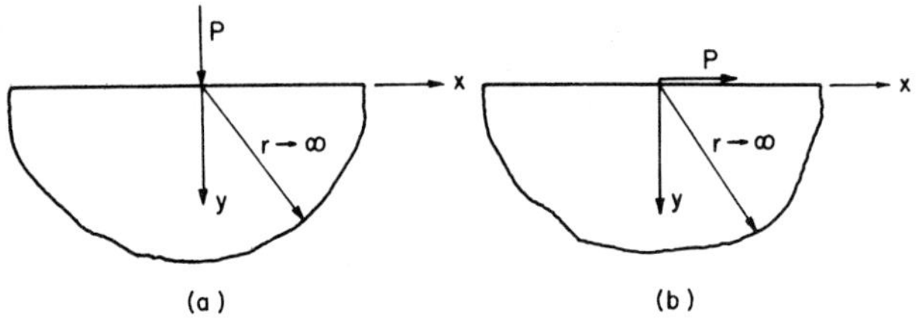

Problem 5–13

5-14 Consider the theorem stated below to show that an Airy stress function exists for each stress distribution which satisfies the two-dimensional equations of equilibrium. For simplicity, let the body forces be zero and the region be simply connected.

"Let P and Q be two functions of x and y, such that P, Q, $\partial P/\partial y$, and $\partial Q/\partial x$ are continuous and single-valued at every point of a simply connected region R. The necessary and sufficient condition that there exist a potential function $F(x,y)$ such that $\partial F/\partial x = P$ and $\partial F/\partial y = Q$ is that $\partial P/\partial y = \partial Q/\partial x$ at all points in R." (This is an extension of Green's theorem, which will be discussed in Chapter 6.)

Hint: Use the theorem to generate a potential function from each equation of equilibrium. Then demonstrate that the relation between these two functions implies the existence of an Airy stress function.

6

Torsion of Cylindrical Bars

6.1. General Solution of the Problem

The shearing stress in a cylinder of circular cross section under torsion is given by the elementary torsion formula in strength of materials. It will be shown that these stresses satisfy the governing equations of elasticity as well as the boundary conditions, therefore they represent the exact solution for a circular cylinder. The behavior of a circular cylinder under torsion is such that all cross sections normal to the axis remain plane. For cylinders of other cross sections subjected to torque, however, this condition does not prevail and warping occurs. We shall now consider the case of a cylinder of any cross section being twisted by couples applied at the ends.

The exact solution for this problem was first formulated by Saint-Venant by using the semi-inverse method. Guided by the deformations which occur in a circular cylinder, Saint-Venant assumed that the displacements which occur in a twisted noncircular cylinder are of this nature: (1) The projection of any deformed cross section on the x-y plane (Fig. 6.1) rotates as a rigid body, the angle of twist per unit length being constant, and (2) each point is displaced in the longitudinal direction (warping occurs). He further assumed that this warping is the same for all cross sections, i.e., the displacement w is independent of z. This behavior may be visualized by considering the displacements in (1) and (2) as occurring separately. The rotation considered in (1) is the same as that which occurs in a circular cylinder. Accompanying this rotation then, we have the longitudinal displacement described in (2).

The loading condition is shown in Fig. 6.1. The x and y axes are taken in the plane of the bottom cross section. The z axis is parallel to the longitudinal axis of the cylinder. The torque T on an end plane is positive if the vector representing T (by the right hand rule) acts in the direction of the outward normal of this plane.

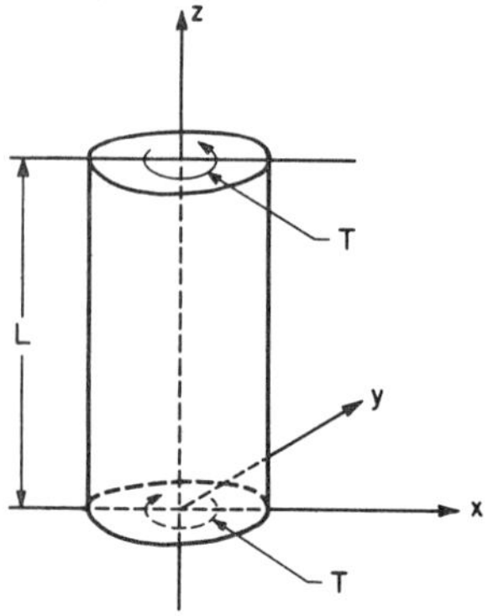

Fig. 6.1 Torsion of a Cylindrical Bar

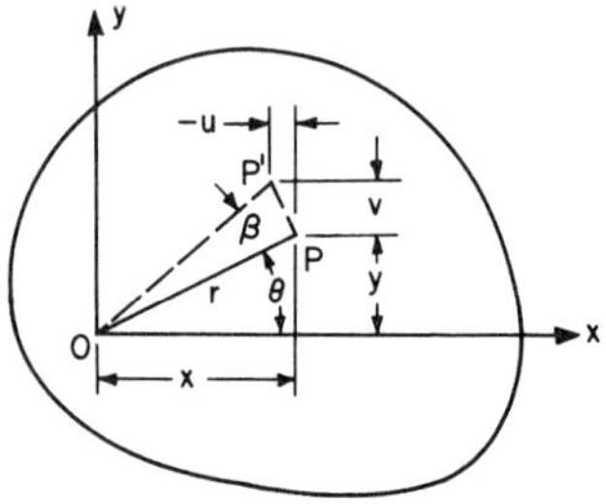

Fig. 6.2 Displacement Due to Rotation

The displacement of any point P due to rotation is shown in Fig. 6.2. The line OP rotates through a small angle β about O, where point O is called the center of twist (the point where $u = v = 0$), the location of which depends upon the shape of the cross section of the bar. The body is positioned such that the center of twist lies on the z axis. Since the angle of twist β is small, arc PP' is assumed to be a straight line normal to OP. The x and y components of the displacement of P, then, are given by

$$u = -r\beta \sin\theta = -\beta y$$

$$v = r\beta \cos\theta = \beta x$$

These are accompanied by a displacement w in the z direction (warping), where

$$w = w(x, y)$$

Since the warping of each cross section is assumed to be the same, w is only a function of x and y, not z. That is, all points on a line parallel to z move through the same distance in the z direction. We notice that Fig. 6.2 is the end view of the cross section showing the deformed position of OP. For convenience, we assume that $u = v = 0$ for the cross section at the origin. Thus if the cross section in Fig. 6.2 is at a distance z from the origin, the angle of twist is given by

$$\beta = \alpha z$$

where α is the angle of twist per unit length along the z direction. Therefore the displacements become

$$
\begin{aligned}
u &= -\alpha yz \\
v &= \alpha xz \\
w &= w(x, y)
\end{aligned}
\tag{6.1}
$$

From the theory of elasticity point of view, Eqs. (6.1) may be considered as the assumed partial solution of the problem in the semi-inverse method. We shall now show that the solution for the stress and strain components—which is derived from Eqs. (6.1) and which also satisfies the governing differential equations and the boundary conditions—may be obtained. By the uniqueness theorem then, this represents the solution to our problem. With the displacements assumed in Eqs. (6.1), the torsion problem may be formulated either in terms of stress (or stress function) or in terms of displacement w. In the following we shall first discuss the stress formulation. The displacement formulation, or the governing equation in terms of w, is presented at the end of this section.

The strain-displacement relations, Eqs. (2.1), when combined with the assumptions of Eqs. (6.1), reduce to

$$
\begin{aligned}
&\epsilon_x = \epsilon_y = \epsilon_z = \gamma_{xy} = 0 \\
&\gamma_{xz} = \frac{\partial w}{\partial x} + \frac{\partial u}{\partial z} = \frac{\partial w}{\partial x} - \alpha y \\
&\gamma_{yz} = \frac{\partial w}{\partial y} + \frac{\partial v}{\partial z} = \frac{\partial w}{\partial y} + \alpha x
\end{aligned}
\tag{6.2}
$$

By applying the stress-strain relations (Hooke's law), the stress-displacement equations become

$$
\begin{aligned}
&\tau_{xz} = G\gamma_{xz} = G\left(\frac{\partial w}{\partial x} - \alpha y\right) \\
&\tau_{yz} = G\gamma_{yz} = G\left(\frac{\partial w}{\partial y} + \alpha x\right) \\
&\sigma_x = \sigma_y = \sigma_z = \tau_{xy} = 0
\end{aligned}
\tag{6.3}
$$

From the first two of Eqs. (6.3), we observe that the nonvanishing stress components, τ_{xz} and τ_{yz}, are functions of x and y only. We may now eliminate w from Eqs. (6.3) by differentiating the first equation with respect to y and the second with respect to x and subtracting. This gives the following compatibility equation in terms of stress.

$$\frac{\partial \tau_{xz}}{\partial y} - \frac{\partial \tau_{yz}}{\partial x} = -2G\alpha \tag{6.4}$$

This equation may also be obtained by integrating the three-dimensional compatibility equations (2.2), and using Eq. (6.2), which gives

$$\frac{\partial \gamma_{xz}}{\partial y} - \frac{\partial \gamma_{yz}}{\partial x} = -2\alpha$$

Combining this with Hooke's law gives Eq. (6.4). The two shear stress components must also satisfy the equations of equilibrium. Writing the third of these with zero body forces, we have

$$\frac{\partial \tau_{xz}}{\partial x} + \frac{\partial \tau_{yz}}{\partial y} = 0 \tag{6.5}$$

the other two equations of equilibrium, with no body forces, being identically satisfied.

If we introduce a stress function ϕ (Prandtl's stress function) where ϕ is a function of x and y, such that

$$\begin{aligned} \tau_{xz} &= \frac{\partial \phi}{\partial y} \\ \tau_{yz} &= -\frac{\partial \phi}{\partial x} \end{aligned} \tag{6.6}$$

we see that Eq. (6.5) is identically satisfied. By substituting Eqs. (6.6) into Eq. (6.4), we get

$$\frac{\partial^2 \phi}{\partial x^2} + \frac{\partial^2 \phi}{\partial y^2} = -2G\alpha$$

or

$$\nabla^2 \phi = -2G\alpha \tag{6.7}$$

where

$$\nabla^2 = \frac{\partial^2}{\partial x^2} + \frac{\partial^2}{\partial y^2}$$

This is the governing differential equation for the variable ϕ.

Now let us consider the boundary conditions as given by Eqs. (4.4). On the lateral surfaces, there are no external forces; therefore, on these surfaces $T_x^{\mu} = T_y^{\mu} = T_z^{\mu} = 0$. Since the normals to these surfaces are perpendicular to the z axis, μ_z must equal zero. The first two of Eqs.

(4.4) are identically satisfied, and the third equation gives the relation (the subscript "o" is omitted from the stress terms)

$$\tau_{xz}\mu_x + \tau_{yz}\mu_y = 0 \tag{6.8a}$$

Keeping in mind that τ_{xz} is the shear stress component on the x plane acting in the z direction, τ_{zx} is that on the z plane in the x direction, and $\tau_{xz} = \tau_{zx}$, we see from Fig. 6.3 that the left side of Eq. (6.8a) is equal to

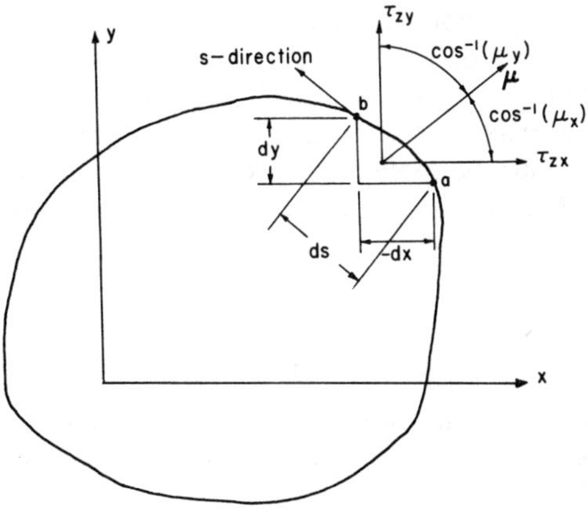

Fig. 6.3 Shear Stresses on a z plane, Near the Boundary

the component of shear stress on a z plane normal to the lateral boundary surface, or

$$\tau_{z\mu} = \tau_{zx}\mu_x + \tau_{zy}\mu_y = 0 \tag{6.8b}$$

Since it is required that this component vanish, the resultant shear stress on the cross section (z plane) is required to be tangent to the boundary. Again referring to Fig. 6.3, we have

$$\mu_x = \frac{dy}{ds} \qquad \mu_y = -\frac{dx}{ds}$$

assuming s is increasing from a toward b. Using these relations and Eqs. (6.6), Eq. (6.8a) becomes

$$\frac{\partial\phi}{\partial y}\frac{dy}{ds} + \frac{\partial\phi}{\partial x}\frac{dx}{ds} = 0$$

and by applying the chain rule of differentiation this becomes

$$\frac{d\phi}{ds} = 0$$

Thus the stress function ϕ is required to be constant along the boundary of the cross section. If this boundary is simply connected, the constant may be chosen arbitrarily (this constant has no effect on the stress components and it only causes rigid body displacement), so that in the following discussion we will assume it is equal to zero, or

$$\phi = 0 \qquad \text{on the boundary} \tag{6.9}$$

Satisfaction of this condition then, implies that Eq. (6.8a) is satisfied.

We have, to this point, established that the stresses defined by ϕ which satisfies Eqs. (6.7) and (6.9), represent the solution of an elasticity problem for this particular cylindrical bar with no surface forces applied on the lateral surfaces. We now proceed to determine what surface force (per unit area) must be applied on the end planes such that these stresses exist. Since the end planes are normal to the z axis, we have

$$\mu_x = \mu_y = 0 \qquad \mu_z = -1$$

for the plane $z = 0$, and

$$\mu_x = \mu_y = 0 \qquad \mu_z = +1$$

for the plane $z = L$. By substituting these values of direction cosines into Eqs. (4.4), we see that surface force components distributed according to

$$T_x^{\mu} = \mp \tau_{xz} \qquad T_y^{\mu} = \mp \tau_{yz} \qquad T_z^{\mu} = 0 \tag{6.10}$$

must be applied on the end planes in order to satisfy the boundary conditions. Here the upper signs refer to the plane $z = 0$, and the lower

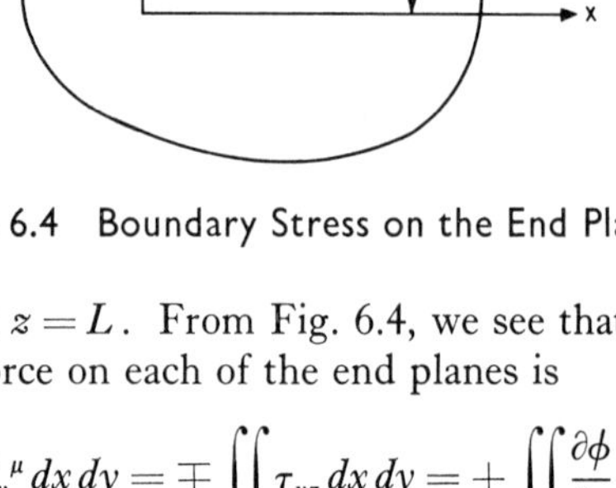

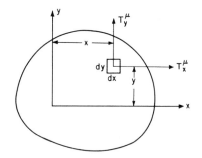

Fig. 6.4 Boundary Stress on the End Planes

signs to the plane $z = L$. From Fig. 6.4, we see that the y component of the resultant force on each of the end planes is

$$\iint T_y^{\mu}\, dx\, dy = \mp \iint \tau_{yz}\, dx\, dy = \pm \iint \frac{\partial \phi}{\partial x}\, dx\, dy \tag{6.10a}$$

The last double integration may be carried out by first performing the inner integration along the x direction as shown in Fig. 6.5, or

$$\begin{aligned}\iint T_y{}^\mu\, dx\, dy &= \pm \int_C \left(\int_{x=x_1}^{x=x_2} \frac{\partial \phi}{\partial x}\, dx \right) dy = \pm \int_C \left(\int_{x=x_1}^{x=x_2} d\phi \right) dy \\ &= \pm \int_C (\phi_2 - \phi_1)\, dy = 0\end{aligned} \tag{6.11}$$

where $x = x_1(y)$ is the equation of the left part of the boundary curve C and $x = x_2(y)$ is the equation of the right portion.[1] Since ϕ_2 and ϕ_1

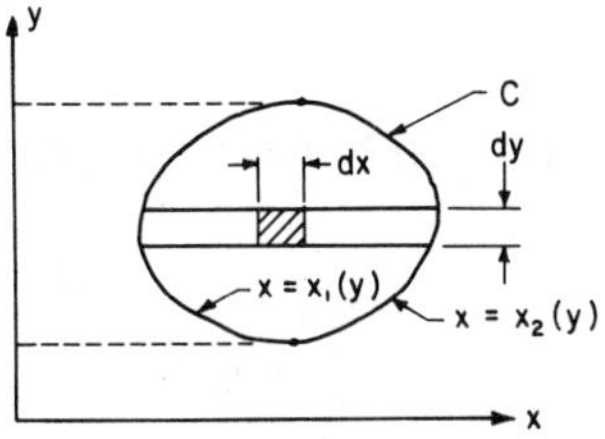

Fig. 6.5 Integration over the End Plane

are evaluated along x_1 and x_2 (on the boundary), they both vanish. Notice that if we assume $\phi = c$ on the boundary where c is a constant different from zero, $\phi_2 - \phi_1$ will also vanish. Similarly, it can be shown that

$$\iint T_x{}^\mu\, dx\, dy = 0 \tag{6.12}$$

Thus, the x and y components of resultant force on each end plane are zero and the applied surface load acting over each end is statically equivalent to a couple (or torque) about the z axis. The torque T, as shown by Fig. 6.4, is

$$T = \mp \iint (T_y{}^\mu x - T_x{}^\mu y)\, dx\, dy \tag{6.13}$$

This may be expressed in terms of the stress function ϕ. Substituting Eqs. (6.6) and (6.10) into Eq. (6.13), we have

$$T = -\iint \frac{\partial \phi}{\partial x} x\, dx\, dy - \iint \frac{\partial \phi}{\partial y} y\, dx\, dy \tag{6.14}$$

[1] The validity of Eq. (6.11) can also be established by applying Green's theorem.

Performing the inner integration of the first term by parts, we obtain

$$\int_{x=x_1}^{x=x_2} \frac{\partial \phi}{\partial x}\, x\, dx = x\phi \Big|_{x=x_1}^{x=x_2} - \int_{x_1}^{x_2} \phi\, dx = x_2 \phi_2 - x_1 \phi_1 - \int_{x_1}^{x_2} \phi\, dx$$

$$= - \int_{x_1}^{x_2} \phi\, dx$$

since $\phi_2 = \phi_1 = 0$ on the boundary.[2] Therefore, the first term of Eq. (6.14) becomes

$$-\iint \frac{\partial \phi}{\partial x}\, x\, dx\, dy = \iint \phi\, dx\, dy$$

Similarly,

$$-\iint \frac{\partial \phi}{\partial y}\, y\, dx\, dy = \iint \phi\, dx\, dy$$

so that

$$T = 2 \iint \phi\, dx\, dy \tag{6.15}$$

Notice that the two integrals in Eq. (6.14) are equal; thus each of the stress components τ_{xz} and τ_{yz} contributes to one-half of the torque.

The applied surface forces per unit area on the end planes as given by Eqs. (6.10) are of a quite restricted form which will seldom be encountered in practical applications. Due to the Saint-Venant's principle, however, the solution for torsion given in this section can be used for any applied end load, as long as the resultant of the end load satisfies the conditions of Eqs. (6.11), (6.12), and (6.15). In other words, if the applied load is a torque with magnitude T, regardless of its detailed distribution over the end planes, at sections a short distance away from the end planes (if the dimensions of the cross section are small compared to the length) the stresses are governed by Eqs. (6.6), (6.7), and (6.9).

Retracing our analysis, we see that our assumed solution of the torsion problem satisfies the 15 governing equations of elasticity. Since it also satisfies the boundary conditions, it is the unique solution. We also see that the governing equations are independent of the origin of coordinates which we select;[3] thus in solving problems, it is not necessary

[2] For the case where ϕ on the boundary is taken as c, where c is a constant different from zero, see Problem 6-3.

[3] If the bar is located at a different position such that the center of twist is at a point (x_1, y_1), the displacement components are

$$u = -\alpha z(y - y_1) \qquad v = \alpha z(x - x_1) \qquad w = w(x, y)$$

Using these displacements, it is easily shown that Eqs. (6.4) to (6.15) remain unchanged.

to place the origin at the center of twist as was done in the derivation.

In the displacement formulation of the torsion problem the stress and strain components will be eliminated and one equation in terms of w will remain, i.e.,

$$\nabla^2 w = 0 \tag{6.16}$$

This is a Laplace equation, which is easier to solve than the Poisson type of equation, Eq. (6.7), in the stress formulation. The boundary condition in the displacement formulation, however, is not as simple as Eq. (6.9). Substituting Eqs. (6.3) into Eq. (6.8a), the boundary condition becomes

$$\left(\frac{\partial w}{\partial x} - y\alpha\right)\mu_x + \left(\frac{\partial w}{\partial y} + x\alpha\right)\mu_y = 0 \qquad \text{on the boundary} \tag{6.17}$$

and the torque is given by

$$T = G\iint\left(x^2\alpha + y^2\alpha + x\frac{\partial w}{\partial y} - y\frac{\partial w}{\partial x}\right)dx\,dy \tag{6.18}$$

The development of the torsion equations is summarized in Table 6.1. We begin with the assumed partial solution on u, v, and w and by considering α as the given applied angle of twist per unit length. The 15 governing equations are then reduced to five equations with five dependent variables, as shown in the first box on top. The left-hand column then gives the stress formulation, which reduces to one governing differential equation and the $\phi = 0$ boundary condition, as shown in the last box. If α is known, the stress function ϕ and thus all the other dependent variables are completely determined. In the displacement formulation, shown on the right-hand column, we again obtain one governing differential equation and one boundary condition equation containing α. For a given shaped bar, w can be found from these equations, and thus all of the other dependent variables can be determined.

In either formulation, the torque T required to produce the applied α is given by Eqs. (6.15) or (6.18). If the torque T, instead of angle α, is given, then the variables ϕ or w may still be determined as indicated, but with α as a parameter appearing in the solution. Then Eqs. (6.15) or (6.18) supply the additional equation to solve for α.

6.2. Solutions Derived from Equations of Boundaries

The governing equations (6.7), (6.9), or (6.16), (6.17), although simple in form, are, in general, not easy to solve in closed form. This is mainly because of the difficulty in satisfying the boundary conditions,

Table 6.1
DEVELOPMENT OF TORSION EQUATIONS

$u = -\alpha yz$
$v = \alpha xz$ $\qquad \alpha =$ applied angle of twist per unit length
$w = w(x, y)$

2 Strain-displacement equations

$$\left.\begin{aligned} \gamma_{xz} &= \frac{\partial w}{\partial x} - \alpha y \\ \gamma_{yz} &= \frac{\partial w}{\partial y} + \alpha x \end{aligned}\right\} \quad (6.2)$$

2 Hooke's law equations

$$\left.\begin{aligned} \tau_{xz} &= G\gamma_{xz} \\ \tau_{yz} &= G\gamma_{yz} \end{aligned}\right\} \quad (6.3)$$

1 Equilibrium equation

$$\frac{\partial \tau_{xz}}{\partial x} + \frac{\partial \tau_{yz}}{\partial y} = 0 \qquad (6.5)$$

5

$\tau_{xz}, \tau_{yz}, \gamma_{xz}, \gamma_{yz}, w$—all functions of x, y only.

1 Compatibility equation in terms of stress

$$\frac{\partial \tau_{xz}}{\partial y} - \frac{\partial \tau_{yz}}{\partial x} = -2G\alpha$$

1 Equilibrium equation

$$\frac{\partial \tau_{xz}}{\partial x} + \frac{\partial \tau_{yz}}{\partial y} = 0$$

τ_{xz}, τ_{yz}

$$\nabla^2 \phi = -2G\alpha \qquad (6.7)$$

where

$$\frac{\partial \phi}{\partial x} = -\tau_{yz}, \quad \frac{\partial \phi}{\partial y} = \tau_{xz} \qquad (6.6)$$

B.C. $\phi = 0$ on the boundary (6.9)

$$T = 2\iint \phi \, dx \, dy \qquad (6.15)$$

2 Stress-displacement equations

$$\tau_{xz} = G\left(\frac{\partial w}{\partial x} - \alpha y\right)$$

$$\tau_{yz} = G\left(\frac{\partial w}{\partial y} + \alpha x\right)$$

1 Equilibrium equation

$$\frac{\partial \tau_{xz}}{\partial x} + \frac{\partial \tau_{yz}}{\partial y} = 0$$

τ_{xz}, τ_{yz}, w

$$\nabla^2 w = 0 \qquad (6.16)$$

B.C.

$$\left(\frac{\partial w}{\partial x} - y\alpha\right)\mu_x + \left(\frac{\partial w}{\partial y} + x\alpha\right)\mu_y = 0 \qquad (6.17)$$

on the boundary where μ_x, μ_y are the direction cosines of the outward normal of the boundary

$$T = G\iint\left(x^2\alpha + y^2\alpha + x\frac{\partial w}{\partial y} - y\frac{\partial w}{\partial x}\right) dx \, dy \qquad (6.18)$$

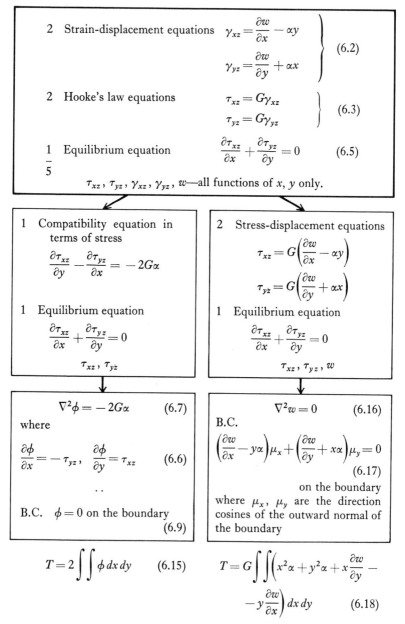

especially when the mathematical equations describing the shape of the boundary are complicated. For a few cases where the boundary is of simple geometrical shape, however, closed form analytical solutions can be obtained directly from the equation of the boundary. For example,

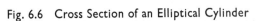

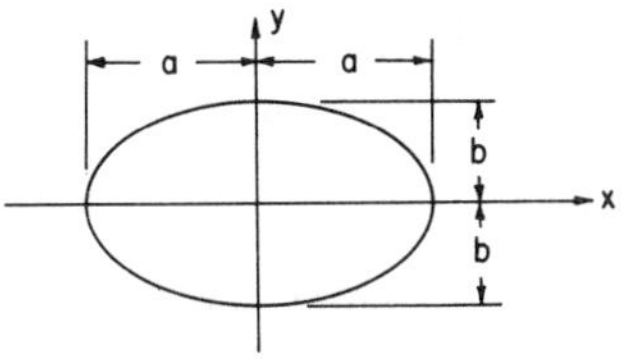

Fig. 6.6 Cross Section of an Elliptical Cylinder

consider the elliptical cylinder, the cross section of which is shown in Fig. 6.6. The equation of the boundary is

$$\frac{x^2}{a^2}+\frac{y^2}{b^2}-1=0 \tag{6.19}$$

Therefore, let us assume the stress function takes the form

$$\phi=k\left(\frac{x^2}{a^2}+\frac{y^2}{b^2}-1\right) \tag{6.20}$$

where k is a constant. We see that this assumed ϕ obviously vanishes on the boundary of the cylinder. Substitution of Eq. (6.20) into the governing equation (6.7) yields

$$k=-\frac{a^2 b^2 G\alpha}{a^2+b^2} \tag{6.21}$$

With this value of k, we see that the governing equations (6.7) and (6.9) are satisfied, and we have found the solution. The torque T may now be determined from Eq. (6.15), which yields

$$T=-\frac{2a^2 b^2}{a^2+b^2}G\alpha\left[\frac{1}{a^2}\iint x^2\,dx\,dy+\frac{1}{b^2}\iint y^2\,dx\,dy-\iint dx\,dy\right]$$

where the integrations are carried out over the area of the cross section. It is interesting to note that the first two integrals represent the moments of inertia of the cross section about the y and x axes, respectively. The third integral is the area of the cross section. For an ellipse, these properties are

$$I_y = \iint x^2 \, dx \, dy = \frac{\pi}{4} ba^3$$
$$I_x = \iint y^2 \, dx \, dy = \frac{\pi}{4} ab^3 \tag{6.22}$$
$$A = \pi ab$$

Substituting, we find that

$$T = \frac{\pi a^3 b^3 G\alpha}{a^2 + b^2} \tag{6.23}$$

therefore the angle of twist per unit length is given by

$$\alpha = \frac{(a^2 + b^2)T}{\pi a^3 b^3 G} \tag{6.24}$$

From Eqs. (6.6), (6.20), (6.21), and (6.24), we have

$$\tau_{xz} = -\frac{2a^2 G\alpha y}{a^2 + b^2} = -\frac{2Ty}{\pi ab^3}$$
$$\tau_{yz} = \frac{2b^2 G\alpha x}{a^2 + b^2} = \frac{2Tx}{\pi ba^3} \tag{6.25}$$

The (shear) stress vector at any point is the vector sum of τ_{xz} and τ_{yz}. It can be shown,[4] by considering the subharmonic nature of τ^2, where τ is the magnitude of the shear stress vector, that for bars of any cross section the maximum value of τ always occurs at points on the boundary. For an ellipse, these are the two points which are closest to the center of the ellipse, thus if $a > b$, we have

$$\tau_{\max} = \frac{2T}{\pi ab^2} \tag{6.26}$$

The same analysis may be carried out for a circular cylinder under torsion. In this case, the stress function is given by

$$\phi = A(x^2 + y^2 - a^2) \tag{6.27}$$

where a is the radius of the cylinder and

$$A = -G\alpha/2$$

[4] See Problem 6-7, which shows that τ^2 is subharmonic. A subharmonic function with continuous second derivatives assumes its maximum value on the boundary. See, for instance, O. D. Kellogg, *Foundations of Potential Theory*, Dover Publications, Inc., New York, 1953, p. 316.

The stress components are found to be

$$\tau_{xz} = -\frac{2Ty}{\pi a^4}$$
$$\tau_{yz} = \frac{2Tx}{\pi a^4} \tag{6.28}$$

which are identical to those given in strength of materials for this problem.

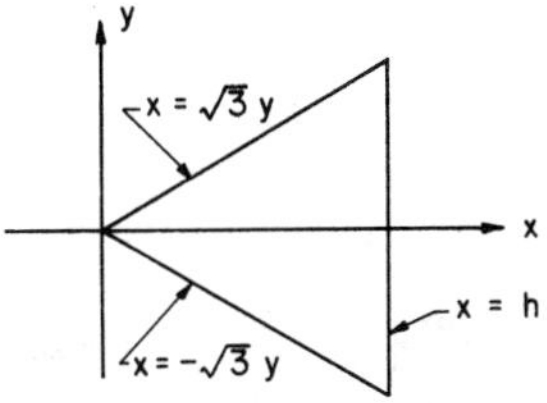

Fig. 6.7 Cross Section of an Equilateral-Triangular Cylinder

The solution for a cylinder with the cross section of an equilateral triangle (Fig. 6.7) is determined from the stress function

$$\phi = B(x - \sqrt{3}y)(x + \sqrt{3}y)(h - x) \tag{6.29}$$

which is also seen to be derived from the equations of the boundary. Here

$$B = \frac{15\sqrt{3}\,T}{2h^5}$$

and

$$\tau_{\max} = \frac{15\sqrt{3}\,T}{2h^3} \tag{6.30}$$

which occurs at the point $(h, 0)$.

6.3. Membrane (Soap Film) Analogy

Since closed form mathematical solutions of the torsion problem are likely to be very difficult to obtain, the membrane analogy has proved to be very valuable. The analogy is based on the fact that the governing elasticity equations for the torsion problem are mathematically identical in form to those describing the deflection of an elastic membrane subjected to uniform pressure. The analogy is restricted to the case where the deflection of the membrane is small. The discussion in this section applies only for simply connected cross sections; the analogy is extended to include multiply connected regions as shown in Problem 6-12.

Consider a thin homogeneous membrane stretched with uniform tension S (lb/in.) over the end of a thin tube and fixed at its edge in the x-y plane. The membrane is then subjected to a uniform lateral pressure p, which causes a small deflection z, where $z = z(x, y)$, as shown in Fig. 6.8. We assume that the application of the small pressure p does not alter the magnitude of the in-plane tension S.

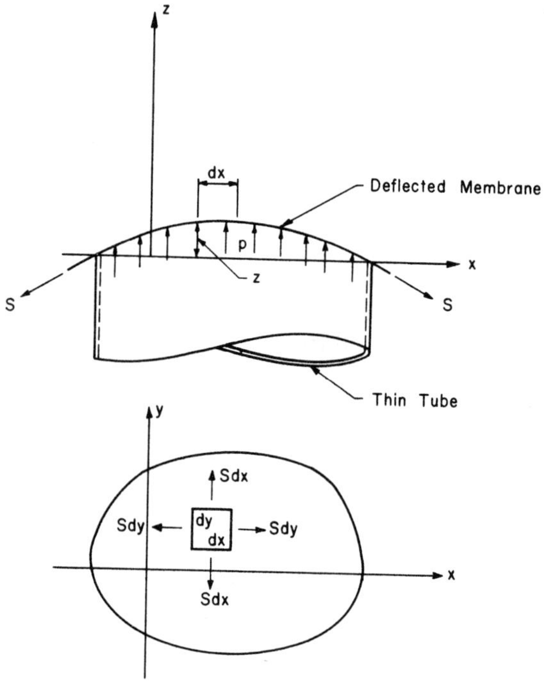

Fig. 6.8 Lateral Deflection of a Stretched Membrane at the End of a Tube

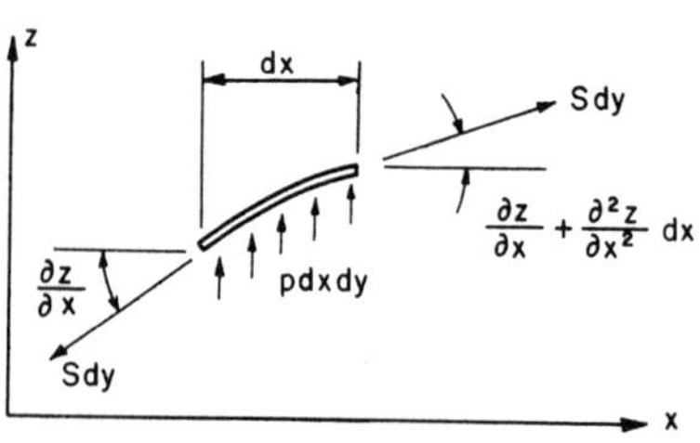

Fig. 6.9 An Element of the Membrane

Consider now the forces acting on the infinitesimal element shown in Fig. 6.8. The cross section of this element cut by a plane parallel to the x and z axes is shown in Fig. 6.9. The slope of the left side of this element measured in the x-z plane is equal to $\partial z/\partial x$, since $z(x, y)$ is the deflection. For small z, the slope is equal to the angle shown in the figure. The slope at the right side of the element, a distance dx away, is $\partial z/\partial x + (\partial/\partial x)(\partial z/\partial x)dx$, where $(\partial/\partial x)(\partial z/\partial x)dx = (\partial^2 z/\partial x^2)dx$ represents the change in $\partial z/\partial x$ over the distance dx. The force acting on the left side of the element is $S\,dy$, and it has a z component $-(S\,dy)(\partial z/\partial x)$. On the right-hand side of the element, the z component of the force is $S\,dy[\partial z/\partial x + (\partial^2 z/\partial x^2)\,dx]$.

Similarly, on the front and back sides of the element, the force is $S\,dx$ and the z components are $-S\,dx(\partial z/\partial y)$ and $S\,dx[\partial z/\partial y + (\partial^2 z/\partial y^2)\,dy]$, respectively. Expressing the equilibrium of the element in the z direction, therefore, yields

$$\frac{\partial^2 z}{\partial x^2} + \frac{\partial^2 z}{\partial y^2} = -\frac{p}{S} \tag{6.31}$$

The boundary condition for the membrane problem is that the deflection at the edge of the tube is zero, or

$$z = 0 \qquad \text{on the boundary} \tag{6.32}$$

The volume enclosed by the deflected membrane and the end plane of the tube (x-y plane) is

$$V = \iint z\,dx\,dy \tag{6.33}$$

A contour line is defined as the line of intersection between the deflected membrane and a horizontal plane ($z = \text{constant}$), as shown in Fig. 6.10. Along a contour line, the deflection z of the membrane is constant; therefore,

$$\frac{dz}{ds} = 0 \qquad \text{along a contour line} \tag{6.34}$$

where s is measured along the contour line from an arbitrary point, counterclockwise as positive.

Let us now compare the membrane problem with the torsion problem. If the shape of the boundary of the membrane (or the shape of the thin tube in Fig. 6.8) is identical with the cross section of a bar under torsion, and if we let

$$\frac{p}{S} = 2G\alpha \tag{6.35}$$

then the governing equations for the membrane problem, Eqs. (6.31) and (6.32), are the same as the governing equations for the stress function in the torsion problem, Eqs. (6.7) and (6.9), with z equivalent to ϕ.

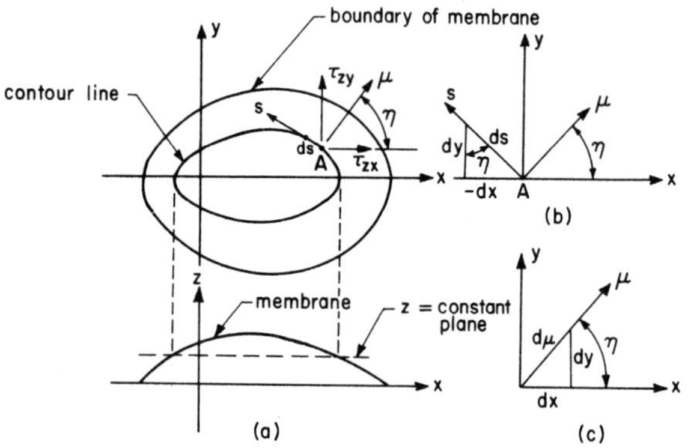

Fig. 6.10 Contour Line of the Membrane

The solution of a membrane problem, therefore, can be used as the solution of a corresponding torsion problem. Some of the characteristics of the stress distribution in a bar under torsion can be analyzed with the help of the corresponding membrane problem.

With $p/S = 2G\alpha$ and $z = \phi$, the shear stress in the torsion problem is related to the slope of the membrane, since

$$\tau_{xz} = \frac{\partial \phi}{\partial y} = \frac{\partial z}{\partial y}$$
$$\tau_{yz} = -\frac{\partial \phi}{\partial x} = -\frac{\partial z}{\partial x} \tag{6.36}$$

That is, the shear stress acting in the x direction is equal to the slope of the membrane measured in the y direction. The shear stress acting in the y direction is equal to the negative of the slope measured in the x direction. This condition may be generalized to determine the shear stress in any direction as follows. Consider a line of constant ϕ in the cross section of the bar under torsion. This is equivalent to the contour line in the membrane; therefore the contour line in Fig. 6.10 may be used as the $\phi = \text{constant}$ line. Along this line we have

$$\frac{d\phi}{ds} = \frac{\partial \phi}{\partial x}\frac{dx}{ds} + \frac{\partial \phi}{\partial y}\frac{dy}{ds} = -\tau_{yz}\frac{dx}{ds} + \tau_{xz}\frac{dy}{ds} = 0 \tag{6.37}$$

Letting μ_x and μ_y be the direction cosines of the outward normal of the contour line, we have, from Fig. 6.10(b) and 6.10(c),

$$\begin{aligned} \mu_x &= \cos\eta = \frac{dx}{d\mu} = \frac{dy}{ds} \\ \mu_y &= \cos(\pi/2 - \eta) = \frac{dy}{d\mu} = -\frac{dx}{ds} \end{aligned} \tag{6.38}$$

Therefore,

$$\tau_{zy}\mu_y + \tau_{zx}\mu_x = \tau_{z\mu} = 0 \tag{6.39}$$

where $\tau_{z\mu}$ is the component of shear stress normal to the contour line. Thus, the shear stress is tangent to the contour line, or the line of constant ϕ. The lines which are everywhere tangent to the shear stress vector (lines of constant ϕ) are sometimes called the *lines of shearing stress.*

The magnitude of the shear stress vector at point A is determined by projecting τ_{zx} and τ_{zy} on the tangent to the contour line at A. Referring to Fig. 6.10, we see that

$$\tau = \tau_{zs} = -\tau_{zx}\cos(\pi/2 - \eta) + \tau_{zy}\cos\eta$$

or

$$\tau = -\frac{\partial\phi}{\partial y}\frac{dy}{d\mu} - \frac{\partial\phi}{\partial x}\frac{dx}{d\mu} = -\frac{d\phi}{d\mu} = -\frac{dz}{d\mu} \tag{6.40}$$

Thus, the magnitude of the shearing stress at point A is given by the magnitude of the slope measured normal to the tangent line, i.e., normal to the contour line at A, of the corresponding membrane.

Comparison of Eqs. (6.15) and (6.33) shows that the torque T in the torsion problem is equal to twice the volume V in the membrane problem.

Summarizing then, we have seen that the torsion problem and the membrane problem are mathematically identical. With identical boundaries and with $p/S = 2G\alpha$, we have the following relations between the variables in the torsion problem and those in the membrane problem

$$\phi = z$$

$$\tau_{zs} = -\frac{dz}{d\mu}$$

$$T = 2\iint z\,dx\,dy$$

i.e., the stress function is equal to the deflection of the membrane; the magnitude of the shear stress at a point in the bar under torsion is equal to the magnitude of the slope of the membrane measured in the vertical plane normal to the contour line at the corresponding point in the membrane; the torque is equal to twice the volume under the membrane.

The membrane analogy is very useful in the experimental determination of stress in bars under torsion. Quantitatively, it may not give very accurate data because of the limitation of small membrane deflection and the difficulty involved in making precise measurements. The true value of the membrane analogy lies in the visual picture it offers. For a bar of irregular cross section, we may readily detect, by the membrane analogy, the point or points where the shearing stress is maximum. In dealing with multiply connected cross sections or inelastic stress distributions in torsion, the visual picture obtained from the membrane analogy is even more valuable.

6.4. Multiply Connected Cross Sections[5]

We have shown that the stress function ϕ must be constant on the boundary of a simply connected cross section. Since the applied surface forces are zero on all lateral surfaces of a bar with a multiply connected cross section, it can be shown by following the derivation of Eq. (6.9) that the stress function ϕ must be constant along each boundary, but these constants are not the same. In this case only one of these constants can be chosen arbitrarily.

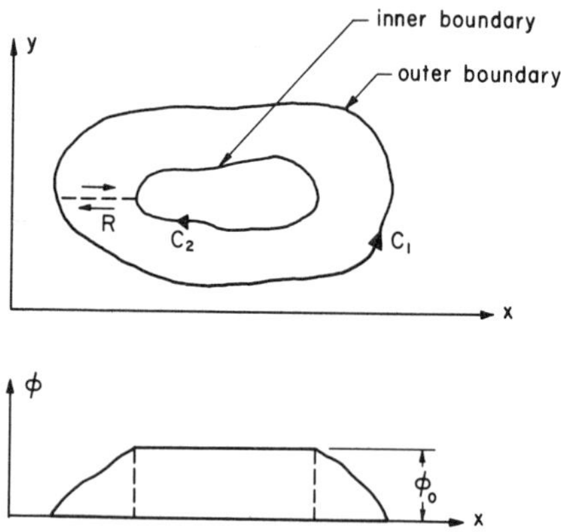

Fig. 6.11 Stress Function Distribution for a Bar with a Doubly Connected Cross Section

[5] For the application of the membrane analogy to multiply connected regions, refer to Problems 6-12 and 6-13.

Consider the doubly connected cross section shown in Fig. 6.11. No generality is lost by again choosing the value of ϕ on the outer boundary to be zero. However, along the inner boundary, the value of ϕ must be another constant. This constant is determined by the condition that the displacement w must be single-valued. If we consider the inner boundary as a contour line as shown in Fig. 6.10, the line integral $\oint dw$ along the contour line is given by

$$\oint dw = \oint \left(\frac{\partial w}{\partial x} dx + \frac{\partial w}{\partial y} dy \right)$$

Upon substitution of Eqs. (6.3), this becomes

$$\oint dw = \frac{1}{G} \oint (\tau_{xz} \, dx + \tau_{yz} \, dy) - \alpha \oint (x \, dy - y \, dx)$$

$$= \frac{1}{G} \oint \tau \, ds - 2\alpha A$$

where A is the area enclosed by the inner boundary. Since w is a single-valued function and the integration is taken around a closed path, $\oint dw$ vanishes. Therefore, we have

$$\oint \tau \, ds = 2G\alpha A \tag{6.41}$$

where, as in Eq. (6.40), $\tau = - d\phi/d\mu$. The constant value of ϕ on the inner boundary must be chosen so that Eq. (6.41) is satisfied. For a bar with a doubly connected cross section, therefore, the stress function ϕ must satisfy, in addition to Eqs. (6.7) and (6.9), the inner boundary condition, Eq. (6.41). The application of Green's theorem, Eq. (6.42), shows that w is single-valued everywhere in the region if Eq. (6.41) is satisfied.

The components of the resultant force on the end planes are again zero, which can be shown either by performing the integration similar to that of Eq. (6.11), or by using Green's theorem, i.e.,

$$\iint_R \left(\frac{\partial M}{\partial x} - \frac{\partial N}{\partial y} \right) dx \, dy = \oint_C (M \, dy + N \, dx) \tag{6.42}$$

where $M(x, y)$, $N(x, y)$, $\partial M/\partial x$, and $\partial N/\partial y$ are single-valued, continuous functions (we limit our discussion to such functions) in region R, which is bounded by the closed curve C. The theorem can be applied for a doubly connected region by traversing the curves C_1 and C_2 in the (positive) directions shown in Fig. 6.11.[6] From Eq. (6.10a), we have

$$\iint_R T_y^{\mu}\,dx\,dy = \pm \iint_R \frac{\partial \phi}{\partial x}\,dx\,dy$$

Transforming the right-hand side of the equation by Green's theorem, we get

$$\iint_R \frac{\partial \phi}{\partial x}\,dx\,dy = \oint_{C_1} \phi\,dy + \oint_{C_2} \phi\,dy = \phi_0 \oint_{C_2} dy = 0$$

so that

$$\iint_R T_y^{\mu}\,dx\,dy = 0 \tag{6.43}$$

Similarly, we get

$$\iint_R T_x^{\mu}\,dx\,dy = 0 \tag{6.44}$$

The torque on the end planes can be evaluated by using Eq. (6.14), i.e.,

$$T = -\iint_R \frac{\partial \phi}{\partial x}\,x\,dx\,dy - \iint_R \frac{\partial \phi}{\partial y}\,y\,dx\,dy$$

which can be written as

$$T = -\iint_R \left[\frac{\partial (x\phi)}{\partial x} + \frac{\partial (y\phi)}{\partial y}\right] dx\,dy + 2\iint_R \phi\,dx\,dy$$

With the aid of Green's theorem, we get

$$T = -\oint_{C_1} (x\phi\,dy - y\phi\,dx) - \oint_{C_2} (x\phi\,dy - y\phi\,dx) + 2\iint_R \phi\,dx\,dy$$

or

$$T = 2\iint_R \phi\,dx\,dy + 2\phi_0 A \tag{6.45}$$

where ϕ_0 is the constant value taken by ϕ on the inner boundary, and

[6] The positive direction of integration is defined such that a person walking on the curve in the positive direction always keeps the region R on his left.

R is the region between the inner and outer boundaries. Equations similar to (6.41) and (6.45) can also be derived for multiply connected cross sections containing more than one inner boundary.

For a bar with an inner boundary of arbitrary shape, it is generally difficult to find a ϕ distribution which satisfies Eq. (6.7), Eq. (6.9) on the outer boundary and Eq. (6.41) on the inner boundary. If the solution for a solid bar is known, however, we can easily obtain the solution of a hollow bar whose inner boundary is a line of shear stress (constant ϕ) of the solid bar.

For example, consider the bar of elliptical cross section with an elliptical area removed as shown in Fig. 6.12. The inner boundary coincides

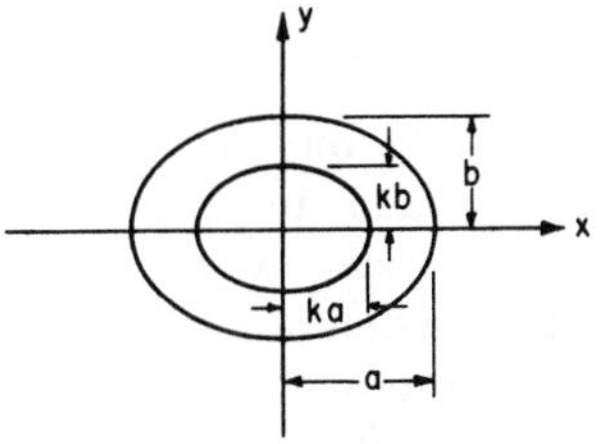

Fig. 6.12 Hollow Elliptic Bar Under Torsion

with a line of shear stress of the solid bar since the lines of shear stress for the latter are concentric ellipses (see Problem 6-5). Thus, in the solid bar, there is no shear stress acting on any internal surface parallel to z axis and whose cross section is a concentric elliptical cylinder. Therefore, if a concentric elliptical cylinder is removed from the bar, the stress distribution in the remaining portion will be the same as in the solid bar. If we make this assumption, the stress function will be given by Eqs. (6.20) and (6.21), i.e.,

$$\phi = -\frac{a^2 b^2 G\alpha}{a^2 + b^2}\left(\frac{x^2}{a^2} + \frac{y^2}{b^2} - 1\right) \tag{6.46}$$

but the torque acting on the hollow bar will be less than that on the solid bar, for a given α. The difference in the two torques is simply the amount carried by the material inside the inner ellipse in the solid bar. Thus, for the bar in Fig. 6.12, the torque is given by

$$T = \frac{\pi a^3 b^3 G\alpha}{a^2 + b^2} - \frac{\pi (ka)^3 (kb)^3 G\alpha}{(ka)^2 + (kb)^2}$$

or

$$T = \frac{\pi G\alpha}{a^2 + b^2} a^3 b^3 (1 - k^4) \tag{6.47}$$

This result may also be obtained from Eq. (6.45). In terms of T, the stress function ϕ is

$$\phi = -\frac{T}{\pi ab(1-k^4)}\left(\frac{x^2}{a^2}+\frac{y^2}{b^2}-1\right) \tag{6.48}$$

and in this case, the maximum magnitude of stress is given by

$$\tau_{\max} = \frac{2T}{\pi ab^2}\frac{1}{1-k^4} \qquad (a > b) \tag{6.49}$$

The solution stated by Eqs. (6.46) to (6.48) is shown to be valid by substitution into Eqs. (6.7), (6.9), and (6.41).

6.5. Solution by Means of Separation of Variables

For certain cross sections of simple geometry, closed form solutions of the torsion problem may be obtained as shown in Section 6.2. For arbitrary cross sections, it is generally not possible to write down the solution directly from the equation of the boundary. For these sections there are numerous other methods available, such as separation of variables, complex variables and conformal mapping,[7] energy or variational methods,[8] and the finite-difference method.[9] In this section, we shall discuss the method of separation of variables applied to a bar of rectangular cross section, as shown in Fig. 6.13.

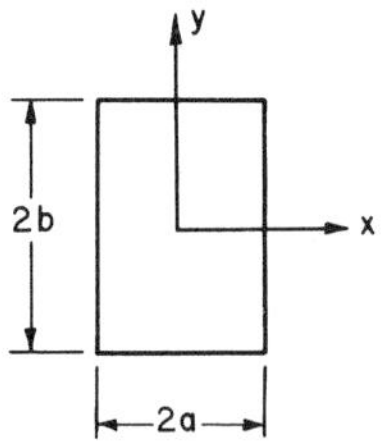

Fig. 6.13 Rectangular Cross Section

We must satisfy the governing equation (6.7),

$$\nabla^2\phi = \frac{\partial^2\phi}{\partial x^2}+\frac{\partial^2\phi}{\partial y^2} = -2G\alpha$$

[7] See S. Timoshenko and J. N. Goodier, *Theory of Elasticity*, Second Edition, McGraw-Hill Book Company, Inc., New York, and I. S. Sokolnikoff, *Mathematical Theory of Elasticity*, Second Edition, McGraw-Hill Book Company, Inc., New York, 1956.

[8] See Chapter 7 and references indicated in footnote 7 of this chapter.

[9] See C. Wang, *Applied Elasticity*, McGraw-Hill Book Company, Inc., New York, 1953, and R. V. Southwell, *Relaxation Methods in Theoretical Physics*, Vol. 1, Oxford University Press, 1946.

and the boundary conditions

$$\begin{aligned}\phi(\pm a, y) &= 0 \\ \phi(x, \pm b) &= 0\end{aligned} \tag{6.50}$$

Now the solution of the governing equation consists of the sum of

(1) the general solution of the homogeneous equation $\nabla^2\phi = 0$ and
(2) a particular solution of the equation $\nabla^2\phi = -2G\alpha$.

We seek a solution of the homogeneous equation in the form of an infinite series of products of functions X_n and Y_n, i.e.,

$$\phi_H = \sum_{n=0}^{\infty} X_n(x) Y_n(y) \tag{6.51a}$$

where X_n is a function of x and Y_n is a function of y and each term of the series satisfies the homogeneous equation. Thus, substituting into the homogeneous equation, we see that

$$\frac{Y_n''}{Y_n} = -\frac{X_n''}{X_n}$$

where the primes denote derivatives with respect to the variable in question. Since the left side of this equation is only a function of y and the right side is only a function of x, the equation can only be satisfied if the quotients are a constant, say λ_n^2. Therefore,

$$\frac{Y_n''}{Y_n} = \lambda_n^2$$

and (6.51b)

$$\frac{X_n''}{X_n} = -\lambda_n^2$$

which are two ordinary differential equations with constant coefficients. The general solutions are

$$\begin{aligned}Y_n &= A_n \sinh \lambda_n y + B_n \cosh \lambda_n y \\ X_n &= C_n \sin \lambda_n x + D_n \cos \lambda_n x\end{aligned} \tag{6.52}$$

where A_n, B_n, C_n, and D_n are arbitrary constants, to be determined by the boundary conditions, and we are taking $\lambda_n > 0$. It will be seen from subsequent work in this solution that it is not necessary to consider any $\lambda_n \leq 0$.

In order to determine the particular solution (ϕ_p), we replace the constant ($-2G\alpha$) by a Fourier series in terms of x, which converges to the function shown in Fig. 6.14.[10] If the signs for λ_n in the two equations (6.51b) are interchanged, a Fourier series in terms of y must be used here.

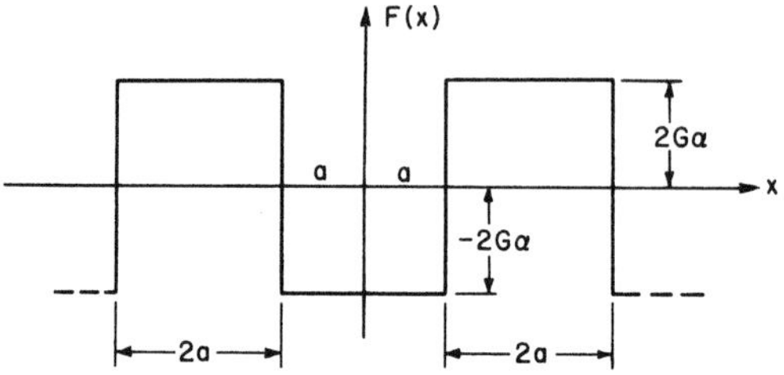

Fig. 6.14 A Periodic Function Equivalent to $-2G\alpha$ in the Range of Interest

Since we are only concerned with the range

$$-a < x < a$$

the fact that the series converges to a different value outside this range is immaterial. For an even function, the Fourier series takes the form

$$F(x) = \tfrac{1}{2} a_0 + \sum a_n \cos \frac{n\pi x}{p} \tag{6.53}$$

where

$$a_n = \frac{2}{p} \int_0^p F(x) \cos \left(\frac{n\pi x}{p} \right) dx \tag{6.54}$$

and $2p$ is the length of the repeating interval. Thus, solving for the constants a_n, we find that

$$\begin{aligned} a_n &= -\frac{8G\alpha}{n\pi} \qquad (n = 1, 5, 9, \ldots) \\ a_n &= +\frac{8G\alpha}{n\pi} \qquad (n = 3, 7, 11, \ldots) \\ a_0 &= 0 \end{aligned} \tag{6.55}$$

and

$$-2G\alpha = -\sum_{n=1,3,5}^{\infty} \frac{8G\alpha}{n\pi} (-1)^{(n-1)/2} \cos \frac{n\pi x}{2a} \qquad (-a < x < a) \tag{6.56}$$

[10] Actually, the series converges to zero at $x = \pm a$, but at points approaching $\pm a$ from $|x| < a$, it converges to the required constant value.

Thus we see that a particular solution of the form

$$\phi_p = \sum_{n=1,3,5}^{\infty} \frac{32G\alpha a^2}{n^3\pi^3}(-1)^{(n-1)/2}\cos\frac{n\pi x}{2a} \tag{6.57}$$

satisfies Eq. (6.7) and the complete solution is

$$\phi = \sum_{n=0}^{\infty} X_n Y_n + \phi_p \tag{6.58}$$

where X_n and Y_n are defined by Eqs. (6.52) and ϕ_p by Eq. (6.57).

Substituting Eqs. (6.52) into Eq. (6.58), we may now determine the arbitrary constants and the permissible values of λ_n (called the *eigenvalues*). To accomplish this, we substitute the boundary conditions (6.50) into Eq. (6.58), and let $E_n = B_n D_n$. We find that

$$A_n = C_n = 0$$

$$\lambda_n = \frac{n\pi}{2a} \qquad (n = 1, 3, 5, \ldots)$$

$$E_n = -\frac{32G\alpha(-1)^{(n-1)/2}a^2}{n^3\pi^3\cosh\dfrac{n\pi b}{2a}} \qquad (n = 1, 3, 5, \ldots)$$

$$\phi = \frac{32G\alpha a^2}{\pi^3}\sum_{n=1,3,5}^{\infty}\frac{(-1)^{(n-1)/2}}{n^3}\left[1 - \frac{\cosh\dfrac{n\pi y}{2a}}{\cosh\dfrac{n\pi b}{2a}}\right]\cos\frac{n\pi x}{2a} \tag{6.59}$$

and the stress components are given by

$$\begin{aligned} \tau_{yz} &= -\frac{\partial\phi}{\partial x} = \frac{16G\alpha a}{\pi^2}\sum_{n=1,3,5}^{\infty}\frac{(-1)^{(n-1)/2}}{n^2}\left[1 - \frac{\cosh\dfrac{n\pi y}{2a}}{\cosh\dfrac{n\pi b}{2a}}\right]\sin\frac{n\pi x}{2a} \\ \tau_{xz} &= \frac{\partial\phi}{\partial y} = -\frac{16G\alpha a}{\pi^2}\sum_{n=1,3,5}^{\infty}\frac{(-1)^{(n-1)/2}}{n^2}\left[\frac{\sinh\dfrac{n\pi y}{2a}}{\cosh\dfrac{n\pi b}{2a}}\right]\cos\frac{n\pi x}{2a} \end{aligned} \tag{6.60}$$

The maximum shear stress will always occur at the center of the longest side, so that if $b > a$,

$$\tau_{\max} = \tau_{yz}\Big|_{\substack{y=0\\x=a}} = 2G\alpha a - \frac{16G\alpha a}{\pi^2} \sum_{n=1,3,5}^{\infty} \frac{1}{n^2 \cosh \dfrac{n\pi b}{2a}} \tag{6.61}$$

since

$$\sum_{n=1,3,5}^{\infty} \frac{1}{n^2} = \frac{\pi^2}{8}$$

If we consider the case of a very thin rectangle ($b/a \to \infty$), the series in Eq. (6.61) vanishes and

$$\tau_{\max} = 2G\alpha a \tag{6.62}$$

The torque T may now be expressed in terms of α, since

$$T = 2 \iint \phi \, dx \, dy$$

so that

$$T = \frac{64G\alpha a^2}{\pi^3} \int_{-b}^{b} \int_{-a}^{a} \sum_{n=1,3,5}^{\infty} \frac{(-1)^{(n-1)/2}}{n^3} \left[1 - \frac{\cosh \dfrac{n\pi y}{2a}}{\cosh \dfrac{n\pi b}{2a}} \right] \cos \frac{n\pi x}{2a} \, dx \, dy \tag{6.63}$$

Performing the integration, and using the fact that

$$\sum_{n=1,3,5}^{\infty} \frac{1}{n^4} = \frac{\pi^4}{96}$$

we get

$$T = \frac{16G\alpha a^3 b}{3} - \frac{1024G\alpha a^4}{\pi^5} \sum_{n=1,3,5}^{\infty} \frac{1}{n^5} \tanh \frac{n\pi b}{2a} \tag{6.64}$$

Thus, if Eq. (6.64) is solved for α and this value is substituted into Eqs. (6.60), the stress components may be expressed in terms of the torque.

If we consider the case of a very thin rectangle (as $b/a \to \infty$ or $a \to 0$), we observe that the second term of Eq. (6.64) approaches zero much faster than the first term, so that the torque approaches the value

$$T = \tfrac{16}{3} G\alpha a^3 b \tag{6.65}$$

while the maximum shear stress becomes

$$\tau_{\max} = \frac{3}{8}\frac{T}{a^2 b} \tag{6.66}$$

This result can be readily applied to the torsion of a bar of simply connected cross section composed of a number of thin rectangles, for example, an angle or a channel. The membrane analogy may be used because over this section the membrane behaves approximately as though it were stretched over the separate rectangles (except in the local regions where the rectangles meet). Thus the torque carried by such a member is approximately given by

$$T = \tfrac{16}{3} G\alpha \sum_{i=1}^{N} a_i^3 b_i$$

where a_i and b_i $(b_i > a_i)$ are the dimensions of the various N rectangles. The maximum shear stress, neglecting the high localized stresses at the re-entrant corners, has the approximate value

$$\tau_{\max} = 2G\alpha a_{\min}$$

where $a_{\min}$ is the minimum width of the cross section. As a practical consideration, the torque capacity of a member with a multiply connected thin cross section is considerably greater than one which has a simply connected cross section of approximately the same dimensions. For an example refer to Problem 6-13.

PROBLEMS

6-1 The stress distribution in a solid bar under pure torsion is given by the stress function

$$\phi = A(a^2 - x^2 + cy^2)(a^2 + cx^2 - y^2)$$

where

$$c = 3 - \sqrt{8}$$

(a) Show that this stress function, with the proper value of the constant A, satisfies the governing differential equation. Find the value of A.

(b) Determine the boundary of the cross section.

(c) Find the torque and the stress distribution.

6-2 Carry out the general formulation of the torsion problem of a cylindrical bar in terms of the displacement w, thus demonstrating that $\nabla^2 w = 0$. Express the boundary conditions in terms of w.

6-3 What does Eq. (6.15) become if $\phi = c$ on the boundary, where c is a constant different from zero?

$$\text{Answer:}\quad T = 2\iint (\phi - c)\,dx\,dy$$

6-4 Plot the lines of constant w for an elliptical cylinder under torsion.

6-5 Show that the shear stress at any point within the cross section of an elliptical cylinder under torsion is tangent to an ellipse which passes through this point and has the same ratio of major to minor axes as that of the boundary ellipse.

6-6 Complete the solution suggested by Section 6.2 for a circular cylinder, i.e., compute the strain and displacement components.

6-7 A function $f(x, y)$ is called a subharmonic function if it satisfies, at every point in the domain, the condition

$$\nabla^2 f \geq 0$$

Utilizing Eqs. (6.6) and (6.7), show that τ^2 is a subharmonic function.

6-8 (a) For a bar of square cross section under torsion find the torque and maximum shear stress in terms of α. Express the maximum shear stress in terms of the torque.

(b) Repeat part (a) if $b/a = 4$.

6-9 The torsional constant J and torsional rigidity GJ are defined by $T = GJ\alpha$. Find the torsional constants for bars of elliptical, circular, equilateral triangular, and narrow rectangular cross sections.

6-10 Show that the strength of materials solution for a hollow circular cylinder under torsion, i.e.,

$$\tau = \frac{Tr}{J} \qquad \alpha = \frac{T}{JG} \qquad J = \tfrac{1}{2}\pi(r_o^4 - r_i^4)$$

where r_o and r_i are the outer and inner radii, respectively, and r is the radial coordinate, is the exact solution.

6-11 Verify the fact that Eq. (6.46) represents the solution to the problem of Fig. 6.12 by showing that it satisfies the governing equation and the boundary conditions.

6-12 Show that the membrane analogy can be applied to a doubly connected region if the arrangement shown in (b) is adopted. The flat plate (assume it is rigid) has the same cross section as the hole in the bar under torsion. Show that

$$T = 2V$$

where V is the volume enclosed by the membrane AB and CD, the plate BC, and the horizontal plane AD. Is this in agreement with Eq. (6.45)?

Also show that the equation

$$\oint \tau \, ds = 2G\alpha A$$

where A is the area enclosed by the inner boundary and the contour integral is taken around this boundary, defines the height h at which the plate must be suspended. Further demonstrate that this equation is a consequence of the fact that the summation of vertical forces acting on the plate is zero.

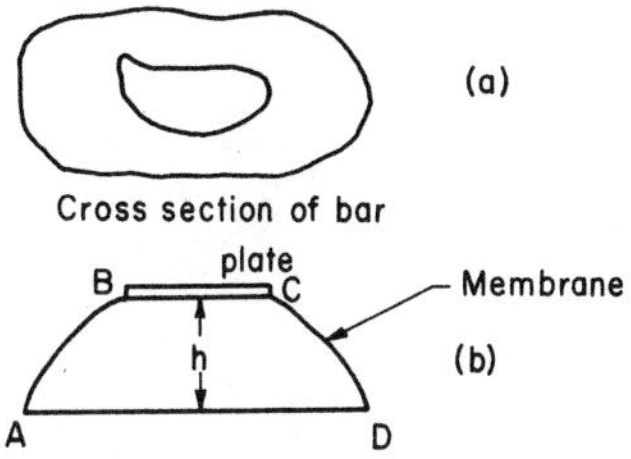

Problem 6–12

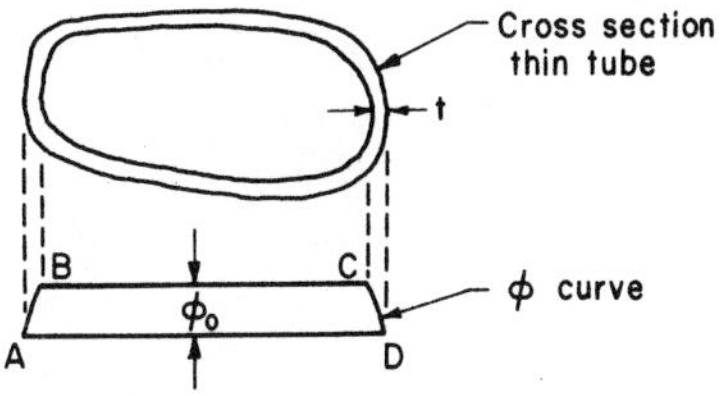

Problem 6–13

6-13 Consider the thin closed tube shown. The thickness is small although not necessarily constant. Since there can be little variation in slope of the ϕ curve shown across the thickness of the section (think of the membrane analogy), assume that the curves AB, CD, etc., are straight lines. Show that

$$T = 2A\phi_o$$

and

$$\tau = \frac{T}{2At}$$

where A is the area enclosed by the center line of the tube, τ is the approximate shear stress at any section, and t is the thickness of the section at the point where τ is to be found. Compare this stress with that of an open tube of the same shape, Eq. (6.66), and show that the closed tube is much stronger.

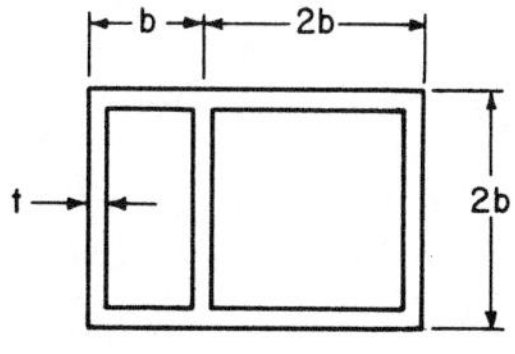

Problem 6–14

6-14 A thin-walled torsion member has the cross section shown. The thickness t is constant. Compute the approximate torque and maximum shear stress in terms of α. Hint: Assume the shear stress does not vary through the thickness of the walls, and remember that ϕ must be constant along each of the inner boundaries. Each of these two constants is determined by the condition $\oint \tau \, ds = 2G\alpha A$.

6-15 Complete the missing steps of the derivation which leads to Eqs. (6.60) and (6.64).

7

Energy Methods

7.1. Introduction

Once an elasticity problem is formulated, it is necessary to find the solution of the resulting differential equations and prescribed boundary conditions. Because of the difficulty encountered in obtaining exact solutions of these equations, it is important to consider various approximate methods. One group of methods is based on the fact that the governing differential equations of an elasticity problem can be obtained as a consqeuence of the minimization of a particular energy expression. In this chapter we shall explore some of these methods and illustrate their application with several examples.

7.2. Strain Energy

When a body is deformed by external forces, work is done by these forces. The energy absorbed in the body due to this external work is called strain energy. If the body behaves elastically, and if we ignore the small amount of energy which is lost in the form of heat, the strain energy can be recovered completely when the body is returned to its unstrained state. We shall first express the strain energy due to a uniaxial stress condition as shown in Fig. 7.1.

In this figure, $ABCD$ is a rectangular prism of dimensions dx, dy, dz. Under a normal stress σ in the x direction, the element is displaced to the position $A'B'C'D'$ as shown. On the plane AB, the stress vector acts in the opposite direction of the displacement u. Thus, the work done by σ on AB is negative, whereas the work done on CD is positive. During the process of deformation, the normal stress σ increases from zero to a final value σ_x. We assume the body force is zero[1] and that the stress is applied very slowly, so that the inertia effect can be neglected and the element is in equilibrium at every instant. The net work done

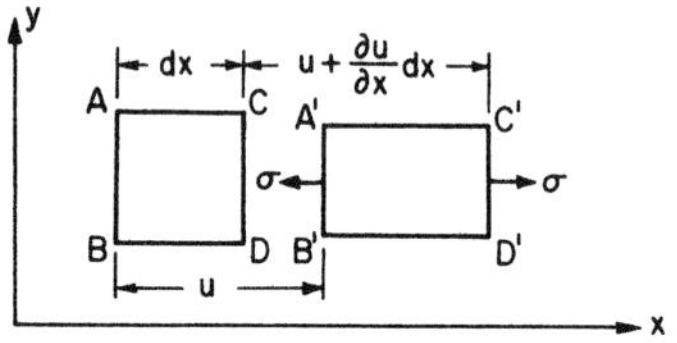

Fig. 7.1 Displacement Under Uniaxial Stress

on the element, which is equal to the strain energy stored, is therefore

$$
\begin{aligned}
dU &= \int_{\sigma=0}^{\sigma=\sigma_x} \sigma d\left(u + \frac{\partial u}{\partial x} dx\right) dy\, dz - \int_{\sigma=0}^{\sigma=\sigma_x} \sigma\, du\, dy\, dz \\
&= \int_{\sigma=0}^{\sigma=\sigma_x} \sigma d\left(\frac{\partial u}{\partial x}\right) dx\, dy\, dz
\end{aligned} \tag{7.1}
$$

Note that σ, u, and $\partial u/\partial x$ are variable and the integration is with respect to the displacement gradient $\partial u/\partial x$; dx, dy, dz are constant for this integration. No contribution is made to the strain energy by v and w because σ_y and σ_z are zero. From the definition of strain and Hooke's law,

$$\frac{\partial u}{\partial x} = \epsilon = \frac{\sigma}{E}$$

we have, therefore,

$$dU = \int_{\sigma=0}^{\sigma=\sigma_x} \sigma \frac{d\sigma}{E} dx\, dy\, dz = \frac{{\sigma_x}^2}{2E} dx\, dy\, dz \tag{7.2}$$

The strain energy per unit volume, or the strain energy density, is therefore given by

$$U_0 = \frac{{\sigma_x}^2}{2E} = \tfrac{1}{2}\sigma_x \epsilon_x = \tfrac{1}{2}E{\epsilon_x}^2 \tag{7.3}$$

The work done by a force is also given by the area under its force-displacement curve. In this case, the final displacement is the relative movement of the stressed faces of the element, i.e., $\epsilon_x\, dx$, and the maximum force is $\sigma_x\, dy\, dz$. For a linear stress-strain relation, the force-displacement curve is a straight line and the work done is equal to the area of the shaded triangle as shown in Fig. 7.2.

[1] Body forces are considered in the next section.

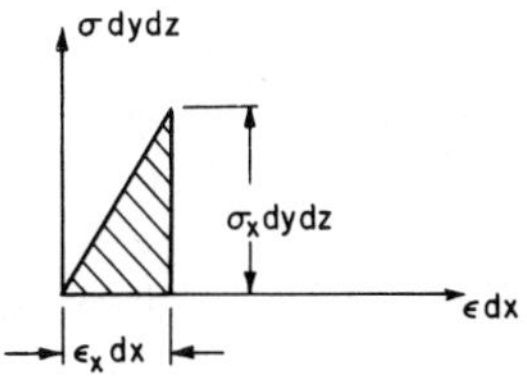

Fig. 7.2 Work Done by Uniaxial Stress σ_x

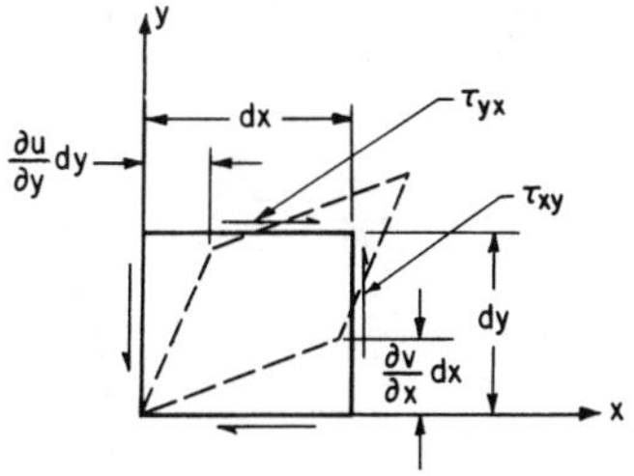

Fig. 7.3 Deformation Due to Simple Shear

By referring to Fig. 7.3, we observe that the strain energy due to the shear stress components τ_{xy} and τ_{yx} is given by

$$dU = \tfrac{1}{2}\left[(\tau_{xy}\,dy\,dz)\left(\frac{\partial v}{\partial x}dx\right) + (\tau_{yx}\,dx\,dz)\left(\frac{\partial u}{\partial y}dy\right)\right]$$

$$= \tfrac{1}{2}\tau_{xy}\left(\frac{\partial v}{\partial x} + \frac{\partial u}{\partial y}\right)dx\,dy\,dz$$

so that

$$U_0 = \tfrac{1}{2}\tau_{xy}\gamma_{xy} = \frac{1}{2}\frac{\tau_{xy}^2}{G} = \tfrac{1}{2}G\gamma_{xy}^2 \tag{7.4}$$

Here, τ_{xy} is the maximum value of the shear stress τ, which varies from zero to τ_{xy}, and γ_{xy} is the maximum value of the shear strain.

Next let us consider the element under the action of both σ_x and σ_y. Since the element is an elastic body, it will return to its original dimensions if σ_x and σ_y are removed. From the principle of conservation of energy we know that the work done will not depend on the order in which the stresses are applied, but only on the final magnitudes of the stresses and strains. If this were not the case, we could load in one order and unload in another order corresponding to a different amount of work; and consequently, a certain amount of work would be done in the complete cycle, in contradiction with the principle of conservation of energy. The strain energy density due to σ_x and σ_y is then[2]

$$U_0 = \tfrac{1}{2}\sigma_x \epsilon_x + \tfrac{1}{2}\sigma_y \epsilon_y \tag{7.5}$$

since the final relative displacements of the x and y faces of the element are $\epsilon_x\, dx$ and $\epsilon_y\, dy$, respectively. Notice that ϵ_x, for example, is the final normal strain in the x direction due to both σ_x and σ_y.

Similarly under a general stress condition, the strain energy density becomes

$$U_0 = \tfrac{1}{2}(\sigma_x \epsilon_x + \sigma_y \epsilon_y + \sigma_z \epsilon_z + \tau_{xy}\gamma_{xy} + \tau_{yz}\gamma_{yz} + \tau_{zx}\gamma_{zx}) \tag{7.6}$$

which, by using generalized Hooke's law, may be expressed in terms of the stress components as

$$\begin{aligned} U_0 &= \frac{1}{2}\left[\frac{1}{E}(\sigma_x{}^2 + \sigma_y{}^2 + \sigma_z{}^2) - \frac{2\nu}{E}(\sigma_x\sigma_y + \sigma_y\sigma_z + \sigma_z\sigma_x) \right. \\ &\qquad \left. + \frac{1}{G}(\tau_{xy}^2 + \tau_{yz}^2 + \tau_{zx}^2)\right] \\ &= \frac{1}{2}\left[\frac{-\nu}{E}(\sigma_x + \sigma_y + \sigma_z)^2 + \frac{1+\nu}{E}(\sigma_x{}^2 + \sigma_y{}^2 + \sigma_z{}^2) \right. \\ &\qquad \left. + \frac{1}{G}(\tau_{xy}^2 + \tau_{yz}^2 + \tau_{xz}^2)\right] \end{aligned} \tag{7.7}$$

or, in terms of the strain components, as

$$U_0 = \tfrac{1}{2}[\lambda\epsilon^2 + 2G(\epsilon_x{}^2 + \epsilon_y{}^2 + \epsilon_z{}^2) + G(\gamma_{xy}^2 + \gamma_{yz}^2 + \gamma_{zx}^2)] \tag{7.8}$$

where $\epsilon = \epsilon_x + \epsilon_y + \epsilon_z$.

We observe from Eq. (7.7) that the derivative of U_0 with respect to any stress component is equal to the corresponding strain component, and from Eq. (7.8) the reverse is true, i.e.,

$$\frac{\partial U_0(\sigma_x, \sigma_y, \ldots, \tau_{xz})}{\partial \sigma_x} = \epsilon_x \quad \text{and} \quad \frac{\partial U_0(\epsilon_x, \epsilon_y, \ldots, \gamma_{xz})}{\partial \epsilon_x} = \sigma_x \tag{7.9}$$

The total strain energy absorbed in an elastic body is found by the integral

$$U = \iiint\limits_V U_0\, dx\, dy\, dz \tag{7.10}$$

throughout the entire volume of the body.

[2] See Problem 7-1.

The strain energy density at a point may be decomposed into two parts, one due to the change in volume (change in size) and the other due to distortion (change in shape). This concept is used in the determination of the yield point of a ductile metal under any stress condition. If the state of stress at a point is designated by the components σ_x, σ_y, σ_z, τ_{xy}, τ_{yz}, τ_{zx}, we can consider this as the sum of two sets of stress components. Let the first set be designated as

$$\sigma'_x = \sigma'_y = \sigma'_z = \sigma_m = \frac{\sigma_x + \sigma_y + \sigma_z}{3}, \quad \tau'_{xy} = \tau'_{xz} = \tau'_{yz} = 0 \qquad (7.11a)$$

Thus the second set is defined by the components

$$\begin{aligned} \sigma''_x &= \sigma_x - \sigma_m & \sigma''_y &= \sigma_y - \sigma_m & \sigma''_z &= \sigma_z - \sigma_m \\ \tau''_{xy} &= \tau_{xy} & \tau''_{xz} &= \tau_{xz} & \tau''_{yz} &= \tau_{yz} \end{aligned} \qquad (7.11b)$$

The first set represents a hydrostatic state of stress, which, by Eq. (3.23), causes the entire change in volume. The second set, the stress deviator components, causes no volume change and is equivalent to a state of three-dimensional pure shear (distortion). The energy of volume change per unit volume (due to the hydrostatic components of stress) is

$$U'_0 = \frac{1}{2}\sigma_m \epsilon = \frac{1}{2}\frac{\sigma_m{}^2}{K} = \frac{1}{18K}(\sigma_x + \sigma_y + \sigma_z)^2 \qquad (7.12)$$

where K is the bulk modulus of elasticity. The strain energy of distortion per unit volume results from the stress deviator components, Eq. (7.11b), and is given by

$$U''_0 = \frac{3}{4}\frac{\tau_{\text{oct}}^2}{G} \qquad (7.13)$$

where τ_{oct} is the octahedral shear stress,

$$\tau_{\text{oct}} = \tfrac{1}{3}\sqrt{(\sigma_x - \sigma_y)^2 + (\sigma_y - \sigma_z)^2 + (\sigma_z - \sigma_x)^2 + 6(\tau_{xy}^2 + \tau_{yz}^2 + \tau_{zx}^2)} \qquad (7.14)$$

7.3. Variable Stress Distribution and Body Forces

In the preceding section we considered the strain energy in a body under a uniform state of stress. We shall next consider the strain energy in the case where the stresses are not uniform, but vary continuously, and we also include the effect of body force components F_x, F_y, and F_z (which may include inertia forces).

Referring to Fig. 7.4, we see that the work done by the stress σ_x on the left-hand surface of the element is $-\frac{1}{2}\sigma_x u\, dy\, dz$. The negative sign is necessary because on this surface the positive σ_x and positive u are in

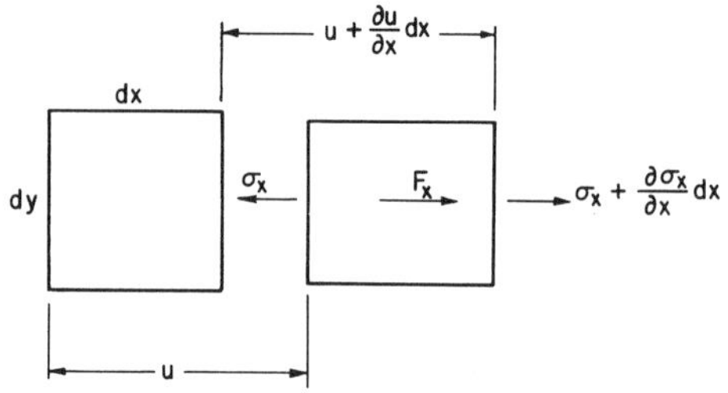

Fig. 7.4 An Element Under a Nonuniform State of Stress

opposite directions. It is implied that σ_x and u are the final values of stress and displacement, both being initially zero. On the right-hand surface the work done is

$$\frac{1}{2}\left(\sigma_x + \frac{\partial \sigma_x}{\partial x} dx\right)\left(u + \frac{\partial u}{\partial x} dx\right) dy\, dz$$

and the net work done by the normal stress on these two x planes is

$$\frac{1}{2}\left(\sigma_x + \frac{\partial \sigma_x}{\partial x} dx\right)\left(u + \frac{\partial u}{\partial x} dx\right) dy\, dz - \frac{1}{2}\sigma_x u\, dy\, dz = \frac{1}{2}\frac{\partial}{\partial x}(\sigma_x u)\, dx\, dy\, dz$$

where the second-order term in dx has been neglected. The stress σ_x and displacement u on the left-hand surface are the mean values of the final stress and displacement over the plane area $dy\, dz$. As shown in Section 1.7 in deriving the equations of equilibrium, the variation of σ_x in the y and z directions has no contribution to the resultant force on the infinitesimal element. It can be shown by a similar procedure that the variation of σ_x and u in the y and z directions has no contribution to the net work done.

Similarly, the net work done by the shear stress components on these planes is

$$\frac{1}{2}\frac{\partial(\tau_{xy} v)}{\partial x} dx\, dy\, dz + \frac{1}{2}\frac{\partial(\tau_{xz} w)}{\partial x} dx\, dy\, dz$$

The total work per unit volume on this element performed by all stresses and body forces may now be written as[3]

$$U_0 = \frac{1}{2}\left[\frac{\partial}{\partial x}(\sigma_x u + \tau_{xy} v + \tau_{xz} w) + \frac{\partial}{\partial y}(\sigma_y v + \tau_{yx} u + \tau_{yz} w)\right.$$
$$\left. + \frac{\partial}{\partial z}(\sigma_z w + \tau_{zx} u + \tau_{zy} v) + F_x u + F_y v + F_z w\right]$$

[3] Strictly speaking, the displacement components multiplying the body forces should be those at the center of the element; however, ignoring this refinement, we have only dropped higher-order terms.

or

$$U_0 = \frac{1}{2}\Big[\Big(\sigma_x \frac{\partial u}{\partial x} + \sigma_y \frac{\partial v}{\partial y} + \sigma_z \frac{\partial w}{\partial z}\Big) + \tau_{xy}\Big(\frac{\partial v}{\partial x} + \frac{\partial u}{\partial y}\Big) + \tau_{yz}\Big(\frac{\partial w}{\partial y} + \frac{\partial v}{\partial z}\Big)$$

$$+ \tau_{zx}\Big(\frac{\partial w}{\partial x} + \frac{\partial u}{\partial z}\Big) + u\Big(\frac{\partial \sigma_x}{\partial x} + \frac{\partial \tau_{xy}}{\partial y} + \frac{\partial \tau_{xz}}{\partial z} + F_x\Big)$$

$$+ v\Big(\frac{\partial \sigma_y}{\partial y} + \frac{\partial \tau_{xy}}{\partial x} + \frac{\partial \tau_{yz}}{\partial z} + F_y\Big) + w\Big(\frac{\partial \sigma_z}{\partial z} + \frac{\partial \tau_{xz}}{\partial x} + \frac{\partial \tau_{yz}}{\partial y} + F_z\Big)\Big]$$

We notice that the expressions in the last three parentheses vanish by reason of the equations of equilibrium. The strain energy density then is given by

$$U_0 = \tfrac{1}{2}[\sigma_x \epsilon_x + \sigma_y \epsilon_y + \sigma_z \epsilon_z + \tau_{xy}\gamma_{xy} + \tau_{yz}\gamma_{yz} + \tau_{zx}\gamma_{zx}]$$

which is identical to Eq. (7.6).

7.4. Principle of Virtual Work and the Theorem of Minimum Potential Energy

We shall introduce the basic concept of virtual work and virtual displacement by considering a small rigid particle which is acted upon by a system of forces. Assume that the particle is at rest: therefore the resultant of all the forces acting on it is zero. If we want to move this particle to a new position (or to give it a small displacement), additional force is required and the original force system must be altered. Now we shall consider a "virtual displacement," defined as an arbitrary displacement which does not affect the force system acting on the particle. In other words, a virtual displacement is a fictitious displacement, and during the process of it each of the forces acting on the particle remains constant in magnitude and direction. For an infinitesimal actual displacement, the actual changes in the forces due to the displacement are small and may be neglected in comparison to the forces themselves; therefore a virtual displacement is sometimes considered as an infinitesimal actual displacement. According to our definition, however, a virtual displacement is not restricted to be infinitesimal.

The work done by the forces acting on the particle during a virtual displacement is called the *virtual work*. The virtual work done is zero if the particle is in equilibrium, since the resultant force vanishes. It is also evident that if the virtual work vanishes, the force system acting on the particle must be in equilibrium. In this case of a single rigid particle, it is easy to write the equilibrium equations governing the forces, and it seems that the principle of virtual work does not contribute much to the problem. For more complicated problems, however, it is sometimes more convenient to require the virtual work corresponding to a certain virtual displacement to vanish than to write down and solve the equilibrium equations.

In discussing the principle of virtual work for an elastic body, we must introduce a virtual displacement field, i.e.,

$$\delta u = \delta u(x, y, z), \quad \delta v = \delta v(x, y, z), \quad \delta w = \delta w(x, y, z)$$

and a virtual strain field

$$\delta \epsilon_x = \delta \epsilon_x(x, y, z), \qquad \text{etc.}$$

all of which take place *after* the body has reached its equilibrium configuration. A virtual displacement is an arbitrary displacement subject to the following restrictions:

1. The components of the virtual displacement, δu, δv, and δw, are single-valued, continuous functions and are of the order of magnitude of displacements admissible in linear elasticity.

2. During a virtual displacement, all forces (surface and body forces) and internal stresses are constant in magnitude and direction.

3. At boundary points where displacement components are prescribed, the corresponding components of virtual displacement must vanish. For example, at the ends of a cylinder under a prescribed angle of twist (torsion problem), we must have $\delta u = \delta v = 0$ since u and v are prescribed on the ends.

We define the virtual strain components in a manner analogous to the definition of real strain components, e.g.,

$$\begin{aligned} \delta \epsilon_x &= \delta\left(\frac{\partial u}{\partial x}\right) = \frac{\partial}{\partial x}(\delta u), \quad \text{etc.} \\ \delta \gamma_{xy} &= \delta\left(\frac{\partial u}{\partial y} + \frac{\partial v}{\partial x}\right) = \frac{\partial}{\partial y}(\delta u) + \frac{\partial}{\partial x}(\delta v) \end{aligned} \tag{7.15}$$

In other words, for every virtual displacement δu, we define $(\partial/\partial x)(\delta u)$ as the virtual normal strain and denote it by $\delta \epsilon_x$.

Except for the three restrictions just mentioned, virtual displacements are perfectly arbitrary. It is not necessary that they be physically possible, for we may simply imagine their occurrence. Once the virtual displacements are specified, the virtual strain components are determined according to Eq. (7.15).

In order to apply the method of virtual work to an elastic body, we shall first develop expressions for δW, the virtual work done by the external (surface and body) forces and the virtual strain energy δU, or the strain energy absorbed in the body during a virtual displacement. Consider an elastic body subjected to a force system causing actual displacements u, v, and w. Then let the body be subjected to a virtual displacement field with components δu, δv, and δw. In order to determine the virtual strain energy, we first consider the virtual work done by σ_x on the element shown in Fig. 7.5. We are not concerned with the

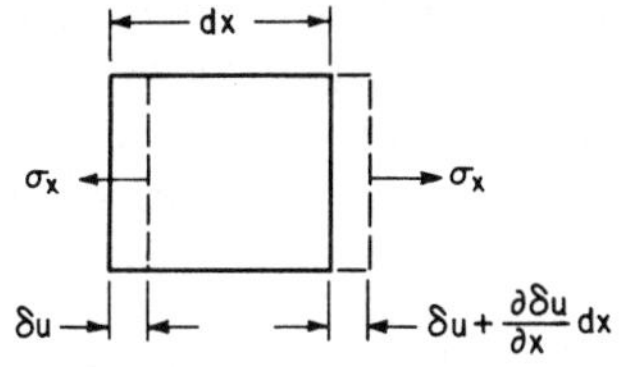

Fig. 7.5 Virtual Work Done on an Element

work done while the actual displacements occur; we assume that u, v, and w occur first and following this we imagine that the virtual displacement components are applied. We then write the work done by the actual stresses in being carried through the virtual displacement field. The virtual work done by σ_x is therefore

$$\sigma_x \left(\delta u + \frac{\partial \delta u}{\partial x} dx \right) dy\, dz - \sigma_x (\delta u)\, dy\, dz \tag{7.16}$$

and the virtual work done per unit volume (δU_0) is

$$\delta U_0 = \sigma_x \frac{\partial \delta u}{\partial x} = \sigma_x\, \delta\epsilon_x = E\epsilon_x\, \delta\epsilon_x \tag{7.17}$$

Since $U_0 = \frac{1}{2} E\epsilon_x^2$ and $\delta U_0 = E\epsilon_x\, \delta\epsilon_x$, the "$\delta$" may be considered as an operator similar to the differentiation operator "d," or

$$\delta U_0 = \frac{dU_0}{d\epsilon_x} \delta\epsilon_x \tag{7.18}$$

where U_0 is expressed as a function of ϵ_x (not σ_x). The function δU_0 is called *the first variation of* U_0, or simply *the variation of* U_0. Now let us consider an actual incremental strain $\Delta\epsilon_x$ and the corresponding increase in strain energy density ΔU. From a Taylor's series expansion, we have

$$U_0(\epsilon_x + \Delta\epsilon_x) = U_0(\epsilon_x) + \frac{dU_0}{d\epsilon_x} \Delta\epsilon_x + \frac{1}{2} \frac{d^2 U_0}{d\epsilon_x^2} (\Delta\epsilon_x)^2 + \cdots \tag{7.19}$$

or

$$\Delta U_0 = U_0(\epsilon_x + \Delta\epsilon_x) - U_0(\epsilon_x) = \frac{dU}{d\epsilon_x} \Delta\epsilon_x + \frac{1}{2} \frac{d^2 U_0}{d\epsilon_x^2} (\Delta\epsilon_x)^2 \tag{7.20}$$

Since the strain energy density is a function of the strain components to the second power, the derivatives of higher than the second order in Eq. (7.19) vanish. If $\Delta\epsilon_x$ is infinitesimal, the term involving $(\Delta\epsilon_x)^2$ may be neglected and Eq. (7.20) reduces to the form of Eq. (7.18). In deriving Eq. (7.18), we did not make the assumption that $\delta\epsilon_x$ is infinitesimal and no second-order terms had been neglected. The fact that the second-order term does not appear in Eq. (7.18) is because $\delta\epsilon_x$ is a virtual strain; during its application the stress remains constant.

Under a general stress condition it can be shown that the virtual strain energy density is given by

$$\delta U_0 = \sigma_x\,\delta\epsilon_x + \sigma_y\,\delta\epsilon_y + \sigma_z\,\delta\epsilon_z + \tau_{xy}\,\delta\gamma_{xy} + \tau_{yz}\,\delta\gamma_{yz} + \tau_{zx}\,\delta\gamma_{zx} \tag{7.21a}$$

which along with Eq. (7.8) and Hooke's law, indicates that the rule for the δ operator is perfectly general. Notice that no factor of $\frac{1}{2}$ is present, since the stresses are constant during the virtual displacement. The total virtual strain energy is therefore

$$\delta U = \iiint_V \delta U_0\,dx\,dy\,dz \tag{7.21b}$$

Consider next the virtual work done by the external forces. Displacement components u, v, and w occur first. Then the virtual components δu, δv, and δw are applied. The virtual work done by the surface forces is therefore

$$\int_A (T_x{}^\mu\,\delta u + T_y{}^\mu\,\delta v + T_z{}^\mu\,\delta w)\,dA$$

where dA is an elemental surface area and the integration is taken over the complete boundary surface of the body. Again, the factor of $\frac{1}{2}$ is not present because the surface forces are constant during the virtual displacement. Of course, the value of this integral taken over that part of the surface where the displacements are prescribed will be identically zero. The virtual work done by the body forces is

$$\int_V (F_x\,\delta u + F_y\,\delta v + F_z\,\delta w)dV$$

where the integration is carried out throughout the entire volume V of the body. The virtual work δW done by external forces is therefore given by

$$\delta W = \int_A (T_x{}^\mu\,\delta u + T_y{}^\mu\,\delta v + T_z{}^\mu\,\delta w)dA + \int_V (F_x\,\delta u + F_y\,\delta v + F_z\,\delta w)dV \tag{7.21c}$$

We shall now develop the governing equation in the method of virtual work. Equations (7.21a) and (7.21b) give

$$\delta U = \int_V [\sigma_x \, \delta\epsilon_x + \sigma_y \, \delta\epsilon_y + \sigma_z \, \delta\epsilon_z + \tau_{xy} \, \delta\gamma_{xy} + \tau_{yz} \, \delta\gamma_{yz} + \tau_{zx} \, \delta\gamma_{zx}] dV$$

Using Eqs. (7.15), this becomes

$$\delta U = \int_V \left\{ \sigma_x \frac{\partial}{\partial x}(\delta u) + \sigma_y \frac{\partial}{\partial y}(\delta v) + \sigma_z \frac{\partial}{\partial z}(\delta w) + \tau_{xy} \left[\frac{\partial}{\partial y}(\delta u) + \frac{\partial}{\partial x}(\delta v) \right] \right.$$

$$\left. + \tau_{yz} \left[\frac{\partial}{\partial z}(\delta v) + \frac{\partial}{\partial y}(\delta w) \right] + \tau_{zx} \left[\frac{\partial}{\partial z}(\delta u) + \frac{\partial}{\partial x}(\delta w) \right] \right\} dV \qquad (7.21d)$$

Denoting the integral of the first term by M, i.e.,

$$M = \int_V \sigma_x \frac{\partial}{\partial x}(\delta u) \, dV = \iiint_V \sigma_x \frac{\partial}{\partial x}(\delta u) \, dx \, dy \, dz$$

integration by parts gives

$$M = \iint_A [\sigma_x \, \delta u]_{x_1(y,z)}^{x_2(y,z)} \, dy \, dz - \int_V \frac{\partial \sigma_x}{\partial x} \, \delta u \, dV$$

where $x_2(y, z)$ and $x_1(y, z)$ are the equations of the right and left surfaces respectively, of A. But on x_2 we have

$$dy \, dz = dA \, \mu_x$$

while on x_1

$$dy \, dz = - \, dA \, \mu_x$$

according to Eqs. (1.23). Thus the expression for M may be written as

$$M = \int_A \sigma_x \, \delta u \, \mu_x \, dA - \int_V \frac{\partial \sigma_x}{\partial x} \, \delta u \, dV$$

Evaluating the integrals of the other terms in Eq. (7.21d) in the same way, the expression for δU becomes

$$\delta U = \int_A [(\sigma_x \mu_x + \tau_{xy} \mu_y + \tau_{zx} \mu_z) \, \delta u + (\sigma_y \mu_y + \tau_{xy} \mu_x + \tau_{yz} \mu_z) \, \delta v$$

$$+ (\sigma_z \mu_z + \tau_{yz} \mu_y + \tau_{zx} \mu_x) \, \delta w] \, dA - \int_V \left[\left(\frac{\partial \sigma_x}{\partial x} + \frac{\partial \tau_{xy}}{\partial y} + \frac{\partial \tau_{zx}}{\partial z} \right) \delta u \right.$$

$$\left. + \left(\frac{\partial \sigma_y}{\partial y} + \frac{\partial \tau_{xy}}{\partial x} + \frac{\partial \tau_{yz}}{\partial z} \right) \delta v + \left(\frac{\partial \sigma_z}{\partial z} + \frac{\partial \tau_{yz}}{\partial y} + \frac{\partial \tau_{zx}}{\partial x} \right) \delta w \right] dV$$

Substituting Eqs. (4.4) and the equations of equilibrium into this equation, we find that

$$\delta U = \int_A (T_x{}^\mu \,\delta u + T_y{}^\mu \,\delta v + T_z{}^\mu \,\delta w) dA + \int_V (F_x \,\delta u + F_y \,\delta v + F_z \,\delta w)\, dV$$

In words, if the displacement components satisfy the equilibrium equations, the virtual strain energy is equal to the virtual work done by the external forces. Since the external forces are unchanged during the virtual displacement and the limits of integration are constant, the operator δ may be placed before the integral signs in the above equation. Defining $\Pi = U - W$, then, we get

$$\delta\Pi = \delta(U - W) = 0 \tag{7.22}$$

where

$$W = \int_A (T_x{}^\mu u + T_y{}^\mu v + T_z{}^\mu w)\, dA + \int_V (F_x u + F_y v + F_z w)\, dV \tag{7.23}$$

The function Π is called the potential energy of the body, since U is the energy of deformation, which is comparable to the energy given up when a compressed spring is released, and $-W$ is proportional to the potential energy of the external forces. The potential energy is taken as zero when the body is in its initial, undeformed position. It must be noted, however, that W, according to its definition by Eq. (7.23), is *not* the work done by the external forces in deflecting the body between its unloaded and equilibrium positions. The work done by the external forces is actually $\frac{1}{2}W$, since the work done by $T_x{}^\mu$ for example, is $\frac{1}{2}\int_A T_x{}^\mu u\, dA$. Thus, according to our definition, the potential energy is $U - W = -U = -\frac{1}{2}W$ since $\frac{1}{2}W = U$ by the principle of conservation of energy.

Equation (7.22) implies that, at the equilibrium configuration of a body, the potential energy assumes a stationary value. The principle of potential energy is then as follows: Of all the displacement distributions satisfying the conditions of continuity and the prescribed displacement boundary conditions, the one which actually takes place (or which satisfies the equilibrium equations) is the one which makes the potential energy assume a stationary value. It can be shown further that this stationary value is a minimum if the body is in stable equilibrium.[4] In order to apply the principle of potential energy, it is necessary to write

[4] The general proof of this can be found from I. S. Sokolnikoff, *Mathematical Theory of Elasticity*, Second Edition, McGraw-Hill Book Company, Inc., New York, 1956, pp. 384–385. The proof for a specific example is given in the next section.

the function U in terms of the strain or displacement components, *not containing any of the stress components*, because of the definition of δU.

From Eqs. (7.9) and (7.21a), we see that

$$\delta U_0 = \frac{\partial U_0}{\partial \epsilon_x}\delta\epsilon_x + \frac{\partial U_0}{\partial \epsilon_y}\delta\epsilon_y + \cdots + \frac{\partial U_0}{\partial \gamma_{xz}}\delta\gamma_{xz} \tag{7.24}$$

In many cases it is possible to express the strain components in terms of a single parameter, such as the deflection w of a beam. Then, $U_0 = U_0(\epsilon_x, \epsilon_y, \ldots, \gamma_{xz})$, where $\epsilon_x, \epsilon_y, \ldots, \gamma_{xz}$ are explicit functions of w. In this case, we have

$$\delta U_0 = \left[\frac{\partial U_0}{\partial \epsilon_x}\frac{d\epsilon_x}{dw} + \frac{\partial U_0}{\partial \epsilon_y}\frac{d\epsilon_y}{dw} + \cdots + \frac{\partial U_0}{\partial \gamma_{xz}}\frac{d\gamma_{xz}}{dw}\right]\delta w \tag{7.25}$$

or

$$\delta U_0 = \frac{dU_0}{dw}\delta w \tag{7.26}$$

We have seen that the actual displacement field which occurs is the one which minimizes the potential energy, if the stresses in a body are held constant while the body is subjected to displacement variations. If we consider the reverse problem, that is, if we vary the stresses while holding the displacements fixed, it can be shown that another minimizing condition results. In this case, among all stresses that satisfy the equilibrium equations, the ones that satisfy also the compatibility equations (i.e., the ones which actually occur) are those which minimize a quantity called the complimentary energy. This quantity Π^* is defined by

$$\Pi^* = U - \int_{A_u} (T_x^{\mu} u + T_y^{\mu} v + T_z^{\mu} w)\, dA \tag{7.27}$$

where U is the strain energy expressed in terms of the stress components and A_u is the surface area over which the displacement is prescribed.

7.5. Illustrative Problems

We shall demonstrate the application of the principle of potential energy by considering a uniformly loaded, homogeneous string as shown in Fig. 7.6. The string is initially under a large tensile force S, (Fig. 7.6(a)). A uniform transverse load q (lb/in.) is then applied (Fig. 7.6(b)), and we assume that the application of q does not change the magnitude of applied force S. Also, the string has no resistance to bending so that, under load q, the internal tensile force in the string at each point is (very nearly) S. We shall limit our discussion to the case where w is small so that the deflection curve is rather flat. This, of course, is a standard assumption of linear elasticity. Considering the stretched

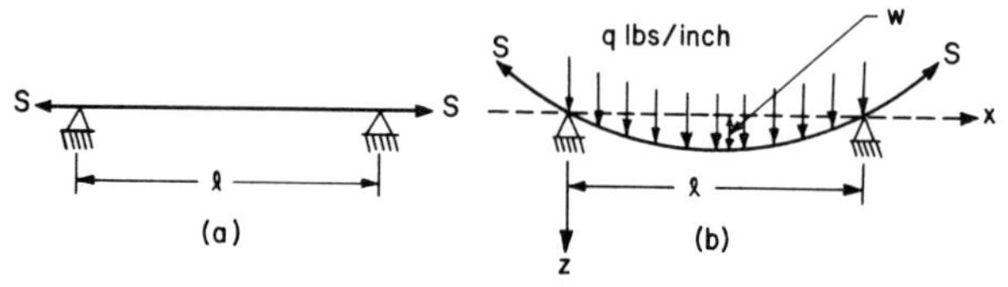

Fig. 7.6 Uniformly Loaded String

string under tension S and $q=0$ as the reference state (zero potential energy), we write

$$\Pi = U - W$$

where

$$W = \int_A (T_x{}^\mu u + T_y{}^\mu v + T_z{}^\mu w)\, dA$$

$$= \int_0^\ell (qw)\, dx$$

since the surface force per unit length is q and there are no body forces.

In order to evaluate U,[5] we must determine the change in length of the string caused by the transverse load q. As can be seen from Fig. 7.7,

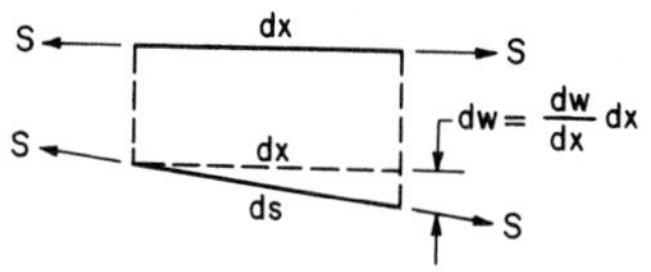

Fig. 7.7 An Element of the Stretched String

the internal work done by S on the element shown is $S(ds - dx)$, where $(ds - dx)$ is the elongation of the element due to the application of q. Notice that no factor of $\frac{1}{2}$ is introduced, since S is constant during the displacement by our assumption. Thus

$$ds = \sqrt{dx^2 + dw^2} = dx\sqrt{1 + (dw/dx)^2} \cong dx[1 + \tfrac{1}{2}(dw/dx)^2] \tag{7.28}$$

[5] We are actually computing the change in strain energy, since the initial strain energy is not zero. However, in taking the variation, the initial value has no effect.

where the higher powers of dw/dx are neglected in the last step. Therefore,

$$U = \int_0^\ell S(ds - dx)$$

$$= \frac{S}{2}\int_0^\ell \left(\frac{dw}{dx}\right)^2 dx$$

and

$$\Pi = U - W = \frac{S}{2}\int_0^\ell \left(\frac{dw}{dx}\right)^2 dx - \int_0^\ell qw\, dx \tag{7.29}$$

The first variation of Π, $\delta\Pi$, may now be found by taking the first variation, i.e., using Eq. (7.26). Thus

$$\delta\Pi = S\int_0^\ell \frac{dw}{dx}\,\delta\left(\frac{dw}{dx}\right) dx - q\int_0^\ell \delta w\, dx$$

The first integral may be rewritten as

$$S\int_0^\ell \frac{dw}{dx}\frac{d\delta w}{dx}\, dx$$

since the operators d and δ may be interchanged (see Eq. (7.15)). Evaluating this integral by parts, we get

$$S\int_0^\ell \frac{dw}{dx}\frac{d\delta w}{dx}\, dx = S\frac{dw}{dx}\,\delta w\Big|_0^\ell - S\int_0^\ell \delta w\,\frac{d^2w}{dx^2}\, dx$$

$$= -S\int_0^\ell \delta w\,\frac{d^2w}{dx^2}\, dx$$

since the actual deflection is prescribed at the pinned ends; therefore, according to the definition of virtual displacement, δw is zero at $x=0$ and $x=\ell$. We now have

$$\delta\Pi = -S\int_0^\ell \delta w\,\frac{d^2w}{dx^2}\, dx - q\int_0^\ell \delta w\, dx = 0$$

or

$$\int_0^\ell \left(S\frac{d^2w}{dx^2} + q\right)\delta w\, dx = 0 \tag{7.30}$$

and since δw is arbitrary, we find that[6]

$$S\frac{d^2w}{dx^2}+q=0 \tag{7.31}$$

which is the equilibrium equation of the string in terms of w.

We see in the example that the condition of $\delta\Pi=0$ leads directly to the equilibrium equation in terms of displacement, which is the governing differential equation as discussed in Chapter 4. In solving this problem, therefore, we can either use Eq. (7.31) as the governing equation, or use $\delta\Pi=0$ as the governing equation. For this particular problem, Eq. (7.31) is a simple ordinary differential equation which can be derived directly from equilibrium and geometry (as in the membrane problem of Chapter 6) and can be integrated easily. For more complicated problems, it is sometimes more convenient to derive the equilibrium equations by first formulating the expression for Π and then requiring $\delta\Pi=0$. If it is difficult to find a solution for the equilibrium equations, we can find an approximate solution which satisfies the equation $\delta\Pi=0$, as will be shown in the next section.

We have previously shown that, of all the continuous displacement fields satisfying the boundary conditions, the one which satisfies the equilibrium equations and which actually occurs is that which gives the potential energy a stationary value. We shall now demonstrate that this stationary value is a minimum for the uniformly loaded string. In order to prove this, we shall show that the quantity[7]

$$\Delta\Pi=\Pi(w+\Delta w)-\Pi(w)$$

is always positive, where Δw is a function of x, equal to zero at $x=0$ and $x=\ell$. This means that if the string is displaced by Δw from its equilibrium position the potential energy is increased. Now by direct

[6] In the calculus of variations, this is called the *fundamental lemma*, a rigorous proof of which may be found, for example, in R. Courant and D. Hilbert, *Methods of Mathematical Physics*, Vol. 1, John Wiley & Sons, New York, 1953. We may explain it as follows. Suppose that the quantity in the parentheses of Eq. (7.30) is larger than zero in the interval $x_1<x<x_2$, where $0<x_1<x_2<\ell$. Then, since δw is arbitrary, we can choose δw to be positive in $x_1<x<x_2$, and zero everywhere else. The integrand is thus always positive and the integral must also be positive. This is in contradiction with Eq. (7.30), therefore we must have Eq. (7.31).

[7] Recall that a function $f(x)$ is stationary at x_1 if $df/dx=0$ at x_1, or $df=0$ for an infinitesimal change in position dx. If the function $f(x)$ is a minimum at x_1, then $\Delta f=f(x_1+\Delta x)-f(x_1)$ must be larger than zero, where Δx is a small, but finite quantity. In Eq. (7.22), we have $\delta\Pi=0$, where $\delta\Pi$ corresponds to *virtual* displacements δu, δv, δw. Referring to actual small increments in displacements Δu, Δv, and Δw, $\Delta\Pi$ is zero if Δu, Δv, and Δw are infinitesimal; $\Delta\Pi>0$ for finite Δu, Δv, and Δw, where the second-order term in the Taylor's expansion for $\Delta\Pi$ must be retained.

substitution into Eq. (7.29) we see that

$$\begin{aligned}\Delta\Pi &= \Pi(w+\Delta w)-\Pi(w)\\ &=\frac{S}{2}\int_0^{\ell}\left(\frac{dw}{dx}+\frac{d\Delta w}{dx}\right)^2 dx - q\int_0^{\ell}(w+\Delta w)\,dx\\ &\quad -\frac{S}{2}\int_0^{\ell}\left(\frac{dw}{dx}\right)^2 dx + q\int_0^{\ell} w\,dx\\ &=\frac{S}{2}\int_0^{\ell} 2\,\frac{dw}{dx}\frac{d\Delta w}{dx}\,dx+\frac{S}{2}\int_0^{\ell}\left(\frac{d\Delta w}{dx}\right)^2 dx - q\int_0^{\ell}\Delta w\,dx\end{aligned}$$

Evaluating the first integral by parts, we have

$$\Delta\Pi = S\frac{dw}{dx}\Delta w\Big|_0^{\ell} - S\int_0^{\ell}\Delta w\,\frac{d^2w}{dx^2}\,dx+\frac{S}{2}\int_0^{\ell}\left(\frac{d\Delta w}{dx}\right)^2 dx - q\int_0^{\ell}\Delta w\,dx$$

The first term is zero, since Δw vanishes at the ends $x=0$ and $x=\ell$, and by combining the first and third integrals we get

$$\Delta\Pi = -\int_0^{\ell}\Delta w\left(S\frac{d^2w}{dx^2}+q\right)dx+\frac{S}{2}\int_0^{\ell}\left(\frac{d\Delta w}{dx}\right)^2 dx$$

But the first integral vanishes because of the equilibrium condition for the string, and we have finally

$$\Delta\Pi = \frac{S}{2}\int_0^{\ell}\left(\frac{d\Delta w}{dx}\right)^2 dx \geqslant 0 \tag{7.32}$$

since the integral cannot be negative. The integral vanishes only in the exceptional case when $d(\Delta w)/dx=0$, which only occurs at the equilibrium position. Thus we have demonstrated the theorem of minimum potential energy for the uniformly loaded string.

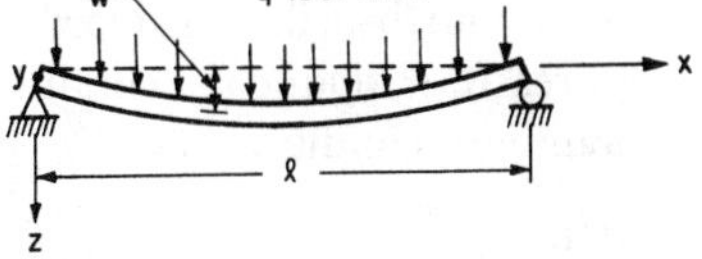

Fig. 7.8 A Simply Supported Beam

As another example of the principle of potential energy, we shall consider a simply supported beam of constant cross section, supporting a uniformly distributed load q, as shown in Fig. 7.8. We shall only consider the strain energy due to pure bending, i.e., due to σ_x. We shall also use the assumption that plane cross sections remain plane after bending; therefore the normal stress is $\sigma_x = Mz/I$, where M is the bending moment and I is the moment of inertia of the cross section with respect to the y axis. According to the Bernoulli-Euler law in beam theory given in most strength of materials textbooks, the moment is given by

$$M = \frac{EI}{R} \cong EI\frac{d^2w}{dx^2}$$

where R is the radius of curvature of the neutral axis of the beam. With these assumptions, we have

$$U_0 = \frac{\sigma_x{}^2}{2E} = \frac{M^2z^2}{2EI^2} = \frac{E}{2}\left(\frac{d^2w}{dx^2}\right)^2 z^2 \tag{7.33}$$

We use this form because the potential energy must be expressed in terms of strain or displacement. The total strain energy is therefore

$$\begin{aligned} U &= \iiint \frac{E}{2}\left(\frac{d^2w}{dx^2}\right)^2 z^2\, dx\, dy\, dz \\ &= \int_0^\ell \frac{EI}{2}\left(\frac{d^2w}{dx^2}\right)^2 dx \end{aligned} \tag{7.34}$$

since $\iint z^2\, dy\, dz = I$. Also

$$W = \int_0^\ell qw\, dx \tag{7.35}$$

So that the potential energy in this case is

$$\Pi = U - W = \int_0^\ell \left[\frac{EI}{2}\left(\frac{d^2w}{dx^2}\right)^2 - qw\right] dx \tag{7.36}$$

Due to the Bernoulli-Euler assumption, the beam problem has only one independent variable, x. The governing equation (equilibrium equation in terms of w) will be a fourth-order ordinary differential equation, Eq. (7.39), which requires four boundary conditions. For the simply supported beam the boundary conditions are

$$w = \frac{d^2w}{dx^2} = 0 \qquad \text{at } x = 0 \text{ and } x = \ell. \tag{7.37}$$

The first variation of Π, or the change in Π due to a virtual displacement is

$$\delta\Pi = \frac{EI}{2}\int_0^\ell 2\frac{d^2w}{dx^2}\frac{d^2\,\delta w}{dx^2}\, dx - \int_0^\ell q\delta w\, dx \tag{7.38}$$

Integrating the first term by parts twice, and utilizing the boundary conditions of Eq. (7.37) and the fundamental lemma in the calculus of variations, it can be shown that (see Problem 7-7)

$$EI\frac{d^4w}{dx^4} - q = 0 \tag{7.39}$$

which is the governing equilibrium equation for the beam. It can also be shown that $\Delta\Pi > 0$ for a finite Δw. (See Problem 7-8.)

Equation (7.39) can be obtained directly by utilizing the Bernoulli-Euler assumption and considering the equilibrium of an infinitesimal element of the beam. However, the relations $\sigma_x = Mz/I$ and $M = EI\, d^2w/dx^2$ must be used in either this approach or the energy approach. The purpose of this example is to demonstrate that the condition $\delta\Pi = 0$ leads directly to the governing equilibrium equation in terms of displacement. The approximation involved in using the Bernoulli-Euler assumption as compared to the exact elasticity solution is not discussed here.

7.6. Rayleigh-Ritz Method

For boundary value problems where an exact solution is not possible, a number of methods have been developed for finding approximate solutions based on the principle of potential energy. One of the more important of these is called the Rayleigh-Ritz method. In this section we shall present a brief treatment of the method as it applies to one-dimensional problems (problems in which the dependent variables are functions of one space coordinate).

Considering a one-dimensional problem such as the case of an elastic string or beam, the potential energy assumes the following form

$$\Pi = \int_{x_0}^{x_1} f(x, w, w', w'', \ldots)\, dx \tag{7.40}$$

where $w', w'', \ldots$ are the derivatives of w with respect to x. We may think of the solution then as the determination of a function $w(x)$ which satisfies the boundary conditions of the problem and which minimizes Π. If we can find such a function $w(x)$, it will automatically satisfy the equilibrium equation and will be the solution of the problem.

Recognizing that any continuous function may be expressed in the form of a series (e.g., power series, Fourier series), we assume the solution for $w(x)$ is of the form

$$w = a_0 w_0 + a_1 w_1 + a_2 w_2 + \cdots = \sum_{i=0}^{n} a_i w_i$$

where $w_0, w_1, \ldots$ are functions of x, each of which satisfies the boundary conditions,[8] and $a_0, a_1, a_2, \ldots$ are parameters to be determined. Substituting this function in Eq. (7.40), we find

$$\Pi = \Pi(a_0, a_1, a_2, \ldots)$$

That is, after carrying out the integration and substituting the limits of integration, the potential energy is a function of the parameters a_0, a_1, $a_2, \ldots$.

The minimizing condition, therefore, is

$$\frac{\partial \Pi}{\partial a_0} = 0 \qquad \frac{\partial \Pi}{\partial a_1} = 0 \qquad \cdots$$

Thus we obtain a system of algebraic equations which may be solved for the parameters. For approximate solutions, we may limit the series to a few terms; however, for an exact solution, the functions in the series must constitute a complete set and we must consider the entire series.

As an example of the application of the Rayleigh-Ritz method, let us consider the simply supported beam discussed in the preceding section, where the potential energy is given by Eq. (7.36) and the boundary conditions by Eqs. (7.37).

We assume the solution for w is the series

$$w(x) = \sum_{n=1}^{\infty} a_n \sin \frac{n\pi x}{\ell} \tag{7.41}$$

where a_n are undetermined coefficients. The second derivative of w is

$$\frac{d^2 w}{dx^2} = -\sum_{n=1}^{\infty} a_n \left(\frac{n\pi}{\ell}\right)^2 \sin \frac{n\pi x}{\ell}$$

It can be seen that each term of the series satisfies all of the boundary conditions. Substituting into the expression for Π, we find

$$\Pi = \int_0^{\ell} \left\{ \frac{EI}{2} \left[\sum_{n=1}^{\infty} a_n \left(\frac{n\pi}{\ell}\right)^2 \sin \frac{n\pi x}{\ell} \right]^2 - q \sum_{n=1}^{\infty} a_n \sin \frac{n\pi x}{\ell} \right\} dx \tag{7.42}$$

Squaring the series, we see that

[8] A boundary condition is homogeneous if every term of the boundary condition equation contains w or its derivatives, i.e., the constant term of the equation is zero. For a nonhomogeneous boundary condition, one of the functions, say $a_0 w_0$, may be selected to satisfy the nonhomogeneous boundary condition equation, while each of the other w_i's satisfies the corresponding homogeneous boundary condition.

$$\left[\sum_{n=1}^{\infty} a_n\left(\frac{n\pi}{\ell}\right)^2 \sin\frac{n\pi x}{\ell}\right]^2 = a_1\left(\frac{1\pi}{\ell}\right)^2 \sin\frac{1\pi x}{\ell}\sum_{n=1}^{\infty} a_n\left(\frac{n\pi}{\ell}\right)^2 \sin\frac{n\pi x}{\ell} +$$

$$a_2\left(\frac{2\pi}{\ell}\right)^2 \sin\frac{2\pi x}{\ell}\sum_{n=1}^{\infty} a_n\left(\frac{n\pi}{\ell}\right)^2 \sin\frac{n\pi x}{\ell} + \dots$$

$$= \sum_{m=1}^{\infty}\sum_{n=1}^{\infty} a_m a_n \left(\frac{m\pi}{\ell}\right)^2\left(\frac{n\pi}{\ell}\right)^2 \sin\frac{m\pi x}{\ell}\sin\frac{n\pi x}{\ell}$$

However,

$$\int_0^{\ell} \sin\frac{m\pi x}{\ell}\sin\frac{n\pi x}{\ell}\,dx = 0 \qquad [m \neq n]$$

$$\int_0^{\ell} \sin\frac{m\pi x}{\ell}\sin\frac{n\pi x}{\ell}\,dx = \frac{\ell}{2} \qquad [m = n]$$

so that

$$\Pi = \int_0^{\ell}\left[\frac{EI}{2}\sum_{n=1}^{\infty} a_n^{\,2}\left(\frac{n\pi}{\ell}\right)^4 \sin^2\frac{n\pi x}{\ell} - q\sum_{n=1}^{\infty} a_n \sin\frac{n\pi x}{\ell}\right] dx$$

$$= \frac{EI}{2}\left(\frac{\ell}{2}\right)\sum_{n=1}^{\infty} a_n^{\,2}\left(\frac{n\pi}{\ell}\right)^4 + q\left(\frac{\ell}{\pi}\right)\sum_{n=1}^{\infty} a_n\frac{1}{n}\cos\frac{n\pi x}{\ell}\Bigg|_0^{\ell}$$

We observe that if n is even the second term vanishes, thus

$$\Pi(a_n) = \frac{\pi^4 EI}{4\ell^3}\sum_{n=1}^{\infty} a_n^{\,2} n^4 - \frac{2q\ell}{\pi}\sum_{n=1,3,5}^{\infty}\frac{a_n}{n} \tag{7.43}$$

The minimizing condition is

$$\frac{\partial\Pi}{\partial a_m} = 0$$

where all coefficients except a_m are taken as constant during the partial differentiation. Therefore,

$$\frac{2\pi^4 EI}{4\ell^3} m^4 a_m = 0 \qquad [\text{for even } m]$$

or

$$a_m = 0 \qquad [\text{for even } m]$$

and

$$\frac{2\pi^4 EIm^4 a_m}{4\ell^3} - \frac{2q\ell}{\pi}\frac{1}{m} = 0 \qquad \text{and} \qquad a_m = \frac{4q\ell^4}{EI(m\pi)^5} \qquad [\text{for odd } m]$$

The expression for w becomes

$$w(x) = \frac{4q\ell^4}{EI\pi^5} \sum_{n=1,3,5}^{\infty} \frac{1}{n^5} \sin \frac{n\pi x}{\ell} \tag{7.44}$$

and since the maximum deflection occurs at $x = \ell/2$, we find that

$$w_{\max} = \frac{4q\ell^4}{EI\pi^5}\left(1 - \frac{1}{3^5} + \frac{1}{5^5} - \dots\right)$$

Taking only one term of the series, we get

$$w_{\max} = \frac{q\ell^4}{76.6EI}$$

whereas the exact solution is

$$w_{\max} = \frac{q\ell^4}{76.8EI}$$

In this example the infinite series for $w(x)$ yields the exact solution. In general, it may not be possible to determine all the coefficients in the series, and only a finite number of terms can be included.

It should be noted that Rayleigh-Ritz solutions based on only one term of the series will not in general produce the remarkable accuracy which was attained in this problem.

PROBLEMS

7-1 Demonstrate that the strain energy due to σ_x and σ_y is that given by Eq. (7.5) by applying σ_x first, from zero to its final value, and then σ_y.

7-2 Derive Eq. (7.13).

7-3 Express the strain energy density in terms of
(a) σ_x, σ_y, and τ_{xy};
(b) ϵ_x, ϵ_y, and γ_{xy};
(c) σ_x, σ_y, τ_{xy}, ϵ_x, ϵ_y, and γ_{xy}
for plane stress problems.

7-4 Repeat Problem **7-3** for plane strain problems.

7-5 A prismatic bar with constant cross-sectional area A is under uniform tensile stress as shown.

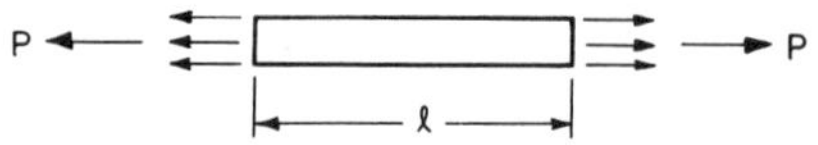

Problem 7–5

(a) Find the total strain energy U of the bar in terms of P, ℓ, A, and E.

(b) What is the total elongation of the bar? Find the work done *by the force P*.

(c) Find the potential energy in terms of ϵ, ℓ, A, and E ($\epsilon = \sigma/E$). The results should show that

$$\Pi = U - W = -U = -\tfrac{1}{2}W$$

where $\frac{1}{2}W$ is the work done by P.

7-6 Interpret the equation

$$\iint U_0\,dx\,dy = \tfrac{1}{2}\iint (F_x u + F_y v)\,dx\,dy + \tfrac{1}{2}\int (T_x^{\mu} u + T_y^{\mu} v)\,ds$$

and justify the factors $\frac{1}{2}$ on the right. What is the potential energy in terms of the stress components, F_x, F_y, T_x^{μ}, and T_y^{μ} for a plane stress problems?

7-7 Derive Eq. (7.39) from Eq. (7.38) for the simple beam problem.

7-8 By determining $\Delta\Pi = \Pi(w + \Delta w) - \Pi(w)$ for the simple beam problem, show that the potential energy for the equilibrium configuration is a minimum.

7-9 Explain the difference between the expressions

$$\delta\Pi = 0 \qquad \text{and} \qquad \Delta\Pi > 0$$

7-10 A beam with both ends fixed is loaded by a concentrated force P at the center of beam as shown.

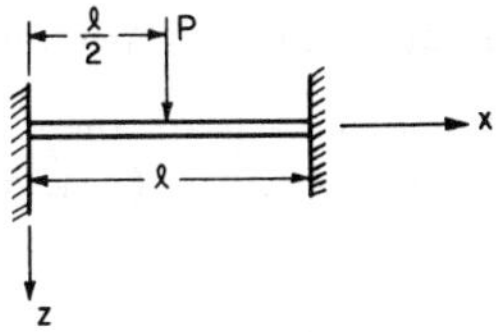

Problem 7–10

Find: (a) the strain energy in terms of w;

(b) W and Π.

(c) The boundary conditions are given as:

$$\text{at } x = 0 \qquad w = 0 \quad dw/dx = 0$$
$$x = \ell \qquad w = 0 \quad dw/dx = 0$$

Show that the following series satisfies the boundary conditions.

$$w = \sum_{n=2}^{\infty} A_n\left(\cos\frac{n\pi x}{\ell} - 1\right) \qquad n = 2, 4, 6, \ldots$$

(d) Take one term from the series above and determine A_2 by the Rayleigh-Ritz method.

7-11 (a) The uniform prism shown is subjected to a uniform normal stress p on all faces. Write the expression for Π in terms of $\epsilon_x, \epsilon_y, \epsilon_z, \gamma_{xy}, \gamma_{yz}, \gamma_{zx}$, p, u, v, and w.

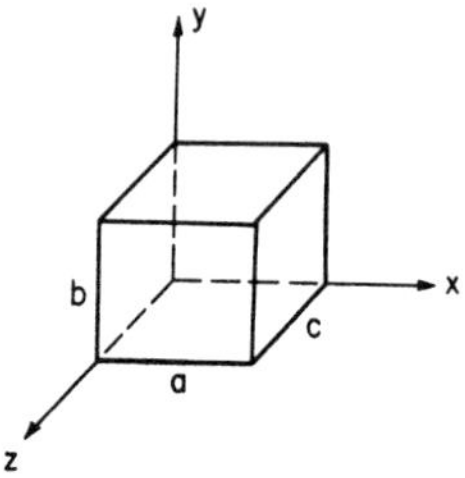

Problem 7–11

(b) Consider the virtual components of displacement δu, δv, and δw. Since these may be arbitrarily prescribed (as long as the assumptions regarding virtual displacements are satisfied), assume that δv and δw, as well as $\partial \delta u/\partial z$ and $\partial \delta u/\partial y$, are zero, i.e., all points move uniformly and only in the x direction. Then show that $\delta \Pi = 0$ results in the following expression (limits of integration are not shown)

$$-\iiint \delta u \left[\lambda \frac{\partial \epsilon}{\partial x} + 2G \frac{\partial^2 u}{\partial x^2} \right] dx\, dy\, dz + \iint \left[\delta u \left(\lambda \epsilon + 2G \frac{\partial u}{\partial x} - p \right) \right]_0^a dy\, dz = 0$$

Hint: Use integration by parts and $\epsilon_x = \partial u/\partial x$.

(c) Since δu and $\delta u\Big|_0^a$ are completely arbitrary, according to the fundamental lemma the quantities in each of the two brackets must vanish. Therefore, show that $\lambda\epsilon + 2G\epsilon_x \equiv \sigma_x = p$; (x, y, z) at every point inside the prism.

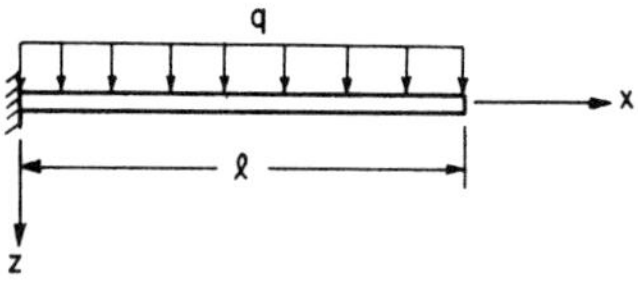

Problem 7–12

7-12 A cantilever beam supports a uniformly distributed load as shown. The boundary conditions are

$$w = \frac{dw}{dx} = 0 \quad \text{at} \quad x = 0$$

$$\frac{d^2 w}{dx^2} = \frac{d^3 w}{dx^3} = 0 \quad \text{at} \quad x = \ell$$

(a) Show that each term of the infinite series

$$w(x)=\sum_{n=0}^{\infty} C_n\left\{\left[\frac{2\ell}{(2n+1)\pi}\right]^3 \sin\frac{(2n+1)\pi x}{2\ell}+\frac{(-1)^n \ell x^2}{(2n+1)\pi}-\left[\frac{2\ell}{(2n+1)\pi}\right]^2 x\right\}$$

where the C_n are constants, satisfies all the boundary conditions.

(b) Determine C_0 by the Rayleigh-Ritz method, using one term of the series.

(c) Compare the maximum deflection obtained from (b) with that given by the exact formula

$$w=\frac{q}{24EI}(x^4-4\ell x^3+6\ell^2x^2)$$

8

Cartesian Tensor Notation

8.1. Introduction

It was pointed out on several occasions in Chapter 1 that a description of physical laws frequently results in expressions which differ from one another only in the subscripts referring to the directions of the coordinate axes (for example, the equations of transformation of stress, equations of equilibrium, etc.). In such cases, if one of the equations is written, the other expressions may be derived by cyclic permutation, which simply corresponds to an orthogonal rotation of the coordinate axes. Furthermore, the various terms of a given equation frequently follow a pattern which may be reproduced once a typical term is written. These facts lead to the so-called "indicial notation" or "tensor notation" (also known as the coordinate method).

In Chapter 4 we introduced one notation which leads to a condensed form of representation of the governing equations. For example, the equilibrium equations were written as

$$\frac{\partial \sigma_x}{\partial x} + \frac{\partial \tau_{xy}}{\partial y} + \frac{\partial \tau_{xz}}{\partial z} + F_x = 0 \qquad (x, y, z) \tag{4.1}$$

which indicates that there are two more equations obtainable by cyclic permutation of x, y, and z. By using indicial notation, we shall further condense equations such as (4.1) in this chapter.

We shall restrict our discussion to rectangular (Cartesian) coordinates. In the more general tensor calculus, variables and equations are expressed in such a manner that they are valid in any coordinate system, including oblique and curvilinear coordinates. By restricting ourselves to Cartesian systems, we greatly simplify the mathematical derivations. Discussion of curvilinear coordinates, using vector operations, will be given in Chapter 12. In this chapter, we presume the student has a grasp of some of the elementary vector operations, such as the dot and cross products.

The indicial notation is only a shorthand for scalar equations written in terms of Cartesian coordinates. Unlike vector notation, no new algebraic rules are required. With a few simple conventions, we can readily write most of our equations in tensor notation. It must be pointed out, however, that the conventions adopted in the Cartesian tensor notation are sometimes not sufficient to represent certain complicated equations. Also, in certain cases, a student may have trouble in digesting the physical meaning of an equation written in indicial notation, or he may have doubts in the mathematical operations. In these cases, one should not hesitate to write the equation in unabridged component form.

In treating mechanics of particles and rigid bodies, vector notation is sufficient and is commonly adopted. Vector and dyadic notations can also be used in the study of mechanics of fluids (ideal or linear viscous) and linear elasticity. In treating finite elastic deformations, plastic flow, or continuum mechanics in general, the symbolic method (vector or dyadic notation) is no longer suitable, because it becomes unmanageable due to the large number of operations which must be defined. In these cases, Cartesian or general tensor notation must be used for conciseness and clarity. The great advantage of the symbolic method is that relations expressed in this way are independent of any coordinate system. Further discussions on the symbolic method will be given in Chapter 11.

8.2. Indicial Notation and Vector Transformations

We shall use x_1, x_2, and x_3 to represent the three Cartesian coordinates, in place of x, y, and z. Consider a position vector with components x_1, x_2, and x_3 along the corresponding axes. In another Cartesian system x'_1, x'_2, x'_3 having the same origin,[1] its components are

$$\begin{aligned} x'_1 &= a_{11}x_1 + a_{21}x_2 + a_{31}x_3 \\ x'_2 &= a_{12}x_1 + a_{22}x_2 + a_{32}x_3 \\ x'_3 &= a_{13}x_1 + a_{23}x_2 + a_{33}x_3 \end{aligned} \tag{8.1}$$

where x'_1, x'_2, and x'_3 are the components of the vector along these axes and the coefficients are the direction cosines, i.e.,

[1] We are assuming a common origin merely for convenience. The components of a vector are unchanged by a translation of the origin.

$$a_{11} = \cos(x_1, x'_1)$$
$$a_{12} = \cos(x_1, x'_2)$$
$$\cdots$$
$$a_{32} = \cos(x_3, x'_2)$$
$$a_{33} = \cos(x_3, x'_3)$$

Table 8.1
DIRECTION COSINES

	x'_1	x'_2	x'_3
x_1	a_{11}	a_{12}	a_{13}
x_2	a_{21}	a_{22}	a_{23}
x_3	a_{31}	a_{32}	a_{33}

These direction cosines can be conveniently displayed as in Table 8.1. In writing the direction cosines, our convention is that the first subscript (or index) refers to the unprimed axes (x_1, x_2, or x_3), while the second refers to the primed system x'_1, x'_2, x'_3. A convention of this sort is necessary because, in general,

$$a_{12} \neq a_{21}$$

although

$$\cos(x_1, x'_2) = \cos(x'_2, x_1)$$

The three equations of (8.1) may now be condensed to

$$x'_1 = \sum_{i=1}^{3} a_{i1} x_i$$

$$x'_2 = \sum_{i=1}^{3} a_{i2} x_i \qquad (8.2)$$

$$x'_3 = \sum_{i=1}^{3} a_{i3} x_i$$

and they may be further consolidated as follows:

$$x'_j = \sum_{i=1}^{3} a_{ij} x_i \qquad j = 1, 2, 3 \qquad (8.3)$$

where, as j takes on values of 1, 2, and 3 in succession, we obtain the three equations of (8.2) or (8.1).

Equations (8.3) may be simplified further by observing that the summation extends over the range of i, which is a repeated subscript on the right-hand side. Thus we may introduce a summation convention, whereby the $\sum$ is redundant, as follows: if a repeated alphabetic subscript appears in one monomial,[2] automatic summation over the range of this

[2] A monomial is defined here as a term in an equation written in indicial notation. An equation in indicial notation represents three or more equations in unabridged notation. We shall call equations like (8.4) either one tensor equation or three scalar equations.

subscript is required. A repeated subscript is called a *dummy* index, whereas a nonrepeated subscript is termed a *free* index. Equation (8.3) now becomes

$$x'_j = a_{ij} x_i \qquad i, j = 1, 2, 3 \tag{8.4}$$

The letter used for the free index is completely arbitrary as long as the same letter is used in every monomial, i.e., Eq. (8.4) can be written as

$$x'_k = a_{ik} x_i \qquad i, k = 1, 2, 3$$

The letter used for the dummy index is also arbitrary, as can be seen (by expansion if necessary) from the relations

$$x'_j = a_{ij} x_i = a_{\ell j} x_\ell \tag{8.5}$$

Both these expressions are identical to Eq. (8.1) in unabridged form. This indicates that the letter used for a dummy index in a given monomial can be changed arbitrarily; indeed, at times this is mandatory (see the derivation of Eq. (8.10)). The only limitation is that the same letter cannot appear more than two times in one monomial since such an operation is not defined. These rules may be summarized as follows: *In a correctly written tensor equation, a subscript may appear not more than twice in each monomial; if a subscript appears only once in a monomial, it must appear just once in each other monomial.*

As an example of the application of the summation convention, one may verify the equation

$$b_{ij} b_{jk} = 0 \qquad i, j, k = 1, 2, 3 \tag{8.6}$$

when expanded, becomes nine equations, i.e.,

$$\begin{aligned}
i = 1, k = 1&: b_{11}b_{11} + b_{12}b_{21} + b_{13}b_{31} = 0\\
k = 2&: b_{11}b_{12} + b_{12}b_{22} + b_{13}b_{32} = 0\\
k = 3& \quad \ldots \quad \ldots \quad \ldots\\
i = 2, k = 1& \quad \ldots \quad \ldots \quad \ldots\\
k = 2& \quad \ldots \quad \ldots \quad \ldots\\
k = 3& \quad \ldots \quad \ldots \quad \ldots\\
i = 3, k = 1& \quad \ldots \quad \ldots \quad \ldots\\
k = 2& \quad \ldots \quad \ldots \quad \ldots\\
k = 3&: b_{31}b_{13} + b_{32}b_{23} + b_{33}b_{33} = 0
\end{aligned} \tag{8.7}$$

where j is a dummy subscript, so that summation over the range of j

is understood. As an aid in deciphering equations such as Eq. (8.6), it may be helpful to first carry out the summation over the dummy index as follows:

$$b_{i1} b_{1k} + b_{i2} b_{2k} + b_{i3} b_{3k} = 0 \tag{8.8}$$

and then assign all possible values to the other subscripts. Thus using $i = 1$, $k = 1$, 2, and 3 results in the first three of Eqs. (8.7); $i = 2$, $k = 1$, 2, 3, in the second three; and $i = 3$, $k = 1$, 2, 3, gives the third set.

One should not hesitate to write an equation in expanded form such as Eq. (8.7) if the details of any operation are not clearly understood. After expanding only a portion of the expressions occurring in the derivations which follow, the general form should become clear.

To save writing, the expression $(i, j = 1, 2, 3)$ will be omitted in the rest of the text, and it is understood that unless otherwise specified all indices assume values 1, 2, and 3. We shall also refer to a vector with components u_i simply as the vector u_i.

Returning to the transformation expressed by Eqs. (8.1), we observe that, given x'_1, x'_2, and x'_3, we may write the components x_1, x_2, x_3 as follows:

$$\begin{aligned} x_1 &= a_{11} x'_1 + a_{12} x'_2 + a_{13} x'_3 \\ x_2 &= a_{21} x'_1 + a_{22} x'_2 + a_{23} x'_3 \\ x_3 &= a_{31} x'_1 + a_{32} x'_2 + a_{33} x'_3 \end{aligned}$$

which are equivalent to

$$x_i = a_{ij} x'_j \tag{8.9}$$

Changing the dummy index[3] in this equation from j to k and substituting it into Eq. (8.4), we have

$$x'_j = a_{ij} a_{ik} x'_k \tag{8.10}$$

Since x'_j and x'_k represent the coordinates in the same coordinate system, and since x'_1, x'_2, and x'_3 are independent of each other, we have

$$\begin{aligned} a_{ij} a_{ik} &= 1 \qquad \text{if} \quad j = k \\ a_{ij} a_{ik} &= 0 \qquad \text{if} \quad j \neq k \end{aligned} \tag{8.11a}$$

[3] Notice that this is necessary in order to avoid a subscript appearing more than twice on the right-hand side of Eq. (8.10). This can also be seen by expanding Eq. (8.10); j and k are not the same in each term. In this equation i and k are dummy indices while j is a free index.

In unabridged form, these are

$$
\begin{aligned}
a_{11}^2 + a_{21}^2 + a_{31}^2 &= 1 \qquad & a_{12}a_{13} + a_{22}a_{23} + a_{32}a_{33} &= 0 \\
a_{12}^2 + a_{22}^2 + a_{32}^2 &= 1 \qquad & a_{13}a_{11} + a_{23}a_{21} + a_{33}a_{31} &= 0 \\
a_{13}^2 + a_{23}^2 + a_{33}^2 &= 1 \qquad & a_{11}a_{12} + a_{21}a_{22} + a_{31}a_{32} &= 0
\end{aligned}
$$

which are the equations governing the direction cosines, Eqs. (1.29). Similarly, it can be shown that

$$a_{ji}a_{ki} = 1 \qquad \text{if} \quad j = k$$

and (8.11b)

$$a_{ji}a_{ki} = 0 \qquad \text{if} \quad j \neq k$$

It is important to note that the use of indicial notation does not require the definition of any new algebraic operations. This is in sharp contrast to vector notation, which requires a redefinition of the operations of addition, subtraction, and multiplication. Indicial notation simply represents a convenient method of condensing algebraic, differential and integral equations.

For any vector quantity **F**, it can be shown that its components transform to a primed coordinate system according to the rule

$$F'_j = a_{ij}F_i \tag{8.12}$$

This equation may be taken as the definition of a vector, or tensor of the first order. That is, a set of three quantities F_i referred to a coordinate system x_i and transformed to another system x'_i by Eq. (8.12) is defined as a vector.

The statement that a certain physical quantity is a vector can not be made arbitrarily. Not any set of three scalar quantities attached to the x_1, x_2, x_3 system can be called a vector. The set must be defined in another coordinate system, and the relationship between these two sets must be known from physical or geometrical principles; finally, the relationship must agree with Eq. (8.12). For instance, let I_1, I_2, and I_3 be the moments of inertia of a body with respect to the x_1, x_2, and x_3 axes, respectively, and I'_1, I'_2, and I'_3 be those with respect to x'_1, x'_2, and x'_3, both coordinate systems having the same origin. These quantities are all defined and may be written in terms of the geometrical dimensions of the body. However, the set of three moments of inertia is not a vector, because

$$I'_j \neq a_{ij}I_i$$

As a matter of fact, moment of inertia is a second-order tensor quantity, the same as stress. Force is a vector because its components in two different coordinate systems are defined and are related by Eq. (8.12).

8.3. Higher-Order Tensors

Consider any two vectors u_i and v_k. In another coordinate system their components, according to Eq. (8.4), are

$$\begin{aligned} u'_j &= a_{ij} u_i \\ v'_\ell &= a_{k\ell} v_k \end{aligned} \tag{8.13}$$

Now suppose that we wish to multiply each component of u_i by each component v_k, i.e., we write $u_i v_k$. The product represents nine terms, since i and k will be independently given all the values 1, 2, and 3. By using Eq. (8.13), the corresponding products in the primed coordinate system become

$$\begin{aligned} u'_j v'_\ell &= (a_{ij} u_i)(a_{k\ell} v_k) \\ &= a_{ij} a_{k\ell} u_i v_k \end{aligned} \tag{8.14}$$

Since i and k are dummy indices, the right-hand side of Eq. (8.14) is seen to represent a double summation. Thus the expression stands for nine equations, each containing nine terms on the right. Each term on the right is the product of two factors: the first factor $a_{ij} a_{k\ell}$, which depends solely on the orientation of the axes in the two coordinate systems, and the second $u_i v_k$, which is the product of the components of the two vectors in the original coordinate system. In this way, the various quantities $u'_j v'_\ell$ may be found in terms of $u_i v_k$. In fact, we might think of the products $u_i v_k$ as representing some quantity, say A_{ik}, which has the nine components given by $A_{ik} = u_i v_k$. If we define nine components of this quantity in the primed system by $A'_{j\ell} = u'_j v'_\ell$, the transformation of this quantity would be expressed by Eq. (8.14), or

$$A'_{j\ell} = a_{ij} a_{k\ell} A_{ik} \tag{8.15}$$

Products of the components of two vectors, however, are not the only quantities which transform according to Eq. (8.15). This equation was derived simply to introduce the concept of a higher-order tensor. Any group of nine scalar quantities w_{ik} referred to a coordinate system x_i and which transforms to a group of nine quantities referred to another coordinate system x'_i by the rule

$$w'_{j\ell} = a_{ij} a_{k\ell} w_{ik} \tag{8.16}$$

is called a tensor of the second order (examples are the stress and strain tensors, and the system of mass moments of inertia and products of inertia). In this text we shall only consider tensors referred to Cartesian coordinate systems.

Similarly, we may proceed to construct and define tensors of higher order, each containing 3^n components, where n is the order of the tensor; for example,

$$w'_{pqr} = a_{ip} a_{jq} a_{kr} w_{ijk} \tag{8.17}$$

expresses the transformation of a third-order tensor (27 components) and

$$w'_{pq\ldots} = a_{ip} a_{jq} \cdots w_{ij\ldots} \tag{8.18}$$

defines the transformation of a tensor of order n, where both w' and w have n subscripts, and the n subscripts of w' appear as second subscripts of the a's on the right-hand side of Eq. (8.18), while the first subscript of the a's are the same as the subscripts of w. This tensor consists of 3^n components, and there are 3^n coefficients (which are products of direction cosines) appearing in each of the expanded equations (8.18). According to this definition, then, vector quantities are first-order tensors and scalars are tensors of order zero.

Similar to the discussion given for the definition of a vector, we must emphasize that not every set of quantities assigned to the coordinate axes can be called a tensor. The set of quantities must be proven to possess all the properties of a tensor. We must have a method, based on physical properties, to establish the existence of two sets of quantities in two different coordinate systems. If we can show then, that these sets are related by Eq. (8.18), we may state that the quantities are components of a tensor of order n. We observe that, in the development of Eq. (8.15), for instance, we did not start by assuming that the products $u_i v_k$ are the components of a tensor. We demonstrated that the products $u_i v_k$ are the components of a second-order tensor by showing that the products may be expressed in any coordinate system and that the relation between the products in any two coordinate systems (the transformation) is given by the transformation equation for a second-order tensor, Eq. (8.16). Now we may state that the products of the components of any two vectors, i.e., $u_i v_k$, are themselves the components of a second-order tensor.

In general the subscripts used for a tensor are completely arbitrary. We may use the notation w_{ij}, $w_{k\ell}$, w_{mn}, or w_{ji} to represent the same tensor. But once a specific alphabet is used in an equation, the rest of the letters used for subscripts in this equation, and preferably even in subsequent equations, must be consistent with it. If used in the same equation, w_{ij} and w_{ji} do not represent the same tensor; rather they are the transpose of each other. That is, $u_{ij} = w_{ji}$ is the transpose of w_{ij}. For example, the equation

$$w_{ij} = w_{ji} + v_{ij}$$

has the following components:

$$w_{12} = w_{21} + v_{12}$$
$$w_{23} = w_{32} + v_{23}$$
$$\cdots$$

We shall prove in this section that if w_{ij} is a tensor, then its transpose w_{ji} is also a tensor.

To show this, we define

$$\begin{aligned} u_{ij} &= w_{ji} \\ u'_{k\ell} &= w'_{\ell k} \\ w'_{k\ell} &= a_{ik} a_{j\ell} w_{ij} \end{aligned} \tag{8.19}$$

Therefore, we must prove that

$$u'_{k\ell} = a_{ik} a_{j\ell} u_{ij} = a_{ik} a_{j\ell} w_{ji} \tag{8.20}$$

But

$$a_{ik} a_{j\ell} w_{ji} = a_{jk} a_{i\ell} w_{ij}$$

This can be shown by expanding both sides of this equation. It follows that

$$a_{ik} a_{j\ell} u_{ij} = a_{jk} a_{i\ell} w_{ij} = a_{i\ell} a_{jk} w_{ij} = w'_{\ell k} = u'_{k\ell}$$

and Eq. (8.20) is proved. Thus if we represent the components of a tensor in the form of a matrix,[4] and, if we *know* that the set of quantities arranged as

$$\begin{pmatrix} w_{11} & w_{12} & w_{13} \\ w_{21} & w_{22} & w_{23} \\ w_{31} & w_{32} & w_{33} \end{pmatrix} \tag{8.21}$$

represents a second-order tensor, then the arrangement

$$\begin{pmatrix} w_{11} & w_{21} & w_{31} \\ w_{12} & w_{22} & w_{32} \\ w_{13} & w_{23} & w_{33} \end{pmatrix} \tag{8.22}$$

is also a tensor of the second order. The fact that a set of terms is

[4] A matrix is simply a set of elements placed in an ordered array. The fact that a group of elements appears in matrix form does not imply that any particular operation need be performed on the elements. The elements of a matrix may or may not have any physical meaning. We shall use matrices only to the extent of displaying the nine components of a tensor and shall not introduce any topics in matrix algebra or matrix calculus.

represented in matrix form, however, does not imply that these terms are the components of a tensor.

It follows that the quantities $(w_{ik}+w_{ki})$ and $(w_{ik}-w_{ki})$ are tensors of the second order. Now if we interchange i and k, the first set is unaltered; a tensor of this type is called a *symmetric tensor*. The elements which are not on the main diagonal (the line through the elements with $i=k$) of the corresponding matrix are seen to occur in equal pairs. The second set $(w_{ik}-w_{ki})$ has all its components reversed in sign when i and k are interchanged and is called an *antisymmetric tensor*. Here the elements on the main diagonal are zero. Also, since

$$w_{ik}=\tfrac{1}{2}(w_{ik}+w_{ki})+\tfrac{1}{2}(w_{ik}-w_{ki}) \tag{8.23}$$

we conclude that any second-order tensor can be expressed as the sum of a symmetric and an antisymmetric tensor. The symmetric part of a second-order tensor (the first term of Eq. (8.23)) will be denoted by w_{ik}^S, and the antisymmetric part (the second term of Eq. (8.23)) by w_{ik}^A.

8.4. Gradient of a Vector

We know from vector analysis that the gradient of a scalar field is a vector field. We can demonstrate this fact by showing that the gradient of a scalar transforms according to Eq. (8.12). Let U be a scalar function of x_1, x_2, and x_3, that is, $U=U(x_1, x_2, x_3)$,[5] and consider its gradient $\partial U/\partial x_i$ and $\partial U/\partial x'_j$ in two coordinate systems, x_i and x'_j, respectively. By the chain rule of differentiation we may write

$$\frac{\partial U}{\partial x'_j}=\frac{\partial U}{\partial x_1}\frac{\partial x_1}{\partial x'_j}+\frac{\partial U}{\partial x_2}\frac{\partial x_2}{\partial x'_j}+\frac{\partial U}{\partial x_3}\frac{\partial x_3}{\partial x'_j}=\frac{\partial U}{\partial x_i}\frac{\partial x_i}{\partial x'_j} \tag{8.24}$$

where we have used the summation convention in differentiation.[6] However, from Eq. (8.9) we see that

$$\frac{\partial x_i}{\partial x'_j}=a_{ij} \tag{8.25}$$

thus Eq. (8.24) takes the form

$$\frac{\partial U}{\partial x'_j}=a_{ij}\frac{\partial U}{\partial x_i} \tag{8.26}$$

so that the gradient transforms according to Eq. (8.12) and is therefore a vector. Similarly, we may show that the gradient of a vector is a second-order tensor, for if u_i and u'_j are the components of a vector with respect

[5] Notice that we do not use the notation $U=U(x_i)$, which is different from $U=U(x_1, x_2, x_3)$. In this case the indicial notation is of no help to us.

[6] Differentiation is sometimes represented by a comma, for example, $\partial U/\partial x_i = U, i$ or by the notation $\partial U/\partial x_i=\partial_i U$. Second derivatives are denoted by $U_{,ij}$ or $\partial_{ij}U$.

to two sets of axes, and its gradient is defined by $\partial u_i/\partial x_k$ and $\partial u'_j/\partial x'_\ell$ in the respective systems, we have

$$\begin{aligned} \frac{\partial u'_j}{\partial x'_\ell} &= \frac{\partial u'_j}{\partial x_k}\frac{\partial x_k}{\partial x'_\ell} \\ &= a_{k\ell}\frac{\partial}{\partial x_k}(a_{ij}u_i) \\ &= a_{ij}a_{k\ell}\frac{\partial u_i}{\partial x_k} \end{aligned} \tag{8.27}$$

since the a_{ij} are constants for two given coordinate systems, or

$$\frac{\partial a_{ij}}{\partial x_k} = \frac{\partial}{\partial x_k}\left(\frac{\partial x_i}{\partial x'_j}\right) = 0$$

Thus the transformation Eq. (8.27) is that of a second-order tensor. Similarly, the gradient of an n-th order tensor is a tensor of order $n + 1$.

8.5. The Kronecker Delta

Consider the gradient $\partial x_i/\partial x_k$ of the position vector x_i. As we have just seen, this will be a second-order tensor. Now x_1, x_2, and x_3 are independent variables, so that

$$\frac{\partial x_1}{\partial x_1} = 1 \qquad \frac{\partial x_1}{\partial x_2} = 0 \qquad \frac{\partial x_1}{\partial x_3} = 0 \qquad \text{etc.}$$

or, in general terms,

$$\begin{aligned} \frac{\partial x_i}{\partial x_k} &= 1 \qquad \text{for} \quad i = k \\ \frac{\partial x_i}{\partial x_k} &= 0 \qquad \text{for} \quad i \neq k \end{aligned} \tag{8.28}$$

The components of the second-order tensor given by Eqs. (8.28) may be conveniently expressed by δ_{ik}, which is called the *Kronecker delta*, and which is defined by

$$\begin{aligned} \delta_{ik} &= 1 \qquad \text{for} \quad i = k \\ \delta_{ik} &= 0 \qquad \text{for} \quad i \neq k \end{aligned} \tag{8.29}$$

Writing out the components in Eq. (8.29), we have

$$\begin{aligned} \delta_{11} = \delta_{22} = \delta_{33} = 1 \\ \delta_{12} = \delta_{13} = \delta_{21} = \delta_{23} = \delta_{31} = \delta_{32} = 0 \end{aligned} \tag{8.30}$$

or, in matrix form,

$$\delta_{ik} = \begin{pmatrix} 1 & 0 & 0 \\ 0 & 1 & 0 \\ 0 & 0 & 1 \end{pmatrix} \tag{8.31}$$

We shall now show that the Kronecker delta has identical components in any coordinate system. By applying the transformation equation for a second-order tensor, Eq. (8.16), we get

$$\delta'_{j\ell} = a_{ij} a_{k\ell} \delta_{ik} \tag{8.32}$$

However, k is a dummy subscript and is therefore given the values 1, 2, and 3 and summed for each value of i, which is also a dummy index. But if $k \neq i$, $\delta_{ik} = 0$, and the corresponding term is zero. If $k = i$, $\delta_{ik} = 1$, thus the only terms which are retained are those for which $i = k$. Therefore δ_{ik} may be eliminated by setting $k = i$; consequently, we get

$$\delta'_{j\ell} = a_{ij} a_{i\ell} \tag{8.33}$$

From Eq. (8.11), however, we have

$$\begin{aligned} a_{ij} a_{i\ell} &= 1 \qquad \text{for} \quad j = \ell \\ a_{ij} a_{i\ell} &= 0 \qquad \text{for} \quad j \neq \ell \end{aligned} \tag{8.34}$$

and from Eqs. (8.34) and (8.33) we see that

$$\begin{aligned} \delta'_{j\ell} &= 1 \qquad \text{for} \quad j = \ell \\ \delta'_{j\ell} &= 0 \qquad \text{for} \quad j \neq \ell \end{aligned}$$

or

$$\delta'_{j\ell} = \delta_{j\ell} \tag{8.35}$$

and the Kronecker delta transforms into itself. Tensors with identical components in any coordinate system are called isotropic tensors. We notice that Eq. (8.33), when expanded, becomes Eqs. (1.29).

If u_i is a vector and we form the product $\delta_{ik} u_m$, we have, from the result of Problem 8-3, a third-order tensor. Consider now the expression $\delta_{ik} u_k$, that is,

$$\delta_{ik} u_k = \delta_{i1} u_1 + \delta_{i2} u_2 + \delta_{i3} u_3 \tag{8.36}$$

For any given value of i, only one term of Eq. (8.36) will be retained due to Eq. (8.29); thus the result is

$$\delta_{ik} u_k = u_i \tag{8.37}$$

This operation replaces the dummy index k by i, so that δ_{ik} is sometimes called the substitution tensor. The result is quite general, as shown, for example, by the following identities:

$$\delta_{ik}\frac{\partial u_j}{\partial x_k}=\frac{\partial u_j}{\partial x_i}=u_{j,i} \tag{8.38}$$

$$\delta_{ik}w_{ik}=w_{kk}=w_{ii} \tag{8.39}$$

$$\delta_{i\ell}\frac{\partial^2 u_j}{\partial x_\ell\,\partial x_k}=\delta_{i\ell}u_{j,\ell k}=\frac{\partial^2 u_j}{\partial x_i\,\partial x_k}=u_{j,ik} \tag{8.40}$$

8.6. Tensor Contraction

Given the second-order tensor w_{ik}, let us form the expression

$$w_{11}+w_{22}+w_{33}=w_{ii} \tag{8.41}$$

The corresponding quantity w'_{jj} is found from Eq. (8.16) to be

$$\begin{aligned} w'_{jj}&=a_{ij}a_{kj}w_{ik}\\ &=\delta_{ik}w_{ik}\end{aligned}$$

and by the substitution property of δ_{ik}, we get

$$w'_{jj}=w_{ii} \tag{8.42}$$

Thus w_{ii} is independent of the coordinate system in which it is expressed and is therefore a scalar (invariant) quantity.

The operation of equating two letter subscripts in a tensor and summing accordingly is known as contraction. In general, contraction gives another tensor of order two less than that of the original tensor. For example, if we contract the second-order tensor $u_i v_k$, we obtain

$$u_i v_i=u_1 v_1+u_2 v_2+u_3 v_3 \tag{8.43}$$

which is the scalar product of u_i and v_k.

Similarly, the tensor $\partial u_k/\partial x_i$ gives, on contraction, the scalar

$$\frac{\partial u_i}{\partial x_i}=\frac{\partial u_1}{\partial x_1}+\frac{\partial u_2}{\partial x_2}+\frac{\partial u_3}{\partial x_3} \tag{8.44}$$

which is the divergence of u_i.

8.7. The Alternating Tensor

We now introduce another important isotropic tensor, the alternating tensor ϵ_{ikm}, which is defined by the conditions

$$\epsilon_{ikm}\begin{cases} =0 & \text{if the numerals taken by any two of the subscripts } i, k, m \text{ are equal} \\ =+1 & \text{if the numerals taken by } i, k, m \text{ are unequal in cyclic order of 1 2 3, i.e., in the order 123, 231, or 312} \\ =-1 & \text{if the numerals taken by } i, k, m \text{ are unequal in noncyclic order of 1 2 3, i.e., in the order 132, 213, or 321.} \end{cases} \tag{8.45}$$

We further specify that $\epsilon'_{j\ell n}=\begin{Bmatrix}0\\+1\\-1\end{Bmatrix}$ where the value depends on the subscript rules of Eq. (8.45).

Let us now demonstrate that ϵ_{ikm} is a third-order tensor. To do this, we must show that it transforms according to the expression

$$\epsilon'_{j\ell n}=a_{ij}\,a_{k\ell}\,a_{mn}\,\epsilon_{ikm} \tag{8.46}$$

We proceed by demonstrating that $\epsilon'_{j\ell n}$ defined by Eq. (8.46) satisfies all of the requirements of Eq. (8.45). Summing over the range of the dummy indices and using Eq. (8.45), we find that

$$\begin{aligned} a_{ij}\,a_{k\ell}\,a_{mn}\,\epsilon_{ikm} &= a_{1j}\,a_{2\ell}\,a_{3n}+a_{2j}\,a_{3\ell}\,a_{1n}+a_{3j}\,a_{1\ell}\,a_{2n} \\ &\quad -a_{2j}\,a_{1\ell}\,a_{3n}-a_{3j}\,a_{2\ell}\,a_{1n}-a_{1j}\,a_{3\ell}\,a_{2n} \end{aligned} \tag{8.47}$$

If the numerals taken by any two of the subscripts j, ℓ, n are equal, we see that Eq. (8.47) reduces to zero, for the expression results in three pairs of terms having the same value, but opposite signs. The right-hand side of Eq. (8.47) is simply the expansion of the determinant

$$\begin{vmatrix} a_{1j} & a_{1\ell} & a_{1n} \\ a_{2j} & a_{2\ell} & a_{2n} \\ a_{3j} & a_{3\ell} & a_{3n} \end{vmatrix} \tag{8.48}$$

If the numerals taken by j, ℓ, and n are unequal in cyclic order, the value of the determinant is 1. For example let $j=1$, $\ell=2$, $n=3$, and $\mathbf{c}_1$, $\mathbf{c}_2$, and $\mathbf{c}_3$ be unit vectors in the x'_1, x'_2, and x'_3 directions, respectively, and $\mathbf{i}, \mathbf{j}, \mathbf{k}$ unit vectors in the x_1, x_2, and x_3 directions, respectively. Then

$$\begin{aligned} \mathbf{c}_1 &= a_{11}\,\mathbf{i}+a_{21}\mathbf{j}+a_{31}\,\mathbf{k} \\ \mathbf{c}_2 &= a_{12}\,\mathbf{i}+a_{22}\mathbf{j}+a_{32}\,\mathbf{k} \\ \mathbf{c}_3 &= a_{13}\,\mathbf{i}+a_{23}\mathbf{j}+a_{33}\,\mathbf{k} \end{aligned} \tag{8.49}$$

and since the unit vectors in the primed coordinate system are orthogonal, we have

$$\mathbf{c}_1 \times \mathbf{c}_2 = \mathbf{c}_3 \tag{8.50}$$

which yields the following relations,[7]

$$\begin{aligned} a_{13} &= a_{21}\,a_{32} - a_{31}\,a_{22} \\ a_{23} &= a_{31}\,a_{12} - a_{11}\,a_{32} \\ a_{33} &= a_{11}\,a_{22} - a_{21}\,a_{12} \end{aligned} \tag{8.51}$$

Similar relations, obtained by cyclic interchange of the subscripts or by writing $\mathbf{c}_2 \times \mathbf{c}_3 = \mathbf{c}_1$, etc., show that each term in the determinant of Eq. (8.48) is equal to its cofactor. Substituting the relations of Eq. (8.51) into Eq. (8.47), we observe that

$$a_{i1}\,a_{k2}\,a_{m3}\,\epsilon_{ikm} = a_{13}^{2} + a_{23}^{2} + a_{33}^{2} \equiv 1 \begin{cases} j = 1 \\ \ell = 2 \\ n = 3 \end{cases}$$

If j, ℓ, and n take on the values of 2, 3, 1 or 3, 1, 2, respectively, the result is the same, since these values result in an even number of interchanges of the columns of the determinant (8.48). If the numerals taken by j, ℓ, and n are unequal in noncyclic order of 123, this amounts to an odd number of interchanges of the columns. This reverses the sign of the result, i.e., the right side of Eq. (8.47) is equal to -1. Therefore ϵ_{ikm} transforms into itself by Eq. (8.17) and is a tensor of the third order.

Now consider the fourth-order tensor $\epsilon_{ikm}u_p$, where u_p is a vector. If we contract this tensor by putting $p = m$ and then summing, we get a second-order tensor, $w_{ik} = \epsilon_{ikm}u_m$. If the numerals taken by i and k are equal, from Eq. (8.45), the resulting term is zero. With these numerals unequal the result is

$$w_{ik} = \epsilon_{ik1}\,u_1 + \epsilon_{ik2}\,u_2 + \epsilon_{ik3}\,u_3 \tag{8.52}$$

Thus with $i = 1$, $k = 2$, the first two terms vanish and

$$w_{12} = u_3 \tag{8.53}$$

If $i = 2$, $k = 1$, the first two terms again vanish, but $\epsilon_{213} = -1$, so that

$$w_{21} = -u_3 \tag{8.54}$$

[7] These relations will hold if the coordinate systems x_i and x'_i are either both right-handed or both left-handed. We shall always employ right-handed axes; therefore the expressions (8.51) and their counterparts will be valid in our discussion.

Or, in general,

$$w_{ik} = +u_m \text{ if the numerals for } i, k, m \text{ are unequal in cyclic order of 1 2 3} \tag{8.55}$$

$$w_{ik} = -u_m \text{ if the numerals for } i, k, m \text{ are unequal in non-cyclic order of 1 2 3}$$

and the elements of w_{ik} in matrix form are

$$\begin{pmatrix} 0 & u_3 & -u_2 \\ -u_3 & 0 & u_1 \\ u_2 & -u_1 & 0 \end{pmatrix} \tag{8.56}$$

Since u_i is a vector, we can always obtain an antisymmetric tensor of the second order by combining the alternating tensor with any vector, as in Eq. (8.52). Conversely, we can produce a vector quantity by combining the alternating tensor with any antisymmetric tensor of the second order. This fact will be demonstrated next.

Suppose that we have a second-order tensor w_{pq}. The operation $\epsilon_{ikm} w_{pq}$ produces a fifth-order tensor. If we contract twice by letting $p = i$ and $q = k$, the result is a vector u_m, i.e.,

$$\begin{aligned} u_m &= \epsilon_{ikm} w_{ik} \\ &= \epsilon_{12m} w_{12} + \epsilon_{21m} w_{21} + \epsilon_{32m} w_{32} \\ &\quad + \epsilon_{13m} w_{13} + \epsilon_{23m} w_{23} + \epsilon_{31m} w_{31} \end{aligned} \tag{8.57}$$

which gives

$$\begin{aligned} u_1 &= w_{23} - w_{32} \\ u_2 &= w_{31} - w_{13} \\ u_3 &= w_{12} - w_{21} \end{aligned} \tag{8.58}$$

We notice that if w_{pq} is symmetrical, $u_m = 0$. If w_{pq} is antisymmetrical, the components of u_m are numerically twice those of w_{pq}, which proves the statement at the end of the previous paragraph.

If we take any three vectors u_i, v_i, and w_i and construct the scalar $\epsilon_{ikm} u_i v_k w_m$, we see that

$$\begin{aligned} \epsilon_{ikm} u_i v_k w_m &= u_1 v_2 w_3 + u_2 v_3 w_1 + u_3 v_1 w_2 \\ &\quad - u_2 v_1 w_3 - u_3 v_2 w_1 - u_1 v_3 w_2 \\ &= \begin{vmatrix} u_1 & u_2 & u_3 \\ v_1 & v_2 & v_3 \\ w_1 & w_2 & w_3 \end{vmatrix} \end{aligned} \tag{8.59}$$

which represents the mixed triple product $\mathbf{u} \cdot \mathbf{v} \times \mathbf{w}$. If two of the vectors are parallel, the elements of one row of the determinant will be proportional to those of another row and this scalar vanishes.

Now consider any two vectors u_i, v_k and the product $\epsilon_{mik} u_i v_k$. This gives a vector w_m, in which

$$w_m = \epsilon_{mik} u_i v_k \tag{8.60}$$

and which has the components

$$\begin{aligned} w_1 &= u_2 v_3 - u_3 v_2 \\ w_2 &= u_3 v_1 - u_1 v_3 \\ w_3 &= u_1 v_2 - u_2 v_1 \end{aligned} \tag{8.61}$$

The vector w_m is the vector (cross) product, $\mathbf{u} \times \mathbf{v}$. This vector is normal to the vectors u_i and v_k, since

$$\begin{aligned} u_1 w_1 + u_2 w_2 + u_3 w_3 &= u_1(u_2 v_3 - u_3 v_2) + u_2(u_3 v_1 - u_1 v_3) \\ &\quad + u_3(u_1 v_2 - u_2 v_1) = 0 \\ v_1 w_1 + v_2 w_2 + v_3 w_3 &= v_1(u_2 v_3 - u_3 v_2) + v_2(u_3 v_1 - u_1 v_3) \\ &\quad + v_3(u_1 v_2 - u_2 v_1) = 0 \end{aligned}$$

That is, the scalar products $u_i w_i$ and $v_i w_i$ vanish, which indicates that w_i is perpendicular to u_i and v_i.

Similarly, we may relate the antisymmetric tensor $w_{ik} = (\partial u_k/\partial x_i - \partial u_i/\partial x_k)/2$ with a vector v_m by Eq. (8.57), or

$$v_m = \epsilon_{mik} w_{ik} = \epsilon_{mik} \left(\frac{\partial u_k}{\partial x_i} - \frac{\partial u_i}{\partial x_k} \right) \frac{1}{2}$$

which has the components

$$\begin{aligned} v_1 &= \frac{\partial u_3}{\partial x_2} - \frac{\partial u_2}{\partial x_3} \\ v_2 &= \frac{\partial u_1}{\partial x_3} - \frac{\partial u_3}{\partial x_1} \\ v_3 &= \frac{\partial u_2}{\partial x_1} - \frac{\partial u_1}{\partial x_2} \end{aligned} \tag{8.62}$$

This is the curl of u_i.

8.8. The Theorem of Gauss

The theorem of Gauss states that

$$\int_V \frac{\partial}{\partial x_i} A_{jk\ell\ldots} \, dV = \int_S \mu_i A_{jk\ell\ldots} \, dS \tag{8.63}$$

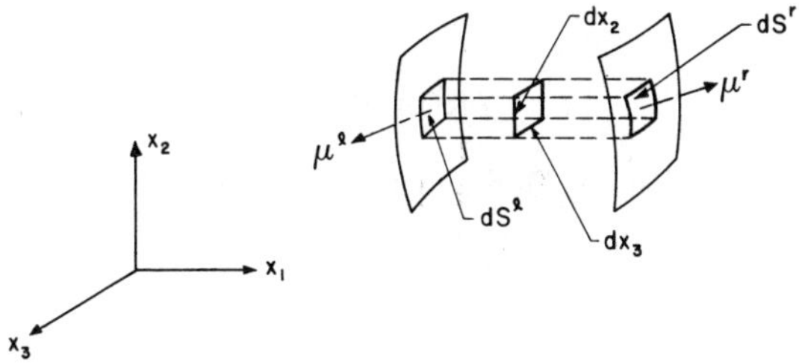

Fig. 8.1 Integration Over a Volume V

where S is the surface area of a region enclosing volume V, $A_{jk\ell\ldots}$ is a tensor of any order, and μ_i is the unit outward vector normal to S. The theorem is used to convert volume integrals to surface integrals, or vice versa.

In order to prove this theorem, we first let $i = 1$, and we integrate in the x_1 direction, i.e.,

$$\iiint_V \frac{\partial}{\partial x_1} A_{jk\ell\ldots}\, dx_1\, dx_2\, dx_3 = \iint_{S^r} A_{jk\ell\ldots}^{r}\, dx_2\, dx_3 - \iint_{S^\ell} A_{jk\ell\ldots}^{\ell}\, dx_2\, dx_3$$

where the superscripts r and ℓ refer to the right and left extremities, respectively, of the element shown in Fig. 8.1. But we note that

$$\begin{aligned} dx_2\, dx_3 &= dS^r \cos(\boldsymbol{\mu}^r, x_1) \\ &= dS^r \mu_1{}^r \end{aligned}$$

and similarly

$$dx_2\, dx_3 = -\, dS^\ell \mu_1{}^\ell$$

Substituting, we find that

$$\int_V \frac{\partial}{\partial x_1} A_{jk\ell\ldots}\, dV = \int_{S^r} A_{jk\ell\ldots}^{r}\, \mu_1{}^r\, dS^r + \int_{S^\ell} A_{jk\ell\ldots}^{\ell}\, \mu_1{}^\ell\, dS^\ell$$

However, the right-hand side of this equation is simply the surface integral

$$\int_S \mu_1\, A_{jk\ell\ldots}\, dS$$

and since analogous expressions can be written for $i = 2$ and $i = 3$, we have

$$\int_V \frac{\partial}{\partial x_i} A_{jk\ell\ldots}\, dV = \int_S \mu_i\, A_{jk\ell\ldots}\, dS$$

proving the theorem.

PROBLEMS

8-1 Demonstrate that the dot product of vectors u_i and v_m may be written in all of the following forms

$$\begin{aligned}\mathbf{u}\cdot\mathbf{v} &= u_i v_i \\ &= a_{ij} a_{mj} u_i v_m \\ &= u'_j v'_j \\ &= a_{ij} a_{ik} u'_j v'_k\end{aligned}$$

8-2 Prove that $u_i v_j w_\ell$ is a third-order tensor, where u_i, v_j, and w_ℓ are arbitrary vectors. Generalize this result for n vectors.

8-3 Prove that $w_{ij} u_k$ is a third-order tensor, where w_{ij} is any second-order tensor and u_k is any vector. Generalize this result for the product between an n-th order tensor and an m-th order tensor.

8-4 Prove that

$$(AB)_{,ii} = AB,_{ii} + 2A,_i B,_i + BA,_{ii}$$

where A and B are scalar functions.

8-5 Using indicial notation, show that Eq. (8.9) can be derived directly from Eq. (8.4). Hint: "Multiply" Eq. (8.4) by a_{kj}.

8-6 Show that the gradient of a tensor of order n is a tensor of order $n+1$.

8-7 Verify the relations stated in Eqs. (8.38), (8.39) and (8.40).

8-8 Establish the identity

$$\epsilon_{iks}\epsilon_{mps} = \delta_{im}\delta_{kp} - \delta_{ip}\delta_{km}$$

by showing that each side of the equation is equal to

$$\begin{array}{rlll} +1 & \text{if } i=m, & k=p & \text{unless } i=k \\ -1 & \text{if } i=p, & k=m & \text{unless } i=k \\ 0 & \text{if } i=k & \text{or } m=p & \end{array}$$

8-9 If τ_{ij} is a symmetric tensor, show that the contracted product $A_{ij}\tau_{ij}$ is independent of the antisymmetric part of A_{ij}. Hint: Form A_{ij} as the sum of a symmetric and an antisymmetric part $A_{ij} = A^S_{ij} + A^A_{ij}$.

8-10 Prove the following formulas:

$$\begin{aligned}\delta_{ii} &= 3 \\ \delta_{ik}\epsilon_{ikm} &= 0 \\ \epsilon_{ijk}\epsilon_{ijk} &= 6 \\ \epsilon_{ijp}\epsilon_{ijq} &= 2\delta_{pq}\end{aligned}$$

8-11 Change the following integral theorems into Cartesian tensor notation. In these equations, dV is the elemental volume; dS the elemental surface area; dL the elemental length; $\boldsymbol{\mu}$ the unit vector along the outward normal direction of S or L.

(a) $$\int_V \text{grad}\,\phi\,dV = \int_S \boldsymbol{\mu}\phi\,dS$$

(b) $$\int_V \text{div}\,\mathbf{U}\,dV = \int_S \boldsymbol{\mu}\cdot\mathbf{U}\,dS$$

(c) $$\int_V \text{curl}\,\mathbf{U}\,dV = \int_S \boldsymbol{\mu}\times\mathbf{U}\,dS$$

(d) $$\int_S \boldsymbol{\mu}\cdot\text{curl}\,\mathbf{U}\,dS = \int_L \mathbf{U}\cdot d\mathbf{L}$$

(e) $$\int_V \nabla^2\phi\,dV = \int_S \boldsymbol{\mu}\cdot\nabla\phi\,dS$$

8-12 Let u_i, u'_j, u''_k be the components of a vector in three coordinate systems, i.e., unprimed, primed, and double-primed coordinates, respectively. Show that the values of u''_k obtained by transforming from u_i are the same as those obtained by first transforming from u_i to u'_j and then from u'_j to u''_k. Hint: Use three different letters for the three sets of direction cosines, such as a_{ik}, b_{ij}, and c_{jk}.

8-13 Are the nine direction cosines a_{ij} components of a second-order tensor? Are a_{ij} components of vectors?

9

The Stress Tensor

9.1. State of Stress at a Point

We shall now show how the theory of Chapter 8 applies in the discussion of elasticity. First it will be shown that the nine components of stress at a point constitute a second-order tensor. In order to prove this, we must show that the stress components transform according to Eq. (8.16). The stress components will be designated by τ_{ik}, where the first subscript indicates the plane on which the component acts and the second subscript gives the direction of the component.

Let us now derive the general transformation equation for stress using indicial notation. As shown in Fig. 9.1, we construct axes x_i $(i = 1, 2, 3)$

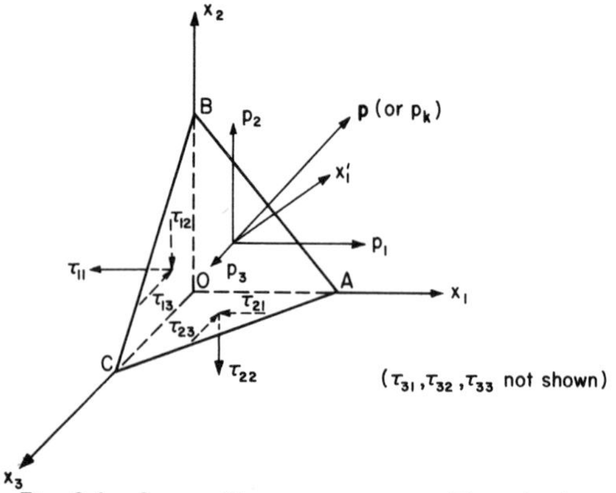

Fig. 9.1 Stress Components on a Tetrahedron

and $x'_n (n = 1, 2, 3)$, where x'_1 is normal to the plane ABC and x'_2, x'_3 are parallel to this plane. The stress components acting on the planes of the unprimed coordinate system are as depicted in the figure. The stress components which act on the x'_1 plane are shown in Fig. 9.1, where **p** (or p_k) is the stress vector on this plane and p_1, p_2, and p_3 are its components parallel to x_i. As shown in Chapter 1, the "stress differences" $(\partial\tau_{11}/\partial x_2)(\Delta x_2/2)$, etc., and the body forces result in higher-order terms as the dimensions of the element approach zero. For this reason they are omitted in Fig. 9.1. Now, expressing the equilibrium condition of the tetrahedron $OABC$ in the x_1 direction, we have (refer to Eqs. (1.24))

$$p_1 = \tau_{i1} a_{i1}$$

This equation automatically accounts for all of the stress components acting in the x_1 direction, due to the repetition of the subscript i. Similarly, expressing the equilibrium conditions in the x_2 and x_3 directions, we have the relations

$$p_2 = \tau_{i2} a_{i1}$$

and

$$p_3 = \tau_{i3} a_{i1}$$

or, in general terms,

$$p_k = \tau_{ik} a_{i1} \tag{9.1}$$

Equation (9.1) is actually identical to Eqs. (1.24) except that Eq. (9.1) is written in indicial notation. We next obtain the component of the stress vector **p** in the x'_1 direction by summing the projections of p_1, p_2, and p_3 in this direction. Thus,

$$\tau'_{11} = \sigma_{x'_1} = p_1 a_{11} + p_2 a_{21} + p_3 a_{31}$$

or

$$\tau'_{11} = p_k a_{k1}$$

and by substituting the value of p_k from Eq. (9.1), we find that

$$\tau'_{11} = \tau_{ik} a_{i1} a_{k1} \tag{9.2}$$

where τ'_{11} represents the stress component on the x'_1 plane in the x'_1 direction (or $\sigma_{x'_1}$). We see that Eq. (9.2) is identical to Eq. (1.25). The components of **p** in the x'_2 and x'_3 directions, i.e., τ'_{12} and τ'_{13} respectively, can now be written as

$$\tau'_{12} = p_k a_{k2} = \tau_{ik} a_{i1} a_{k2}$$

and

$$\tau'_{13} = p_k a_{k3} = \tau_{ik} a_{i1} a_{k3}$$

so that

$$\tau'_{1n} = \tau_{ik} a_{i1} a_{kn} \tag{9.3}$$

In general, then, we have

$$\begin{aligned} \tau'_{jn} &= \tau_{ik} a_{ij} a_{kn} \\ &= a_{ij} a_{kn} \tau_{ik} \end{aligned} \tag{9.4}$$

We observe that the stress components transform according to Eq. (8.16), and therefore the components of stress at a point constitute a second-order tensor. This tensor is called the *stress tensor.* And since $\tau_{12} = \tau_{21}$, etc., or in general $\tau_{ik} = \tau_{ki}$ as will be shown in Section 9.3, the stress tensor is symmetrical. When displayed in matrix form, the stress tensor is

$$\tau_{ik} = \begin{pmatrix} \tau_{11} & \tau_{12} & \tau_{13} \\ \tau_{21} & \tau_{22} & \tau_{23} \\ \tau_{31} & \tau_{32} & \tau_{33} \end{pmatrix} = \begin{pmatrix} \sigma_x & \tau_{xy} & \tau_{xz} \\ \tau_{yx} & \sigma_y & \tau_{yz} \\ \tau_{zx} & \tau_{zy} & \sigma_z \end{pmatrix}$$

We now introduce a superscript symbol (as in Chapter 1) which supplements the indicial notation. We shall represent the normal to the plane on which the stress acts by a superscript. We shall also utilize vector notation as a convenient supplement to our tensor notation. Thus, referring to Fig. 9.1, we let the outward normal of plane *ABC* lie in the μ direction. The *unit* vector in the μ direction is $\boldsymbol{\mu}$. The components of $\boldsymbol{\mu}$ are μ_j where we note that μ_j are the direction cosines of $\boldsymbol{\mu}$ with respect to x_j. The stress vector on this surface is denoted by $\boldsymbol{\tau}^\mu$ and its components by τ_i^μ. In other words, τ_i^μ is the component of $\boldsymbol{\tau}^\mu$ in the x_i direction. Following Eq. (9.1) then, the stress vector on the μ plane may be expressed as

$$\tau_i^\mu = \tau_{ji} \mu_j \tag{9.5}$$

The preceding discussion will now be used in the proof of a theorem which can be stated as follows: At a given point the component of stress on the μ plane in the ν direction is equal to the component of stress on the ν plane in the μ direction, where μ and ν are two arbitrary directions. This theorem is expressed by the relation

$$\boldsymbol{\tau}^\mu \cdot \boldsymbol{\nu} = \boldsymbol{\tau}^\nu \cdot \boldsymbol{\mu} \tag{9.6}$$

where $\boldsymbol{\nu}$ and $\boldsymbol{\mu}$ are unit vectors in the ν and μ directions, respectively. The fact that $\tau_{ik} = \tau_{ki}$ is but a special case of this general theorem.

From the symmetry condition $\tau_{ij}=\tau_{ji}$ and Eq. (9.5), we find that

$$\begin{aligned}\boldsymbol{\tau}^{\mu}\cdot\boldsymbol{\nu} &= \tau_i^{\mu}\nu_i=\tau_{ji}\mu_j\nu_i\\ &=\tau_{ij}\nu_i\mu_j\\ &=\tau_j^{\nu}\mu_j\\ &=\boldsymbol{\tau}^{\nu}\cdot\boldsymbol{\mu}\end{aligned}$$

which proves the theorem.

9.2. Principal Axes of the Stress Tensor

We define the principal axes of the stress tensor as follows: The direction specified by unit vector $\boldsymbol{\mu}$ is called a *principal axis of* τ_{ik} if the stress vector of τ_{ik} acting on the surface defined by $\boldsymbol{\mu}$ (this plane is called a *principal plane*) is parallel to $\boldsymbol{\mu}$. We shall speak specifically of the stress tensor in our discussion; however, the results which we establish will hold for any symmetrical tensor of the second order.

The existence of three principal axes will be proved later, but for the meantime, let us assume that at least one exists. We see that a principal direction along unit vector $\boldsymbol{\mu}$ is characterized by

$$\tau_j^{\mu}=\tau_{ij}\mu_i=\tau\mu_j \tag{9.7}$$

where τ is the magnitude of $\boldsymbol{\tau}^{\mu}$. Equation (9.7) follows from the fact that the stress vector on a principal plane acts normal to the plane (no shear stress). Rewriting Eq. (9.7) then, we have

$$\tau_{ij}\mu_i-\tau\mu_j=0$$

or

$$\tau_{ij}\mu_i-\tau\delta_{ij}\mu_i=0 \tag{9.8}$$

which, in unabridged form, is

$$\begin{aligned}(\tau_{11}-\tau)\mu_1+\tau_{21}\mu_2+\tau_{31}\mu_3&=0\\ \tau_{12}\mu_1+(\tau_{22}-\tau)\mu_2+\tau_{32}\mu_3&=0\\ \tau_{13}\mu_1+\tau_{23}\mu_2+(\tau_{33}-\tau)\mu_3&=0\end{aligned} \tag{9.9}$$

The trivial solution of Eqs. (9.9) is $\mu_1=\mu_2=\mu_3=0$. However, this solution is not acceptable, since μ_1, μ_2, and μ_3 cannot all vanish because $\mu_i\mu_i=1$. Using Cramer's rule for simultaneous linear equations, the solution for one of the μ_i results in the quotient of two determinants, where the numerator is zero. Therefore, if the system of equations is to admit a nontrivial solution, the denominator must also vanish. Indicating this determinant by a typical element, we express this condition in the form

$$|\tau_{ij} - \tau\delta_{ij}| = 0 \tag{9.10}$$

or

$$\begin{vmatrix} (\tau_{11} - \tau) & \tau_{21} & \tau_{31} \\ \tau_{12} & (\tau_{22} - \tau) & \tau_{32} \\ \tau_{13} & \tau_{23} & (\tau_{33} - \tau) \end{vmatrix} = 0 \tag{9.11}$$

Expanding the determinant and noting that $\tau_{ik} = \tau_{ki}$, we obtain the cubic equation

$$\begin{aligned} \tau^3 - (\tau_{11} + \tau_{22} + \tau_{33})\tau^2 - (\tau_{12}^2 + \tau_{23}^2 + \tau_{31}^2 \\ - \tau_{11}\tau_{22} - \tau_{22}\tau_{33} - \tau_{33}\tau_{11})\tau - (\tau_{11}\tau_{22}\tau_{33} \\ + 2\tau_{12}\tau_{23}\tau_{31} - \tau_{11}\tau_{23}^2 - \tau_{22}\tau_{13}^2 - \tau_{33}\tau_{12}^2) = 0 \end{aligned}$$

which is called the *characteristic equation of* τ_{ik}. This equation may be written in the form

$$\tau^3 - I_1\tau^2 - I_2\tau - I_3 = 0 \tag{9.12}$$

where I_1, I_2, and I_3 depend only upon the given stress components, i.e., τ_{ij}. The roots of this equation are the three principal stresses. Since Eq. (9.12) must be the same when referred to any coordinate system (the principal stresses have definite values so that Eq. (9.12) must result from any choice of axes x_i), the coefficients I_1, I_2, and I_3 must be invariant quantities, i.e., independent of the coordinate system selected. Thus I_1, I_2, and I_3 are called the first, second, and third invariants, respectively. We note that

$$\begin{aligned} I_1 &= \tau_{ii} \\ I_2 &= \tfrac{1}{2}(\tau_{ik}\tau_{ki} - \tau_{ii}\tau_{kk}) \\ I_3 &= \tfrac{1}{6}(\epsilon_{ijk}\epsilon_{pqr}\tau_{ip}\tau_{jq}\tau_{kr}) \\ &= \tfrac{1}{6}(2\tau_{ij}\tau_{jk}\tau_{ki} - 3\tau_{ij}\tau_{ji}\tau_{kk} + \tau_{ii}\tau_{jj}\tau_{kk}) \end{aligned} \tag{9.13}$$

The fact that only dummy subscripts appear in Eqs. (9.13) further indicates the scalar character of the invariants, and as such, they are independent of the coordinate system in which they are expressed.

Since Eq. (9.12) is a cubic equation in τ, at least one of its roots must be real. This root will be designated by τ_I. Corresponding to τ_I, one principal direction, defined by unit vector $\boldsymbol{\mu}_I$ may be obtained from Eqs. (9.9) and the condition $\mu_i^I\mu_i^I = 1$. We let the x'_1 axis of a primed coordinate system lie in this direction. The x'_2 direction must be orthogonal to x'_1 but is otherwise arbitrary. Consider then, the prism formed by planes taken normal to x'_1, x'_2, and x'_3. Here we have the known stress components

$$\tau'_{11} = \tau_I$$
$$\tau'_{12} = \tau'_{21} = 0$$
$$\tau'_{13} = \tau'_{31} = 0$$

Thus the characteristic equation in the primed coordinate system is given by

$$\begin{vmatrix} (\tau_I - \tau) & 0 & 0 \\ 0 & (\tau'_{22} - \tau) & \tau'_{32} \\ 0 & \tau'_{23} & (\tau'_{33} - \tau) \end{vmatrix} = 0 \tag{9.14}$$

which, on expansion, becomes

$$(\tau_I - \tau)[\tau^2 - (\tau'_{22} + \tau'_{33})\tau + \tau'_{22}\tau'_{33} - \tau'^2_{23}] = 0 \tag{9.15}$$

The other two principal stresses, say τ_{II} and τ_{III}, are the roots of the quantity in the brackets of Eq. (9.15) set equal to zero, i.e.,

$$\left.\begin{matrix}\tau_{II} \\ \tau_{III}\end{matrix}\right\} = \frac{(\tau'_{22} + \tau'_{33}) \pm \sqrt{(\tau'_{22} + \tau'_{33})^2 - 4(\tau'_{22}\tau'_{33} - \tau'^2_{23})}}{2} \tag{9.16}$$

Now the term under the radical (the discriminant) may be written as

$$(\tau'_{22} + \tau'_{33})^2 - 4(\tau'_{22}\tau'_{33} - \tau'^2_{23}) = (\tau'_{22} - \tau'_{33})^2 + 4\tau'^2_{23}$$

which clearly cannot be negative; therefore, the three principal values of the stress tensor are real.

We now let $\boldsymbol{\mu}_{II}$ and $\boldsymbol{\mu}_{III}$ (components μ_i^{II} and μ_j^{III}, respectively) be the unit vectors in the principal directions of τ_{II} and τ_{III}, respectively, so that from Eq. (9.8) we get

$$\begin{aligned} \tau_{ij}\mu_i^{II} - \tau_{II}\delta_{ij}\mu_i^{II} &= 0 \\ \tau_{ij}\mu_i^{III} - \tau_{III}\delta_{ij}\mu_i^{III} &= 0 \end{aligned} \tag{9.17}$$

If we multiply the first of Eqs. (9.17) by μ_j^{III} and the second by μ_j^{II} and subtract, we find, using the fact that the dummy indices may be interchanged and that $\tau_{ij} = \tau_{ji}$

$$(\tau_{II} - \tau_{III})\delta_{ij}\mu_i^{II}\mu_j^{III} = 0$$

or

$$(\tau_{II} - \tau_{III})\mu_i^{II}\mu_i^{III} = 0 \tag{9.18}$$

We observe that if τ_{II} and τ_{III} are unequal, then

$$\mu_i^{II}\mu_i^{III} = 0 \qquad (\tau_{II} \neq \tau_{III}) \tag{9.19}$$

which indicates that $\boldsymbol{\mu}_{II}$ is normal to $\boldsymbol{\mu}_{III}$. Similarly, it can be shown that $\boldsymbol{\mu}_I$ is perpendicular to $\boldsymbol{\mu}_{II}$ and $\boldsymbol{\mu}_{III}$. Thus we have demonstrated that the principal directions corresponding to distinct principal stresses are orthogonal. The fact that the principal stresses are also the stationary values of normal stress is shown by considering the Mohr circle construction of Fig. 1.21, where we see that the stationary values of normal stress occur on planes where the shear stress vanishes.

If two of the principal stresses are equal in magnitude, for example, $\tau_{II} = \tau_{III} \neq \tau_I$, the direction of τ_I is still normal to the plane defined by the directions of τ_{II} and τ_{III}. In this case the discriminant in Eq. (9.16) vanishes so that

$$\tau'_{23} = 0$$

therefore,

$$\tau_{II} = \tau_{III} = \tau'_{22} = \tau'_{33}$$

and all directions in the plane of τ_{II} and τ_{III} are principal directions.

In terms of the principal stresses, the transformation equation (9.4) may be expressed in the form

$$\tau_{ij} = \tau_I \mu_i^I \mu_j^I + \tau_{II} \mu_i^{II} \mu_j^{II} + \tau_{III} \mu_i^{III} \mu_j^{III} \tag{9.20}$$

and if the coordinate axes are taken in the principal directions, the matrix of the stress tensor takes the form

$$\begin{pmatrix} \tau_I & 0 & 0 \\ 0 & \tau_{II} & 0 \\ 0 & 0 & \tau_{III} \end{pmatrix} \tag{9.21}$$

9.3. Equations of Equilibrium

We now leave our discussion of the state of stress at a point and consider a continuously varying stress distribution acting throughout a body in equilibrium. In Section 1.7 we derived the equations of equilibrium by considering an infinitesimal element with stress components such as τ_{1k} acting on one face and $\tau_{1k} + (\partial \tau_{1k}/\partial x_1)\, dx_1$ on the opposite face. In this section we shall derive the same equilibrium equations by considering an arbitrary finite body and utilizing the divergence theorem (or Gauss theorem).

Consider a finite volume V of a continuous medium that is bounded by a closed surface S. Let the body force intensity be specified by the vector field F_k and the stresses by the tensor field τ_{ik}. On S, the surface stress is designated by $\tau_k{}^\mu$, where $\boldsymbol{\mu}$ is a unit outward vector normal to the surface. Equilibrium requires that the vector sum of all forces acting on the body vanish, or

$$\int_V F_k \, dV + \int_S \tau_k{}^\mu \, dS = 0 \tag{9.22}$$

where dV is the elementary volume, and dS the area of an infinitesimal surface element. According to Eq. (9.5), Eq. (9.22) may be written as

$$\int_V F_k \, dV + \int_S \tau_{ik} \mu_i \, dS = 0 \tag{9.23}$$

The surface integral in Eq. (9.23) may be transformed into a volume integral by applying the divergence theorem as follows (see Eq. (8.63)):

$$\int_S \tau_{ik} \mu_i \, dS = \int_V \frac{\partial \tau_{ik}}{\partial x_i} \, dV$$

and Eq. (9.23) becomes

$$\int_V \left(F_k + \frac{\partial \tau_{ik}}{\partial x_i} \right) dV = 0 \tag{9.24}$$

We observe that if S is an arbitrary closed surface within the medium (or the boundary surface of the medium itself) and V the volume enclosed by S, Eq. (9.24) must be satisfied. This is due to the fact that *every part* of the medium must be in equilibrium. The integral can vanish under this condition only if the integrand is zero[1] (if the integrand is continuous), so that

$$\frac{\partial \tau_{ik}}{\partial x_i} + F_k = 0 \tag{9.25}$$

Equation (9.25) represents the three equations of equilibrium, i.e., Eqs. (1.21). Setting the resultant moment about the origin equal to zero, we get (refer to Eq. (8.60)),

$$M_i = \int_V \epsilon_{ijk} x_j F_k \, dV + \int_S \epsilon_{ijk} x_j \tau_k{}^\mu \, dS = 0 \tag{9.26}$$

[1] This fact may be demonstrated as follows. For convenience let $(F_k + \partial \tau_{ik}/\partial x_i) \equiv G_k$; thus we must show that $G_k \equiv 0$. Since Eq. (9.24) must hold for any volume V within the body, if G_k is continuous but nonvanishing, we can always choose a small volume V within which G_k is positive (or negative). The integral would therefore be positive (or negative), which contradicts Eq. (9.24). Therefore $G_k \equiv 0$.

and by using Eq. (9.5) and the divergence theorem, the surface integral becomes

$$\int_S \epsilon_{ijk} x_j \tau_k{}^\mu dS = \int_S \epsilon_{ijk} x_j \mu_\ell \tau_{\ell k} dS = \int_V \frac{\partial(\epsilon_{ijk} x_j \tau_{\ell k})}{\partial x_\ell} dV$$

Performing the differentiation, and using the definition of the Kronecker delta and Eq. (9.25), we get

$$\int_S \epsilon_{ijk} x_j \tau_k{}^\mu dS = \int_V \epsilon_{ijk}\left(\delta_{j\ell}\tau_{\ell k} + x_j \frac{\partial \tau_{\ell k}}{\partial x_\ell}\right) dV$$

$$= \int_V \epsilon_{ijk}(\tau_{jk} - x_j F_k)\, dV$$

Substituting into Eq. (9.26), then, we have

$$\int_V \epsilon_{ijk} \tau_{jk} dV = 0$$

Again, V is an arbitrary volume and $\epsilon_{ijk}\tau_{jk}$ is continuous, so that

$$\epsilon_{ijk}\tau_{jk} = 0$$

This can be expanded to give

$$\tau_{12} - \tau_{21} = 0$$

$$\tau_{13} - \tau_{31} = 0$$

$$\tau_{23} - \tau_{32} = 0$$

or, in general,

$$\tau_{ik} = \tau_{ki} \tag{9.27}$$

The stress boundary conditions are expressed by considering surface *ABC* of Fig. 9.1 as lying on the boundary. Thus Eq. (9.5) will give the components of the surface force at a point on the boundary. We observe then, that on the boundary of a body, the components of stress must satisfy the relation

$$\tau_{ij}\mu_i = T_j{}^\mu \tag{9.28}$$

where $\boldsymbol{\mu}$ is the unit outward normal to the boundary surface, $\mathbf{T}^\mu$ the applied surface force vector, and τ_{ij} is evaluated on the boundary. (In Chapter 4, $T_j{}^\mu$ were represented by $T_x{}^\mu$, $T_y{}^\mu$, and $T_z{}^\mu$. See Eq. (4.4).)

9.4. The Stress Ellipsoid

At a given point, the stress vectors on all planes of different inclinations may be conveniently represented graphically by the surface of an ellipsoid (the ellipsoid of Lamé, or the stress ellipsoid). This surface offers a graphic model of the three-dimensional state of stress at a point.

In order to write the equation of the stress ellipsoid, we consider a prism with planes normal to the principal directions at a certain point as shown in Fig. 9.2. No generality is lost by using this coordinate

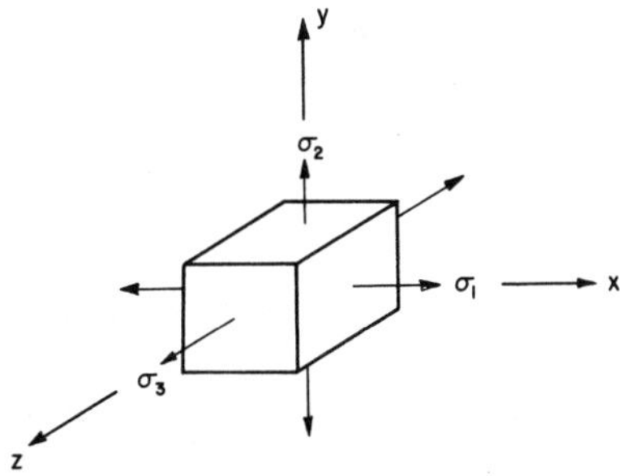

Fig. 9.2 Prism Normal to Principal Directions

system, since we have seen that the principal directions may be determined for any state of stress. From Fig. 9.2, we observe that the given stresses are

$$\sigma_x = \sigma_1, \qquad \sigma_y = \sigma_2, \qquad \sigma_z = \sigma_3$$
$$\tau_{xy} = \tau_{yz} = \tau_{zx} = 0$$

so that the transformation equations (1.24) or (9.1) are expressed by

$$\begin{aligned} p_x &= \sigma_1 a_{11} \\ p_y &= \sigma_2 a_{21} \\ p_z &= \sigma_3 a_{31} \end{aligned} \tag{9.29}$$

where p_x, p_y, and p_z are the components of the stress vector on the x'_1 plane. Using the first of Eqs. (8.11a) and Eqs. (9.29) then, we find that

$$\frac{p_x^2}{\sigma_1^2} + \frac{p_y^2}{\sigma_2^2} + \frac{p_z^2}{\sigma_3^2} = 1$$

which is the equation of an ellipsoid in the p_x, p_y, p_z coordinate system. This surface (the stress ellipsoid) is centered at the origin of the coordinate system, and its axes of symmetry are the coordinate axes. The ellipsoid is as shown in Fig. 9.3, where its intercepts on the coordinate axes are $\pm\sigma_1$, $\pm\sigma_2$, and $\pm\sigma_3$.

The stress ellipsoid has the property that the radius vector from the origin to a point on its surface is equal to the stress vector on some plane through the prism of Fig. 9.2. Since the lengths of the semiaxes of the ellipsoid are equal to the magnitudes of the principal stresses, the two

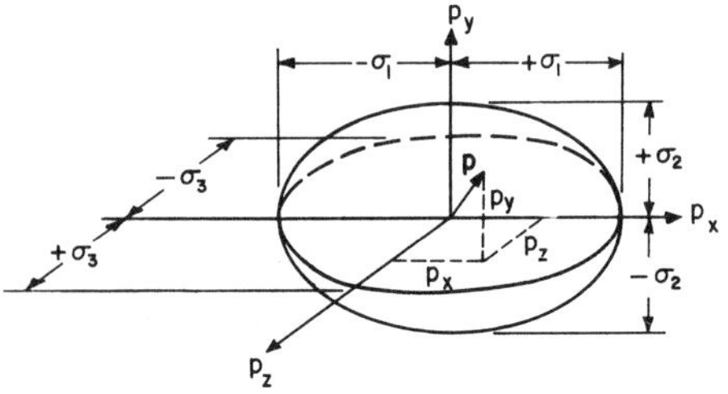

Fig. 9.3 The Stress Ellipsoid of Lamé

principal stresses having the largest and smallest *magnitudes* represent the stress vectors having the largest and smallest magnitudes for any plane at a given point. The stress ellipsoid by itself, however, may not be used to determine the plane on which a given stress vector acts except in certain special cases.

In order to determine the plane on which a given stress vector acts, we may use in addition to the stress ellipsoid, the stress-director surface defined by

$$\frac{p_x^{\ 2}}{\sigma_1} + \frac{p_y^{\ 2}}{\sigma_2} + \frac{p_z^{\ 2}}{\sigma_3} = \pm 1 \tag{9.30}$$

The stress corresponding to a radius vector of the stress ellipsoid acts on a plane parallel to the tangent plane of the stress-director surface at the point where the radius vector intersects the stress-director surface.[2]

9.5. Body Moment and Couple Stress

We shall now present some of the principles based on a more general hypothesis of linear elasticity, which has received considerable attention in recent years.[3] Thus far in our work, we have assumed that infinitesi-

[2] See A. Durelli, Phillips, and Tsao, *Introduction to the Theoretical and Experimental Analysis of Stress and Strain*, McGraw-Hill Book Company, Inc., New York, 1958, pp. 65–68; or S. Timoshenko and J. N. Goodier, *Theory of Elasticity*, Second Edition, McGraw-Hill Book Company, Inc., New York, 1951, p. 216.

[3] R. D. Mindlin and H. F. Tiersten, "Effects of Couple-Stress in Linear Elasticity," *Archive for Rational Mechanics and Analysis*, Vol. II, No. 5, Berlin, Dec. 20, 1962.

mal elements of material are subjected only to body force (force per unit volume) and surface stress (force per unit area). Under certain circumstances, however, the body forces can produce moments per unit volume M_i, as in the case of a polarized dielectric material under the action of an electric field. This moment (per unit volume) can be of finite value and must be included in the moment equilibrium equations.

We shall next hypothesize that moments per unit area, as well as forces per unit area, exist and act on the surfaces of an infinitesimal element. We shall define the couple stress vector on a plane at a point as the limiting value of the ratio of the couple vector acting on this plane divided by the area on which it acts, as the area approaches zero. Since on a given plane the couple stress is a vector, it can be shown easily that the components of couple stress m_{ij} at a point constitute a second-order tensor. In summing up the moments on an infinitesimal element, the moments due to the "conventional" stress components yield higher-order terms, and the couple stress transformation equation, similar to Eq. (9.4), becomes

$$m'_{\ell k} = a_{i\ell} a_{jk} m_{ij} \tag{9.31}$$

Thus, under this hypothesis, the state of stress at a point consists of the components of couple stress m_{ij}, in addition to the components of "conventional" stress τ_{ij}. It should be noted that when couple stresses are included, the stress tensor τ_{ij} (as well as the couple stress tensor m_{ij}) is no longer symmetric.

Let us now consider the equations of equilibrium as governed by the hypothesis under discussion. We thus conceive of a region bounded by a surface S, enclosing volume V, subjected to a vector field on S consisting of surface stress $\tau_i{}^\mu$ and surface couples per unit area $m_i{}^\mu$. In addition to the body force field F_i, there will be body moments per unit volume M_i, acting on the mass points of the region. Equilibrium of the forces acting on the region requires that

$$\int_S \tau_i{}^\mu \, dS + \int_V F_i \, dV = 0$$

which is precisely Eq. (9.22), and as before, this reduces to

$$\frac{\partial \tau_{ji}}{\partial x_j} + F_i = 0 \tag{9.32}$$

except that in this case

$$\tau_{ji} \neq \tau_{ij}$$

Equilibrium of the region also requires that the resultant moment vanish. Thus, taking moments about the origin, we get

$$\int_S \epsilon_{ijk} x_j \tau_k^{\mu} \, dS + \int_S m_i^{\mu} \, dS + \int_V \epsilon_{ijk} x_j F_k \, dV + \int_V M_i \, dV = 0$$

Proceeding as in the development of Eq. (9.27), and using the fact that

$$m_i^{\mu} = m_{ji} \mu_j \tag{9.33}$$

we find that

$$\frac{\partial m_{ji}}{\partial x_j} + M_i + \epsilon_{ijk} \tau_{jk} = 0 \tag{9.34}$$

In the event that the couple stresses and body moments vanish, we see that Eq. (9.34) reduces to Eq. (9.27). It is noted that the equations of equilibrium in the couple stress theory, Eqs. (9.32) and (9.34), are linear. The remainder of the governing equations of this theory will not be presented in this text.

The use of the couple stress hypothesis requires the definition of a new elastic constant, which relates couple stress and curvature. As yet, however, this constant has not been measured experimentally for any material. The effect of couple stress is most pronounced in regions where the strain gradients are relatively large. In this text, unless otherwise noted, couple stresses, as well as body moments are ignored.

PROBLEMS

9-1 Show that Eqs. (9.4) can be generated from the equation $\tau_i^{\mu} \nu_i = \tau_{ij} \mu_j \nu_i$, where $\boldsymbol{\mu}$ and $\boldsymbol{\nu}$ are unit vectors in arbitrary directions.

9-2 Verify Eqs. (9.13).

9-3 Verify Eqs. (9.20).

9-4 Derive Eq. (9.3) by considering the equilibrium of a tetrahedron $OABC$ and by projecting all forces in the x'_n direction. The outward normal to face ABC is in the x'_1 direction.

9-5 Fill in the missing steps of the development which shows that all directions in the plane of τ_{II} and τ_{III} are principal axes and τ_I is normal to this plane for the state of stress at a point where $\tau_{II} = \tau_{III}$.

9-6 Show that all directions are principal axes for the state of stress at a point where

$$\tau_I = \tau_{II} = \tau_{III}$$

9-7 The mean normal stress, σ_m, is defined as $\sigma_m = \frac{1}{3}\tau_{ii}$, and the stress deviator tensor τ_{ij}^* is defined as

$$\tau_{ij}^* = \tau_{ij} - \sigma_m \delta_{ij}$$

Show that the second invariant $I_2{}^*$ of τ_{ij}^* is equal to $\frac{2}{3}$ times the sum of the squares of the three maximum shear stresses.

9-8 Is Eq. (9.4) true if the unprimed coordinate system is right-handed and the primed system is left-handed?

10

Strain, Displacement, and the Governing Equations of Elasticity

10.1. Introduction

In this Chapter we shall continue to explore the tensor character of the various quantities of elasticity. We shall encounter two more second-order tensors, namely, the strain and rotation tensors, and a fourth-order tensor, which is the array of constants characterizing the behavior of elastic media. The material presented in this Chapter parallels the coverage of Chapters 2, 3, and 4, in which the physical concepts were stressed. Consequently, in Chapter 10, there will be no detailed explanations of a physical nature; rather, we shall stress the mathematical structure of tensor quantities.

10.2. Displacement and Strain

In Fig. 10.1, points P_0 and P represent the initial locations of two particles of a body under load. Point P_0 is located at $(x_1{}^0, x_2{}^0, x_3{}^0)$ and P is at (x_1, x_2, x_3). When the body is strained, P_0 moves to $(x_1{}^0+u_1{}^0, x_2{}^0+u_2{}^0, x_3{}^0+u_3{}^0)$, i.e., its displacement vector is $u_i{}^0$. Similarly, particle P is displaced to $(x_1+u_1, x_2+u_2, x_3+u_3)$. We assume that $x_i-x_i{}^0$ is infinitesimal, i.e., we confine our attention to the material in the vicinity of P_0. We also make the assumption of continuous displacement components, which implies that the relative movement of P with respect to P_0 is infinitesimal.

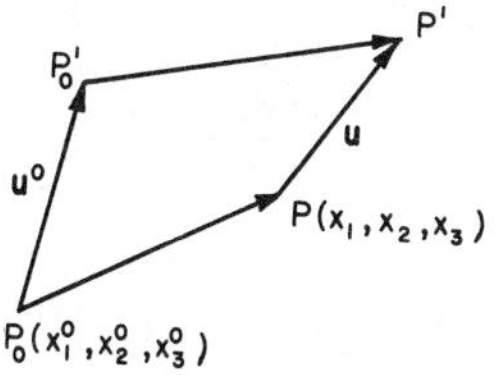

Fig. 10.1 Displacement of Neighboring Points

We note that the displacement is a function of position, or

$$u_i = u_i(x_1, x_2, x_3)$$

Thus, we may expand u_i into a Taylor series around P_0,

$$u_i = u_i^0 + \frac{\partial u_i}{\partial x_j}(x_j - x_j^0) + \frac{1}{2!}\frac{\partial^2 u_i}{\partial x_j \partial x_k}(x_j - x_j^0)(x_k - x_k^0) + \cdots$$

and since $x_j - x_j^0$ are infinitesimal quantities, we may drop the second and higher powers (higher-order terms) in the series. Rewriting, we see that

$$u_i = u_i^0 + \frac{\partial u_i}{\partial x_j} dx_j \tag{10.1}$$

where $dx_j = x_j - x_j^0$. Now since u_i is a vector, its gradient $\partial u_i/\partial x_j$ (see Section 8.4) is a second-order tensor, which may be written as the sum of its symmetric and antisymmetric parts. Thus Eq. (10.1) becomes

$$u_i = u_i^0 + \frac{1}{2}\left(\frac{\partial u_i}{\partial x_j} + \frac{\partial u_j}{\partial x_i}\right) dx_j + \frac{1}{2}\left(\frac{\partial u_i}{\partial x_j} - \frac{\partial u_j}{\partial x_i}\right) dx_j$$

or

$$u_i = u_i^0 + e_{ij}\, dx_j + \omega_{ij}\, dx_j \tag{10.2}$$

where

$$e_{ij} = \frac{1}{2}\left(\frac{\partial u_i}{\partial x_j} + \frac{\partial u_j}{\partial x_i}\right) \tag{10.3a}$$

and

$$\omega_{ij} = \frac{1}{2}\left(\frac{\partial u_i}{\partial x_j} - \frac{\partial u_j}{\partial x_i}\right) \tag{10.3b}$$

The components of e_{ij} are

$$
\begin{aligned}
e_{11} &= \frac{1}{2}\left(\frac{\partial u_1}{\partial x_1} + \frac{\partial u_1}{\partial x_1}\right) = \frac{\partial u_1}{\partial x_1} = \epsilon_x \\
e_{12} = e_{21} &= \frac{1}{2}\left(\frac{\partial u_1}{\partial x_2} + \frac{\partial u_2}{\partial x_1}\right) = \frac{1}{2}\gamma_{xy} \\
e_{13} = e_{31} &= \frac{1}{2}\left(\frac{\partial u_1}{\partial x_3} + \frac{\partial u_3}{\partial x_1}\right) = \frac{1}{2}\gamma_{xz} \\
&\cdots \\
&\cdots
\end{aligned}
\tag{10.4}
$$

where the last step in each expression results from associating x_1, x_2, and x_3 with x, y, and z, respectively. Equations (10.4) indicate that e_{11}, e_{22}, and e_{33} are the normal strains and $e_{12}, \ldots, e_{31}$ are one-half the engineering shear strains as defined in Chapter 2. Since e_{ij} is the symmetric part of the second-order tensor $\partial u_i/\partial x_j$, it is itself a symmetric tensor of the second order. (See Section 8.3.) Similarly, ω_{ij} is an antisymmetric tensor of the second order. We shall call e_{ij} the strain tensor and ω_{ij} the rotation tensor.

The vector associated with the antisymmetric tensor ω_{ij} is the curl of u_i, that is, the only nonvanishing components of the rotation tensor ω_{ij} are

$$
\begin{aligned}
\omega_{32} &= -\omega_{23} = \frac{1}{2}\left(\frac{\partial u_3}{\partial x_2} - \frac{\partial u_2}{\partial x_3}\right) \\
\omega_{13} &= -\omega_{31} = \frac{1}{2}\left(\frac{\partial u_1}{\partial x_3} - \frac{\partial u_3}{\partial x_1}\right) \\
\omega_{21} &= -\omega_{12} = \frac{1}{2}\left(\frac{\partial u_2}{\partial x_1} - \frac{\partial u_1}{\partial x_2}\right)
\end{aligned}
$$

Therefore, in the notation of Chapter 2, we see that

$$
\begin{aligned}
\omega_{32} &= \omega_x \\
\omega_{13} &= \omega_y \\
\omega_{21} &= \omega_z
\end{aligned}
$$

and the components of ω_{ij} are the small angles of rotation about axes parallel to x, y, and z at P_0. It is apparent then that the first term in Eq. (10.2) gives the displacement components of the point P_0, the second term defines the relative movement of P due to deformation (strain), and the third term represents the relative movement of P due to a rotation around P_0.

10.3. Generalized Hooke's Law

It was shown in Chapter 3 that the stress strain relations for a homogeneous elastic medium are assumed to be of the form

$$\begin{aligned}
\tau_{11} &= K_{11}\epsilon_x + K_{12}\epsilon_y + K_{13}\epsilon_z + K_{14}\gamma_{yz} + K_{15}\gamma_{zx} + K_{16}\gamma_{xy} \\
\tau_{22} &= K_{21}\epsilon_x + \cdots \\
\tau_{33} &= K_{31}\epsilon_x + \cdots \\
\tau_{23} &= K_{41}\epsilon_x + \cdots \\
\tau_{31} &= K_{51}\epsilon_x + \cdots \\
\tau_{12} &= K_{61}\epsilon_x + K_{62}\epsilon_y + K_{63}\epsilon_z + K_{64}\gamma_{yz} + K_{65}\gamma_{zx} + K_{66}\gamma_{xy}
\end{aligned} \tag{10.5}$$

that is, a linear relationship is assumed between each stress component and the strain components. Since the symmetry properties of τ_{ij} and e_{ij} are utilized in writing Eqs. (10.5), we have assumed 36 independent elastic constants.

In terms of another set of constants $C_{ijk\ell}$, Eqs. (10.5) may be written in the form

$$\begin{aligned}
\tau_{ij} &= C_{ijk\ell}e_{k\ell} \\
&= C_{ij11}e_{11} + C_{ij12}e_{12} + C_{ij13}e_{13} \\
&\quad + C_{ij21}e_{21} + C_{ij22}e_{22} + C_{ij23}e_{23} \\
&\quad + C_{ij31}e_{31} + C_{ij32}e_{32} + C_{ij33}e_{33}
\end{aligned} \tag{10.6}$$

We must, however, place some restrictions on $C_{ijk\ell}$ because the stress and strain tensors are symmetrical. Without symmetry, $C_{ijk\ell}$ represents 81 constants rather than 36. Because of the symmetry in the strain components, we let

$$C_{ijk\ell} = C_{ij\ell k} \tag{10.7}$$

which reduces the number of independent constants to 54. No generality is lost by this assumption, since, for example,

$$C_{ij21}e_{21} + C_{ij12}e_{12} = (C_{ij21} + C_{ij12})e_{12} = C_{ij12}\gamma_{xy}$$

and the coefficients of the normal strains are unaffected by this assumption. Furthermore, since

$$\tau_{ij} = \tau_{ji}$$

we must have

$$C_{ijk\ell}=C_{jik\ell} \tag{10.8}$$

and we now have 36 independent constants. The restrictions placed on the constants make Eqs. (10.6) equivalent to Eqs. (10.5). Strain energy considerations can be used to reduce the number of constants to 21, however it will not be necessary to employ this reduction here.[1]

We shall first demonstrate that $C_{ijk\ell}$ is a fourth-order tensor. Since

$$\tau_{ij}=C_{ijk\ell}\,e_{k\ell}$$

and

$$\tau_{ij}=a_{im}\,a_{jp}\,\tau'_{mp}$$

we have

$$C_{ijk\ell}\,e_{k\ell}=a_{im}\,a_{jp}\,\tau'_{mp}$$

Let us call the elastic constants in the primed coordinate system C'_{mpqr}, thus

$$\tau'_{mp}=C'_{mpqr}\,e'_{qr}$$

Since strain is a second-order tensor,

$$e'_{qr}=a_{kq}\,a_{\ell r}\,e_{k\ell}$$

so that

$$C_{ijk\ell}\,e_{k\ell}=a_{im}\,a_{jp}\,C'_{mpqr}\,a_{kq}\,a_{\ell r}\,e_{k\ell}$$

or

$$(C_{ijk\ell}-a_{im}\,a_{jp}\,a_{kq}\,a_{\ell r}\,C'_{mpqr})e_{k\ell}=0$$

We note that the quantity in parenthesis is independent of any component of strain. Now the six components of strain may be arbitrarily specified since they are independent of each other; therefore the quantity in the parenthesis must vanish,[2] or

$$C_{ijk\ell}=a_{im}\,a_{jp}\,a_{kq}\,a_{\ell r}\,C'_{mpqr} \tag{10.9}$$

so that $C_{ijk\ell}$ is a fourth-order tensor.

[1] For a discussion of the elastic properties of various classes of crystals, see A. E. H. Love, *A Treatise on the Mathematical Theory of Elasticity*, Dover Publications, Inc., New York, 1944, Chapter 6.

[2] This may be shown by letting all e_{kl} be zero except e_{11}, which results in $C_{ij11}-a_{im}\,a_{jp}\,a_{1q}\,a_{1r}\,C'_{mpqr}=0$. By similar processes, Eq. (10.9) is obtained.

We shall now restrict our discussion to an isotropic medium. The condition of isotropy is expressed by the relation

$$C_{ijk\ell} = C'_{ijk\ell}$$

or

$$\tau'_{ij} = C_{ijk\ell}\, e'_{k\ell} \tag{10.10}$$

since the elastic constants must be the same for any orientation of coordinate axes. For an isotropic material, the principal axes of the stress tensor coincide with the principal axes of the strain tensor as shown by the following analysis. Let x_1, x_2, and x_3 coincide with the principal axes of strain. Thus

$$e_{12} = e_{13} = e_{23} = 0 \tag{10.11}$$

We shall show that $\tau_{12} = \tau_{23} = \tau_{31} = 0$, so that x_1, x_2, and x_3 are also the principal axes for stress. Considering the component τ_{23}, from Eqs. (10.6) and (10.11) we have

$$\tau_{23} = C_{2311}\, e_{11} + C_{2322}\, e_{22} + C_{2333}\, e_{33} \tag{10.12}$$

Now let x'_1, x'_2, x'_3 be obtained by rotating the x_1, x_2, x_3 axes through 180° about x_3. The direction cosines of the primed coordinate axes with respect to the unprimed axes are

	x'_1	x'_2	x'_3
x_1	−1	0	0
x_2	0	−1	0
x_3	0	0	1

$$\text{or} \quad a_{ij} = \begin{pmatrix} -1 & 0 & 0 \\ 0 & -1 & 0 \\ 0 & 0 & 1 \end{pmatrix} \tag{10.13}$$

Combining Eqs. (9.4) and (10.13), we find that

$$\tau'_{23} = a_{k2}\, a_{\ell 3}\, \tau_{k\ell} = -\tau_{23} \tag{10.14}$$

and since $e'_{ij} = a_{ki}\, a_{\ell j}\, e_{k\ell}$, we have, from Eqs. (10.13) and (10.11),

$$e'_{ij} = \begin{pmatrix} e_{11} & 0 & 0 \\ 0 & e_{22} & 0 \\ 0 & 0 & e_{33} \end{pmatrix} \tag{10.15}$$

or the corresponding strain components in these two coordinate systems are the same. Substituting Eqs. (10.14) and (10.15) into Eq. (10.10) with $i = 2$, $j = 3$, we have

$$-\tau_{23} = C_{2311}\, e_{11} + C_{2322}\, e_{22} + C_{2333}\, e_{33} \tag{10.16}$$

Comparing Eqs. (10.12) and (10.16), then, we conclude that

$$\tau_{23} = 0$$

Similarly, it can be shown that τ_{12} and τ_{13} vanish, so that

$$\tau_{23} = \tau_{12} = \tau_{13} = 0 \tag{10.17}$$

Therefore, x_1, x_2, and x_3 are also the principal axes of stress. For an isotropic medium, then, the principal axes of the stress and strain tensors coincide.

We shall next demonstrate that for an isotropic medium only two of the 36 constants $C_{ijk\ell}$ are independent. Consider an infinitesimal prism with its faces normal to the principal axes x_1, x_2, x_3. The stress-strain relations become

$$\begin{aligned}
\tau_{11} &= C_{1111} e_{11} + C_{1122} e_{22} + C_{1133} e_{33} \\
\tau_{22} &= C_{2211} e_{11} + C_{2222} e_{22} + C_{2233} e_{33} \\
\tau_{33} &= C_{3311} e_{11} + C_{3322} e_{22} + C_{3333} e_{33}
\end{aligned} \tag{10.18}$$

Since C_{mpqr} is an isotropic fourth-order tensor, it transforms according to

$$C_{mpqr} = a_{mi} a_{pj} a_{qk} a_{r\ell} C'_{ijk\ell} = a_{mi} a_{pj} a_{qk} a_{r\ell} C_{ijk\ell}$$

For a coordinate system x'_i obtained by a rotation of the old coordinate system through 90° about x_1, the direction cosines are

$$a_{ij} = \begin{pmatrix} 1 & 0 & 0 \\ 0 & 0 & -1 \\ 0 & 1 & 0 \end{pmatrix}$$

and we have

$$\begin{aligned}
C_{11qr} &= a_{qk} a_{r\ell} C_{11k\ell} \\
C_{22qr} &= a_{qk} a_{r\ell} C_{33k\ell} \\
C_{33qr} &= a_{qk} a_{r\ell} C_{22k\ell}
\end{aligned}$$

which yield

$$C_{1133} = C_{1122},\ C_{2211} = C_{3311},\ C_{2222} = C_{3333},\ C_{2233} = C_{3322} \tag{10.19}$$

Furthermore, a 90° rotation about the x_2 axis with

$$a_{ij} = \begin{pmatrix} 0 & 0 & -1 \\ 0 & 1 & 0 \\ 1 & 0 & 0 \end{pmatrix}$$

leads to

$$C_{2211} = C_{2233},\ C_{3322} = C_{1122},\ C_{1111} = C_{3333},\ C_{1133} = C_{3311} \tag{10.20}$$

From Eqs. (10.19) and (10.20), we see that

$$C_{1111} = C_{2222} = C_{3333} = C$$

$$C_{1122} = C_{1133} = C_{2211} = C_{2233} = C_{3311} = C_{3322} = \lambda$$

and the principal stress and principal strain relations (10.18) become

$$\begin{aligned} \tau_{11} &= Ce_{11} + \lambda(e_{22} + e_{33}) \\ \tau_{22} &= Ce_{22} + \lambda(e_{11} + e_{33}) \\ \tau_{33} &= Ce_{33} + \lambda(e_{11} + e_{22}) \end{aligned} \tag{10.21}$$

or, in slightly different form,

$$\begin{aligned} \tau_{11} &= \lambda\epsilon + 2Ge_{11} \\ \tau_{22} &= \lambda\epsilon + 2Ge_{22} \\ \tau_{33} &= \lambda\epsilon + 2Ge_{33} \end{aligned} \tag{10.22}$$

where

$$\epsilon = e_{11} + e_{22} + e_{33} = e_{ii} = e'_{ii}$$

and

$$2G = C - \lambda$$

Using the transformation equation (9.4), we may now determine the stress components referred to any coordinate system x'_i, i.e.,

$$\tau'_{ij} = a_{ki}a_{\ell j}\tau_{k\ell}$$

or

$$\begin{aligned} \tau'_{ij} &= a_{1i}a_{1j}(\lambda\epsilon + 2Ge_{11}) + a_{2i}a_{2j}(\lambda\epsilon + 2Ge_{22}) \\ &\quad + a_{3i}a_{3j}(\lambda\epsilon + 2Ge_{33}) \\ &= a_{mi}a_{mj}\lambda\epsilon + 2Ge'_{ij} \end{aligned}$$

The second term on the right results from the transformation of strain equation. Thus,

$$\tau'_{ij} = \lambda\epsilon\delta_{ij} + 2Ge'_{ij}$$

and since ϵ is an invariant quantity and δ_{ij} is an isotropic tensor, we may write

$$\tau'_{ij} = \lambda\epsilon'\,\delta'_{ij} + 2Ge'_{ij}$$

Since the expression is valid for any coordinate system, we may drop the primes and write

$$\tau_{ij} = \lambda\epsilon\delta_{ij} + 2Ge_{ij} \tag{10.23}$$

for any set of coordinate axes. Equation (10.23) i the same as Eqs. (3.18), which are the generalized Hooke's law expressions for an isotropic, linear elastic material. The two independent constants are the Lamé constants.

Equation (10.23) may be inverted to get the strain components in terms of the stresses by noting that

$$\begin{aligned} \tau_{kk} &= 3\lambda\epsilon + 2Ge_{kk} \\ &= (3\lambda + 2G)\epsilon \end{aligned} \tag{10.24}$$

by contracting Eq. (10.23). Solving for ϵ and substituting into Eq. (10.23), we have

$$e_{ij} = \frac{1}{2G}\left(\tau_{ij} - \lambda\delta_{ij}\frac{\tau_{kk}}{3\lambda + 2G}\right)$$

which can assume the alternative form

$$e_{ij} = \frac{1+\nu}{E}\tau_{ij} - \frac{\nu}{E}\delta_{ij}\Theta \tag{10.25}$$

where

$$\Theta = \tau_{kk}$$

$$E = \frac{G(3\lambda + 2G)}{\lambda + G}$$

$$\nu = \frac{\lambda}{2(\lambda + G)}$$

and E is the modulus of elasticity and ν is Poisson's ratio. Equation (10.25) is identical to Eqs. (3.17).

10.4. Equations of Compatibility

It was shown in Section 2.3 that the strain components must satisfy the compatibility equations as a necessary condition for the existence of single-valued, continuous displacement functions. We shall now demonstrate that this is also a *sufficient* condition in simply connected regions. To prove this, consider the displacement of a point P as shown in Fig. 10.2. The origin is placed at P_0 and the displacement components of P_0 and P (x_1^P, x_2^P, x_3^P) are denoted by u_i^0 and u_i^P, respectively. We shall determine the displacement components at point P by integration, or

$$u_i^P = u_i^0 + \int_C \frac{\partial u_i}{\partial x_j}\,dx_j$$

where C is any continuous curve connecting P_0 and P, as shown in Fig. 10.2. Using Eqs. (10.3), we have

$$u_i^P = u_i^0 + \int_C e_{ij}\,dx_j + \int_C \omega_{ij}\,dx_j \tag{10.26}$$

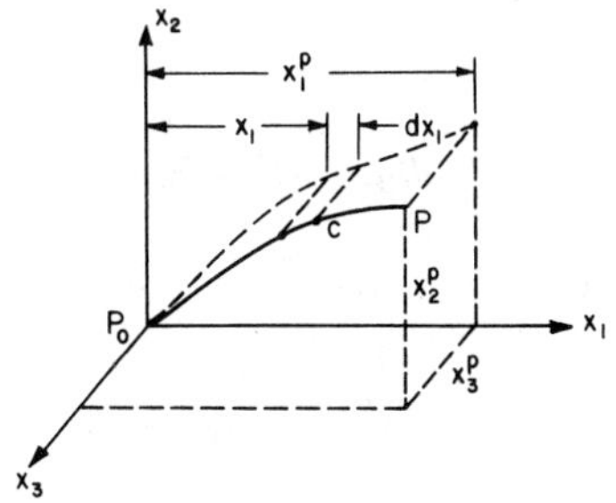

Fig. 10.2 Path of Integration

and by integrating the last term by parts, we get

$$\int_C \omega_{ij}\,dx_j = \omega_{ij}^P x_j^P - \int_C x_j \omega_{ij,k}\,dx_k \tag{10.27}$$

where ω_{ij}^P are the rotation components at P. In order to express the integral in terms of the strain components, we note that

$$\omega_{ij} = \tfrac{1}{2}(u_{i,j} - u_{j,i})$$

so that

$$\begin{aligned}\omega_{ij,k} &= \frac{1}{2}\frac{\partial}{\partial x_k}(u_{i,j} - u_{j,i}) \\ &= \frac{1}{2}\frac{\partial}{\partial x_k}(u_{i,j} - u_{j,i}) + \frac{1}{2}\frac{\partial}{\partial x_i}(u_{k,j} - u_{k,j})\end{aligned}$$

but the order of differentiating continuous functions is immaterial, thus,

$$\omega_{ij,k} = \frac{1}{2}\frac{\partial}{\partial x_j}(u_{i,k} + u_{k,i}) - \frac{1}{2}\frac{\partial}{\partial x_i}(u_{j,k} + u_{k,j})$$

or

$$\omega_{ij,k} = e_{ik,j} - e_{jk,i} \tag{10.28a}$$

Substituting Eqs. (10.27) and (10.28a) into (10.26), we find that

$$u_i^P = u_i^0 + \omega_{ij}^P x_j^P + \int_C \{e_{ik} - x_j(e_{ik,j} - e_{jk,i})\}\,dx_k \tag{10.28b}$$

but the value of u_i^P must be the same for any continuous curve joining P_0 and P. This means that the integral in Eq. (10.28b) must be independent of the path C connecting the two points; therefore, the integrand must be an exact differential. That is, if we let

$$Q_{ik} = e_{ik} - x_j(e_{ik,j} - e_{jk,i}) \tag{10.29}$$

we must have

$$Q_{ik}\,dx_k = dV_i = \frac{\partial V_i}{\partial x_k}\,dx_k$$

if the integrand is an exact differential. Thus we have

$$Q_{ik} = \frac{\partial V_i}{\partial x_k}$$

so that

$$Q_{ik,\ell} = V_{i,k\ell}$$

and

$$Q_{i\ell,k} = V_{i,\ell k}$$

from which it follows that

$$Q_{ik,\ell} = Q_{i\ell,k} \tag{10.30}$$

Substituting Eq. (10.29) into Eq. (10.30), we get

$$e_{ik,\ell} - \delta_{j\ell}(e_{ik,j} - e_{jk,i}) - x_j(e_{ik,j\ell} - e_{jk,i\ell})$$
$$= e_{i\ell,k} - \delta_{jk}(e_{i\ell,j} - e_{j\ell,i}) - x_j(e_{i\ell,jk} - e_{j\ell,ik})$$

which reduces to

$$x_j(e_{ik,j\ell} - e_{jk,i\ell} - e_{i\ell,jk} + e_{j\ell,ik}) = 0$$

Since this equation must be satisfied for all values of x_j in the given region, we have

$$e_{i\ell,jk} - e_{j\ell,ik} + e_{jk,i\ell} - e_{ik,j\ell} = 0 \tag{10.31}$$

Equation (10.31) actually represents a set of 81 equations, since the free indices i, j, k, and ℓ can independently take on values of 1, 2, and 3. Many of these, however, vanish identically (for example, if the numerical values taken by all the subscripts are equal) and others are repetitions. The system thereby reduces to six independent equations,[3] which are identical to Eqs. (2.2).

[3] Here we use the term "independent" in the sense that there are six different equations. As we have noted in Chapter 2, these six equations are equivalent to three independent fourth-order equations.

We can obtain a system of equations which is equivalent (but not identical) to Eq. (10.31) by contracting Eq. (10.31) with respect to j and k, which gives

$$e_{i\ell,jj} - e_{j\ell,ij} + e_{jj,i\ell} - e_{ij,j\ell} = 0 \tag{10.32}$$

Due to the symmetry of the strain tensor, the nine equations of (10.32) represent only six independent equations. When written out in unabridged notation, Eq. (10.32) becomes equivalent to Eqs. (2.2), which are given below for easy reference. (For $i = 1$, $j = k = 3$, $\ell = 2$, Eqs. (10.31) reduce to the last equation of (2.2). From the three equations obtained from (10.32) with $i = \ell$, the first three equations of (2.2) may be derived.)

$$\begin{aligned}
\frac{\partial^2 \epsilon_x}{\partial y^2} + \frac{\partial^2 \epsilon_y}{\partial x^2} &= \frac{\partial^2 \gamma_{xy}}{\partial x \partial y} \\
\frac{\partial^2 \epsilon_y}{\partial z^2} + \frac{\partial^2 \epsilon_z}{\partial y^2} &= \frac{\partial^2 \gamma_{yz}}{\partial y \partial z} \\
\frac{\partial^2 \epsilon_z}{\partial x^2} + \frac{\partial^2 \epsilon_x}{\partial z^2} &= \frac{\partial^2 \gamma_{zx}}{\partial z \partial x} \\
2\frac{\partial^2 \epsilon_x}{\partial y \partial z} &= \frac{\partial}{\partial x}\left(-\frac{\partial \gamma_{yz}}{\partial x} + \frac{\partial \gamma_{xz}}{\partial y} + \frac{\partial \gamma_{xy}}{\partial z}\right) \\
2\frac{\partial^2 \epsilon_y}{\partial z \partial x} &= \frac{\partial}{\partial y}\left(\frac{\partial \gamma_{yz}}{\partial x} - \frac{\partial \gamma_{xz}}{\partial y} + \frac{\partial \gamma_{xy}}{\partial z}\right) \\
2\frac{\partial^2 \epsilon_z}{\partial x \partial y} &= \frac{\partial}{\partial z}\left(\frac{\partial \gamma_{yz}}{\partial x} + \frac{\partial \gamma_{xz}}{\partial y} - \frac{\partial \gamma_{xy}}{\partial z}\right)
\end{aligned} \tag{2.2}$$

We observe that if the strain components satisfy the compatibility equations (10.32), the line integral of Eq. (10.28b) will be independent of the path. Furthermore, the corresponding line integral between *any* two points in a region where Eqs. (10.32) are satisfied must be independent of the path, so that the displacement components in the region must be single-valued, continuous functions if the region is simply connected. In multiply connected regions, the compatibility relations are necessary, but not sufficient, to guarantee single-valued, continuous displacements. Multiply connected regions, however, can be made simply connected by assuming the body is severed along one or more lines, as shown in Fig. 10.3. This imposes additional boundary conditions on the problem in that the displacement components must be the same on either side of the cut (or cuts). A simpler solution may sometimes be obtained without assuming the body is cut by simply inspecting the displacements to determine if they are single valued, continuous functions. If they meet these requirements, the solution is correct: if not, it may be possible to revise the displacements so that they are single-valued and continuous. More commonly, however, single-valued displacements are guaranteed through the medium of contour integrals, as was done in deriving Eq. (6.41).

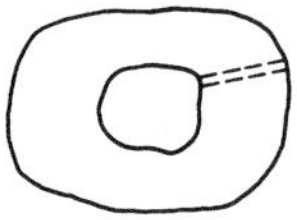

Fig. 10.3 Multiply Connected Region

Following the procedure used in Section 4.4, we shall next derive the compatibility equations in terms of stress by first substituting Eq. (10.25) into Eq. (10.31), which yields

$$\tau_{i\ell,jk} - \tau_{j\ell,ik} + \tau_{jk,i\ell} - \tau_{ik,j\ell} = \frac{\nu}{1+\nu}\{\delta_{i\ell}\Theta_{,jk} - \delta_{j\ell}\Theta_{,ik} + \delta_{jk}\Theta_{,i\ell} - \delta_{ik}\Theta_{,j\ell}\}$$

Multiplying by $\delta_{j\ell}$ (and summing), we get

$$\tau_{ij,jk} - \Theta_{,ik} + \tau_{jk,ij} - \tau_{ik,jj} = \frac{\nu}{1+\nu}\{\delta_{ij}\Theta_{,jk} - \delta_{jj}\Theta_{,ik} + \delta_{jk}\Theta_{,ij} - \delta_{ik}\Theta_{,jj}\} \quad (10.33)$$

where we have used the relation

$$\tau_{jj} = \Theta$$

Since

$$\tau_{ik,jj} = \nabla^2\tau_{ik}$$

and

$$\Theta_{,jj} = \nabla^2\Theta$$

and

$$\delta_{jj} = 3$$

we have

$$\tau_{ij,jk} + \tau_{jk,ij} - \Theta_{,ik} - \nabla^2\tau_{ik} = \frac{-\nu}{1+\nu}\{\delta_{ik}\nabla^2\Theta + \Theta_{,ik}\} \quad (10.34)$$

We notice that Eq. (10.34) represents nine equations since i and k are free indices; however, interchanging i and k yields the same expressions, so that Eq. (10.34) represents six independent equations. Now it was known in advance that six equations would result, since we started with six independent equations, (10.31), and substituted Eqs. (10.25), which are linear in e_{ij} and τ_{ij}. This means that our operation of applying $\delta_{j\ell}$, although resulting in elimination of 72 equations, did not cause any loss in generality.

We shall now present Eq. (10.34) in a more useful form by utilizing the equations of equilibrium, i.e.,

$$\tau_{ij,j} + F_i = 0$$

so that

$$\tau_{ij,jk} = -F_{i,k}$$

and also

$$\tau_{jk,ij} = \tau_{kj,ji} = -F_{k,i}$$

Substituting these relations into Eq. (10.34), we get

$$\nabla^2 \tau_{ik} + \frac{1}{1+\nu}\Theta_{,ik} - \frac{\nu}{1+\nu}\delta_{ik}\nabla^2\Theta = -(F_{i,k} + F_{k,i}) \quad (10.35)$$

Let us now multiply Eq. (10.34) by δ_{ik} and sum which gives

$$\tau_{ij,ji} + \tau_{ji,ij} - \Theta_{,ii} - \nabla^2 \tau_{ii} = -\frac{\nu}{1+\nu}[3\nabla^2\Theta + \Theta_{,ii}]$$

This can be reduced to

$$2\tau_{ij,ij} - 2\nabla^2\Theta = -\frac{\nu}{1+\nu}[4\nabla^2\Theta]$$

or

$$\tau_{ij,ij} = \left(\frac{1-\nu}{1+\nu}\right)\nabla^2\Theta \quad (10.36)$$

By differentiating the equations of equilibrium, we see that

$$\tau_{ij,ji} = \tau_{ij,ij} = -F_{i,i}$$

so that Eq. (10.36) becomes

$$\nabla^2\Theta = -\left(\frac{1+\nu}{1-\nu}\right)F_{i,i} \quad (10.37)$$

With this identity, Eq. (10.35) takes the form

$$\nabla^2 \tau_{ik} + \frac{1}{1+\nu} \Theta_{,ik} = -\frac{\nu}{1-\nu} \delta_{ik} F_{j,j} - (F_{i,k} + F_{k,i}) \qquad (10.38)$$

which represents the six compatibility equations in terms of stress. The system of equations consisting of Eq. (10.38) and the equations of equilibrium are the most convenient set to use in the solution of first boundary-value problems. The boundary conditions are specified by Eq. (9.28). Note that Eqs. (10.38) are the Beltrami-Michell compatibility equations, first derived in Chapter 4 as Eqs. (4.25).

10.5. Governing Equations in Terms of Displacement

We shall now develop the equations of equilibrium in terms of displacement. By substituting the relations

$$\tau_{ij} = \lambda \epsilon \delta_{ij} + 2G e_{ij}$$

where

$$e_{ij} = \tfrac{1}{2}(u_{i,j} + u_{j,i})$$

and

$$\epsilon = u_{k,k}$$

into the equations of equilibrium, we obtain

$$\lambda\, \delta_{ij} u_{k,kj} + G(u_{i,jj} + u_{j,ij}) + F_i = 0$$

or

$$\lambda u_{k,ki} + G\nabla^2 u_i + G \frac{\partial}{\partial x_i} (u_{j,j}) + F_i = 0$$

which reduces to

$$G\nabla^2 u_i + (\lambda + G)\epsilon_{,i} + F_i = 0 \qquad (10.39)$$

Equation (10.39) represents the equations of equilibrium in terms of displacement, or the Navier equations. Of course, these are most effective in the solution of second boundary-value problems, where the boundary conditions are specified by

$$u_i(x^0) \equiv u_i(x_1{}^0, x_2{}^0, x_3{}^0) = u_i{}^b \qquad (10.40)$$

where $u_i(x^0)$ are the displacement components evaluated on the boundary and $u_i{}^b$ are the prescribed displacements on the boundary. A given second boundary-value problem is solved if Eqs. (10.39) and (10.40) are satisfied.

10.6. Strain Energy

In Cartesian tensor notation, the strain energy density function U_0 may be expressed as follows. In terms of stress and strain, corresponding to Eq. (7.6), we have

$$U_0 = \tfrac{1}{2}\tau_{ij} e_{ij} \tag{10.41}$$

In terms of stress components only, we have

$$U_0 = -\frac{\nu}{2E}\tau_{kk}^2 + \frac{1}{4G}\tau_{ij}\tau_{ij} \tag{10.42}$$

which is equivalent to Eq. (7.7). In terms of strain components only, we get

$$U_0 = \frac{\lambda}{2} e_{kk}^2 + G e_{ij} e_{ij} \tag{10.43}$$

which is the same as Eq. (7.8).

10.7. Governing Equations of Elasticity

The governing equations in elasticity are summarized in the following three tables—Table 10.1 for three-dimensional problems, Table 10.2 for plane strain, and Table 10.3 for plane stress problems. These three tables correspond to Tables 4.2, 4.1, and 5.1, which were written in engineering notation. The equations in each box represent the field equations governing the variables indicated. The box on the top of each table contains the equations for stress, strain, and displacement. Boxes in the left-hand columns represent the stress formulation suitable for the solution of the first boundary-value problems; the right-hand columns give the displacement formulation suitable for the solution of the second boundary-value problems in elasticity. The equations for plane stress problems in Table 10.3 are approximate in nature. These are sometimes called the *generalized plane stress equations.*

A summary of the development of equations similar to that of Table 10.1, but in dyadic notation, may be found in Chapter 11, Table 11.1. Further development of the equations in elasticity in terms of stress functions and displacement functions will be given in Chapter 13.

In addition to the field equations which must be satisfied at every interior point of the body, on the surface of the body the boundary conditions must also be satisfied. The boundary conditions can be prescribed either in terms of surface stress T_i^μ, or in terms of surface displacements u_i^b. On those parts of the boundary surface where T_i^μ are specified, the stresses on the surface must fulfill the conditions

$$\tau_{ji}\mu_j = T_i^\mu$$

Table 10.1

THREE-DIMENSIONAL PROBLEMS (CORRESPONDS TO TABLE 4.2)

$(i, j, k, \ell = 1, 2, 3)$

$$\tau_{ij,j} + F_i = 0$$
$$e_{ij} = \tfrac{1}{2}(u_{i,j} + u_{j,i})$$
$$\tau_{ij} = \lambda \delta_{ij} e_{kk} + 2G e_{ij}$$
$$e_{ij}, \quad \tau_{ij}, \quad u_i$$

$$\tau_{ij,j} + F_i = 0$$
$$\tau_{ij} = \lambda \delta_{ij} u_{k,k} + G(u_{i,j} + u_{j,i})$$
$$\tau_{ij}, \quad u_i$$

$$\tau_{ij,j} + F_i = 0$$
$$e_{i\ell,jj} - e_{j\ell,ij} + e_{jj,i\ell} - e_{ij,j\ell} = 0$$
$$e_{ij} = \frac{1+\nu}{E}\tau_{ij} - \frac{\nu}{E}\delta_{ij}\tau_{kk}$$
$$\tau_{ij}, e_{ij}$$

$$G\nabla^2 u_i + (\lambda + G)u_{j,ji} + F_i = 0$$
$$u_i$$

$$\tau_{ij,j} + F_i = 0$$
$$\nabla^2 \tau_{ik} + \frac{1}{1+\nu}\tau_{jj,ik} = -\frac{\nu}{1-\nu}\delta_{ik}F_{j,j} - (F_{i,k} + F_{k,i})$$
$$\tau_{ij}$$

$$\lambda = \frac{E\nu}{(1+\nu)(1-2\nu)}$$
$$G = \frac{E}{2(1+\nu)}$$

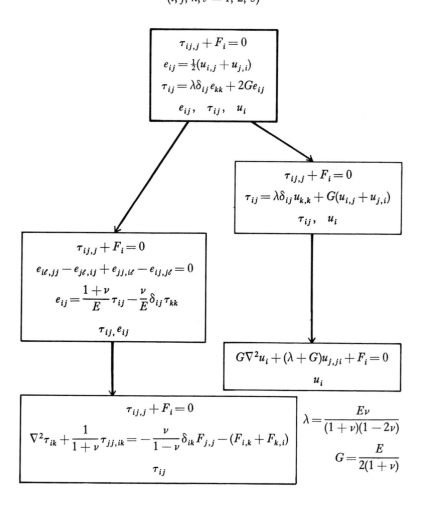

Table 10.2

PLANE STRAIN PROBLEMS (CORRESPONDS TO TABLE 4.1)

$(\alpha, \beta, \gamma = 1, 2)$

The equations for plane strain may be transformed into those for plane stress by replacing E and ν in the following equations by E_2 and ν_2, respectively,

$$E_2 = \frac{E(1+2\nu)}{(1+\nu)^2} \qquad \nu_2 = \frac{\nu}{1+\nu}$$

$$u_1 = u_1(x_1, x_2) \qquad u_2 = u_2(x_1, x_2) \qquad u_3 = 0$$

$$\tau_{\alpha\beta,\beta} + F_\alpha = 0$$
$$e_{\alpha\beta} = \tfrac{1}{2}(u_{\alpha,\beta} + u_{\beta,\alpha})$$
$$\tau_{\alpha\beta} = \lambda\delta_{\alpha\beta} e_{\gamma\gamma} + 2Ge_{\alpha\beta}$$
$$e_{\alpha\beta}, \quad \tau_{\alpha\beta}, \quad u_\alpha$$

↓ (left branch)

$$\tau_{\alpha\beta,\beta} + F_\alpha = 0$$
$$e_{11,22} + e_{22,11} = 2e_{12,12}$$
$$e_{\alpha\beta} = \frac{1}{2G}(\tau_{\alpha\beta} - \delta_{\alpha\beta}\nu\tau_{\gamma\gamma})$$
$$\tau_{\alpha\beta}, e_{\alpha\beta}$$

↓

$$\tau_{\alpha\beta,\beta} + F_\alpha = 0$$
$$\nabla^2\tau_{\alpha\alpha} = \frac{1}{\nu - 1}F_{\alpha,\alpha}$$
$$\tau_{\alpha\beta}$$

↓

$$\nabla^4\phi = \frac{1-2\nu}{1-\nu}\nabla^2 V$$

where

$$V_{,\alpha} = F_\alpha \qquad \phi_{,11} = \tau_{22} + V$$
$$\phi_{,12} = -\tau_{12} \qquad \phi_{,22} = \tau_{11} + V$$

↓ (right branch)

$$\tau_{\alpha\beta,\beta} + F_\alpha = 0$$
$$\tau_{\alpha\beta} = \lambda\delta_{\alpha\beta}u_{\gamma,\gamma} + G(u_{\alpha,\beta} + u_{\beta,\alpha})$$
$$\tau_{\alpha\beta}, u_\alpha$$

↓

$$G\nabla^2 u_\alpha + (\lambda + G)u_{\beta,\beta\alpha} + F_\alpha = 0$$
$$u_\alpha$$

$$\tau_{33} = \nu\tau_{\alpha\alpha}$$

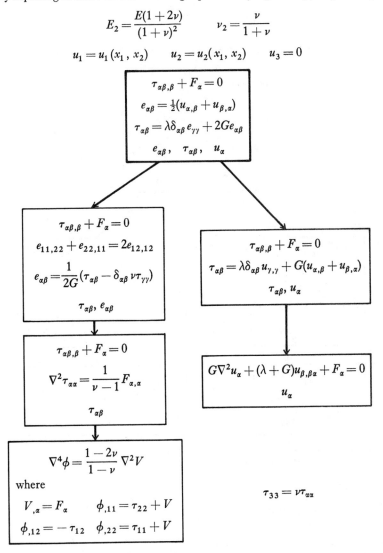

Table 10.3
PLANE STRESS PROBLEMS (CORRESPONDS TO TABLE 5.1)
$(\alpha, \beta, \gamma = 1, 2)$

The equations for plane stress may be transformed into those for plane strain by replacing E and ν in the following equations by E_1 and ν_1, respectively.

$$E_1 = \frac{E}{1-\nu^2} \qquad \nu_1 = \frac{\nu}{1-\nu}$$

$$\tau_{11} = \tau_{11}(x_1, x_2) \qquad \tau_{22} = \tau_{22}(x_1, x_2) \qquad \tau_{12} = \tau_{12}(x_1, x_2)$$
$$\tau_{33} = \tau_{13} = \tau_{23} = 0$$

$$\tau_{\alpha\beta,\beta} + F_\alpha = 0$$
$$e_{\alpha\beta} = \tfrac{1}{2}(u_{\alpha,\beta} + u_{\beta,\alpha})$$
$$e_{\alpha\beta} = \frac{1+\nu}{E}\tau_{\alpha\beta} - \frac{\nu}{E}\delta_{\alpha\beta}\tau_{\gamma\gamma}$$
$\tau_{\alpha\beta}$, $e_{\alpha\beta}$, u_α

↓ (left branch)

$$\tau_{\alpha\beta,\beta} + F_\alpha = 0$$
$$e_{11,22} + e_{22,11} = 2e_{12,12}$$
$$e_{\alpha\beta} = \frac{1+\nu}{E}\tau_{\alpha\beta} - \frac{\nu}{E}\delta_{\alpha\beta}\tau_{\gamma\gamma}$$
$\tau_{\alpha\beta}$, $e_{\alpha\beta}$

↓

$$\tau_{\alpha\beta,\beta} + F_\alpha = 0$$
$$\nabla^2\tau_{\alpha\alpha} = -(\nu+1)F_{\alpha,\alpha}$$
$\tau_{\alpha\beta}$

↓

$$\nabla^4\phi = (1-\nu)\nabla^2 V$$
where
$$V_{,\alpha} = F_\alpha \qquad \phi_{,11} = \tau_{22} + V$$
$$\phi_{,22} = \tau_{11} + V \quad \phi_{,12} = -\tau_{12}$$

↓ (right branch)

$$\tau_{\alpha\beta,\beta} + F_\alpha = 0$$
$$\tau_{\alpha\beta} = \frac{\nu E}{1-\nu^2}\delta_{\alpha\beta}u_{\gamma,\gamma} + G(u_{\alpha,\beta} + u_{\beta,\alpha})$$
$\tau_{\alpha\beta}$, u_α

↓

$$G\nabla^2 u_\alpha + \frac{E}{2(1-\nu)}u_{\beta,\beta\alpha} + F_\alpha = 0$$
u_α

$$e_{33} = -\frac{\nu}{E}\tau_{\alpha\alpha}$$

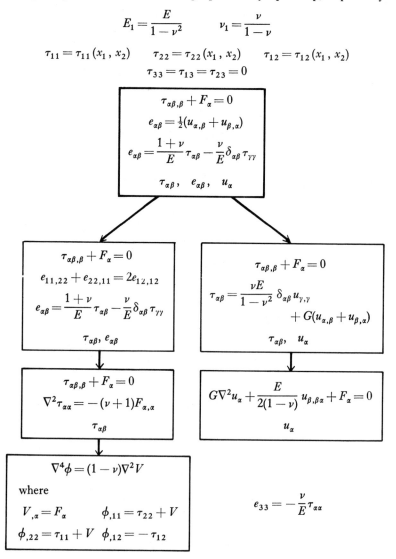

where the μ_j are the direction cosines of the outward normal of the surface. On those parts of the surface where u_i^b are specified, we have the condition

$$u_i(x^0) = u_i^b$$

PROBLEMS

10-1 Extract the six compatibility equations from Eq. (10.31).

10-2 Show that if the body force field **F** is conservative, so that

$$\mathbf{F} = \nabla\phi$$

or

$$F_i = \phi_{,i}$$

Eq. (10.38) takes the form

$$\nabla^2\tau_{ik} + \frac{1}{1+\nu}\Theta_{,ik} = -\frac{\nu}{1-\nu}\delta_{ik}\nabla^2\phi - 2\phi_{,ik}$$

10-3 In the case where the body force field F_i is constant, show that Θ and ϵ are harmonic functions, i.e.,

$$\nabla^2\Theta = 0$$

and

$$\nabla^2\epsilon = 0$$

Furthermore, show that τ_{ij} and e_{ij} are biharmonic functions, i.e.,

$$\nabla^4\tau_{ij} = \nabla^2\nabla^2\tau_{ij} = 0$$

and

$$\nabla^4 e_{ij} = 0$$

Hint: Show first that in this case Eq. (10.38) reduces to

$$\nabla^2\tau_{ik} + \frac{1}{1+\nu}\Theta_{,ik} = 0$$

10-4 Show that the governing equations in elasticity reduce to

$$\nabla^4 V_i + \frac{F_i}{1-\nu} = 0$$

if

$$2Gu_i = 2(1-\nu)\nabla^2 V_i - V_{j,ij}$$

The vector function V_i is called the Galerkin vector, which will be discussed further in Chapter 13.

10-5 Establish the relation (3.23), i.e.,

$$e_{ii} = \frac{1}{K}\frac{\tau_{ii}}{3} = \frac{1}{K}\frac{\Theta}{3}$$

using tensor notation (K is the bulk modulus of elasticity).

10-6 If a material is not isotropic but its elastic properties are symmetrical with respect to one plane, using the method of Section 10.3, show that out of the 36 elastic constants in Eq. (10.6) only 20 are independent and different from zero. If there is symmetry with respect to three mutually perpendicular planes (orthotropic), show that only 12 independent coefficients exist.

10-7 Navier's equation

$$u_{i,jj} + \frac{1}{1-2\nu} u_{j,ij} + \frac{1}{G} F_i = 0$$

becomes indeterminate when the material is incompressible ($\nu = \frac{1}{2}$). Show that in this case

$$\epsilon = e_{ii} = u_{j,j} = 0 \quad \text{and} \quad \nabla^2 u_i + \frac{1}{G}\left(\frac{1}{3}\Theta,_i + F_i\right) = 0$$

Hint: Consider the Hooke's law equations and the equations of equilibrium in terms of stress.

11

Vector and Dyadic Notation in Elasticity

11.1. Introduction

A scalar is a quantity with magnitude only. A vector is a quantity with magnitude and direction, and which obeys the transformation Eq. (8.12). For example, the density and temperature of a body are scalars, whereas force and velocity are vectors. A vector is completely determined if its three components in the x, y, and z directions are known. The algebraic and differential operations involving vectors can be presented in scalar form in terms of the components of the vectors. Thus, if A_x, A_y, A_z are the components of a force **A**, and B_x, B_y, B_z those of **B**, then force **C**, the resultant of **A** and **B**, has the components

$$C_x = A_x + B_x \qquad C_y = A_y + B_y \qquad C_z = A_z + B_z \tag{11.1}$$

The concept of work in elementary mechanics is presented as the product of the magnitudes of the force and displacement vectors and the cosine of the angle between them. The work W done by a constant conservative force **F** moving through a displacement **S** can be expressed as

$$W = F_x S_x + F_y S_y + F_z S_z \tag{11.2}$$

The vector operations of addition and scalar multiplication are thus accomplished by Eqs. (11.1) and (11.2), which are scalar equations. For more complicated operations between vectors, the use of scalar equations is still possible, but the expressions can become very lengthy.

One method of condensing the expressions is through the use of indicial notation as discussed in Chapter 8. This method is simply a shorthand for condensing equations written in component form. Another method of accomplishing this is to use vector notation (or vector symbols). Thus, Eqs. (11.1) and (11.2) are represented by

$$\mathbf{C} = \mathbf{A} + \mathbf{B} \tag{11.3}$$

$$W = \mathbf{F} \cdot \mathbf{S} \tag{11.4}$$

Each vector notation and operation, however, must be defined explicitly, since the rules of scalar algebra, in general, are not applicable for vector algebra. For instance, the commutative law in addition is true for vectors, but the commutative property does not hold for the cross product (vector product),

$$A + B = B + A, \qquad \mathbf{A} + \mathbf{B} = \mathbf{B} + \mathbf{A}$$

$$AB = BA, \qquad \text{but} \qquad \mathbf{A} \times \mathbf{B} \neq \mathbf{B} \times \mathbf{A} \qquad (\mathbf{A} \times \mathbf{B} = -\mathbf{B} \times \mathbf{A})$$

The vector notations are symbolic in nature; vectors and scalars are represented by different symbols, and each mathematical operation is represented by a symbol. These symbols and their mathematical operations are independent of coordinate system, i.e., in using this notation, we need not construct any coordinate system. In order to *evaluate* a specific vector, however, it is most convenient to determine its components referred to a coordinate system.

There are also quantities such as stress and strain which are neither vectors nor scalars, but are tensor quantities. When dealing with tensors, all quantities and their mathematical operations can again be represented in component form, as in the previous two chapters. In the symbolic method of representation, dyadic notation is introduced for second-order tensors, just as vector notation is used for vectors. Dyadic notation, as vector notation, has the advantage of being independent of coordinate systems.

In the symbolic representation of tensors many symbols must be defined for the numerous operations, as a result its utility diminishes as the order of the tensors increases. However, the use of the symbolic method in the study of linear elasticity does not present any major difficulties. As we shall see in this chapter, by introducing a few symbols in addition to conventional vector symbols, we can treat the majority of the equations of linear elasticity. Since equations expressed in the symbolic method of notation are independent of any coordinate system, if the governing equations are expressed in this notational system, they are valid for any coordinate system, e.g., polar coordinates, elliptic coordinates, etc. It is only necessary to express the operations, such as divergence, curl, and gradient, in the particular coordinate system utilized. In this chapter, we shall discuss the equations of elasticity in vector and dyadic notations. We shall assume the students have already had some background in vector analysis. Whenever a vector or dyadic is expressed in component form, the Cartesian coordinates are used in this chapter. A discussion of curvilinear coordinates will be given in Chapter 12.

11.2. Review of Basic Notations and Relations in Vector Analysis

A vector is represented by a boldfaced letter in this text; for example, the symbol **A** stands for the vector with components A_x, A_y, A_z. The magnitude (or length) of vector **A** is denoted by $|\mathbf{A}|$, or simply A. In handwritten work, a vector is commonly represented by a letter with a single bar above it.

The symbols **i**, **j**, and **k** represent unit vectors in the positive x, y, and z directions, respectively. Thus we have

$$|\mathbf{i}| = |\mathbf{j}| = |\mathbf{k}| = 1 \tag{11.5}$$

and by applying the law of vector addition, we may express **A** in the convenient form

$$\mathbf{A} = A_x\mathbf{i} + A_y\mathbf{j} + A_z\mathbf{k} \tag{11.6}$$

where A_x, A_y, and A_z are the components (scalars) or projections of **A** in the x, y, and z directions.

The scalar (dot) product of vectors **A** and **B** is indicated by $\mathbf{A}\cdot\mathbf{B}$. By definition, this scalar quantity is

$$\mathbf{A}\cdot\mathbf{B} = |\mathbf{A}|\,|\mathbf{B}| \cos(\mathbf{A}, \mathbf{B}) \tag{11.7}$$

This quantity is equivalent to the product of the magnitude of **A** and the component of **B** in the direction of **A** or vice versa. Also

$$\mathbf{A}\cdot\mathbf{B} = (A_x\mathbf{i} + A_y\mathbf{j} + A_z\mathbf{k})\cdot(B_x\mathbf{i} + B_y\mathbf{j} + B_z\mathbf{k})$$

and since

$$\mathbf{i}\cdot\mathbf{i} = \mathbf{j}\cdot\mathbf{j} = \mathbf{k}\cdot\mathbf{k} = 1 \tag{11.8}$$

and

$$\mathbf{i}\cdot\mathbf{j} = \mathbf{j}\cdot\mathbf{k} = \mathbf{i}\cdot\mathbf{k} = 0 \tag{11.9}$$

we have

$$\mathbf{A}\cdot\mathbf{B} = A_xB_x + A_yB_y + A_zB_z \tag{11.10}$$

Also

$$\mathbf{A}\cdot\mathbf{B} = \mathbf{B}\cdot\mathbf{A} \tag{11.11}$$

The vector (cross) product of vectors **A** and **B** is indicated by **A** × **B**. This vector quantity is equivalent to

$$\mathbf{A} \times \mathbf{B} = |\mathbf{A}|\,|\mathbf{B}| \sin (\mathbf{A}, \mathbf{B})\mathbf{n} \tag{11.12}$$

where **n** is a unit vector normal to the plane containing **A** and **B**. The direction of **n** is defined as shown in Fig. 11.1; if we turn a right-hand screw from **A** to **B**, the vector **n**, or **A** × **B**, is in the direction of the advancement of the screw (right-hand screw rule).

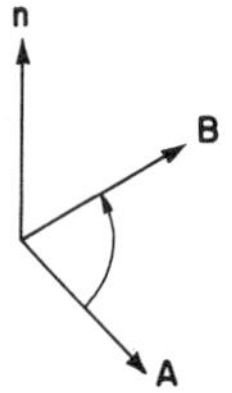

Fig. 11.1 Vector Product

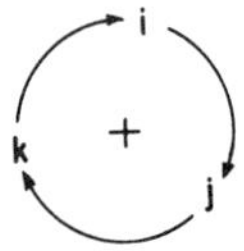

Fig. 11.2 Vector Products Between Unit Vectors

For the cross products between unit vectors **i**, **j**, and **k**, we have

$$\mathbf{i} \times \mathbf{i} = \mathbf{j} \times \mathbf{j} = \mathbf{k} \times \mathbf{k} = 0 \tag{11.13}$$

$$\begin{array}{lll} \mathbf{i} \times \mathbf{j} = \mathbf{k} & \mathbf{j} \times \mathbf{k} = \mathbf{i} & \mathbf{k} \times \mathbf{i} = \mathbf{j} \\ \mathbf{j} \times \mathbf{i} = -\mathbf{k} & \mathbf{k} \times \mathbf{j} = -\mathbf{i} & \mathbf{i} \times \mathbf{k} = -\mathbf{j} \end{array} \tag{11.14}$$

The signs of the cross products in Eqs. (11.14) follow the cyclic exchange rule as indicated in Fig. 11.2. The cross product may also be represented by

$$\mathbf{A} \times \mathbf{B} = (A_x\mathbf{i} + A_y\mathbf{j} + A_z\mathbf{k}) \times (B_x\mathbf{i} + B_y\mathbf{j} + B_z\mathbf{k})$$

or

$$\mathbf{A} \times \mathbf{B} = \begin{vmatrix} \mathbf{i} & \mathbf{j} & \mathbf{k} \\ A_x & A_y & A_z \\ B_x & B_y & B_z \end{vmatrix} \tag{11.15}$$

Also

$$\mathbf{A} \times \mathbf{B} = -\mathbf{B} \times \mathbf{A} \tag{11.16}$$

The vector operator ∇ (del or nabla), or gradient operator, in Cartesian coordinates is defined by

$$\nabla = \mathbf{i}\frac{\partial}{\partial x} + \mathbf{j}\frac{\partial}{\partial y} + \mathbf{k}\frac{\partial}{\partial z} \tag{11.17}$$

When applied to a scalar function, it gives

$$\text{grad } A = \nabla A = \mathbf{i}\frac{\partial A}{\partial x} + \mathbf{j}\frac{\partial A}{\partial y} + \mathbf{k}\frac{\partial A}{\partial z}$$

where A is a scalar function, $A = A(x, y, z)$. The gradient gives the maximum rate of change of A at any point. In addition, the gradient of A is a vector which is normal to the surface $A(x, y, z) = \text{const.}$ at each point on the surface.

The divergence of a vector function is a scalar and is defined by

$$\text{div } \mathbf{A} = \nabla \cdot \mathbf{A} = \frac{\partial A_x}{\partial x} + \frac{\partial A_y}{\partial y} + \frac{\partial A_z}{\partial z} \tag{11.18}$$

The curl of a vector function is expressed by

$$\text{curl } \mathbf{A} = \nabla \times \mathbf{A} = \begin{vmatrix} \mathbf{i} & \mathbf{j} & \mathbf{k} \\ \frac{\partial}{\partial x} & \frac{\partial}{\partial y} & \frac{\partial}{\partial z} \\ A_x & A_y & A_z \end{vmatrix} \tag{11.19}$$

Another useful operator is

$$\text{div } \nabla = \nabla \cdot \nabla = \frac{\partial^2}{\partial x^2} + \frac{\partial^2}{\partial y^2} + \frac{\partial^2}{\partial z^2} = \nabla^2 \tag{11.20}$$

which is the Laplace operator.

11.3. Dyadic Notation

In the symbolic representation, vector notation is not sufficient to treat second-order or higher-order tensor quantities, such as stress and strain. In this section we shall introduce a symbol, called a dyadic symbol, to represent a second-order tensor quantity. To treat third-order tensors, triadics must be introduced; and for fourth-order tensors, tetradics are needed. In this text, we shall limit our discussion in the symbolic approach to dyadics. It will be shown that all of the governing elasticity equations in Table 4.2 (or Table 10.1) can be represented by vector and dyadic notations. Generalized Hooke's law, Eq. (10.6), which contains a fourth-order tensor quantity, cannot be written in dyadic notation.

Dyadic symbols represent second-order tensor quantities, just as vector symbols are for first-order tensor quantities. In this text, a dyadic symbol is indicated by a letter in capital script. (In handwritten work one may use two bars over a letter to designate a dyadic symbol.)

Thus, the stress tensor may be represented by the stress dyadic symbol $\mathscr{T}$.

At this point, it is appropriate to review some of our discussion relating to vector quantities. A quantity with magnitude and direction and which adds according to the parallelogram law, or, a quantity whose components transform according to Eq. (8.12), is defined as a vector quantity, or simply, a vector. A vector quantity is also called a first-order tensor quantity, or simply, a first-order tensor. A vector quantity (or first-order tensor quantity) is represented in tensor notation by its three components, e.g., u_i. This same quantity, in symbolic notation, is represented by a vector symbol $\mathbf{u}$. Certain differential operators may also be called vector operators (in symbolic notation), or first-order tensor operators (in tensor notation). Equations involving vector quantities and vector operators are called vector equations, or equations in vector notation. For brevity, the word "vector" has been used to designate, on different occasions, vector quantity, vector symbol, and vector notation.

In a similar way, the word "dyadic" is used to designate a dyadic symbol, a dyadic quantity (i.e., a second-order tensor quantity), or dyadic notation. Dyadic notations, which are symbolic representations of various second-order tensor quantities and operations, will be defined subsequently in this chapter.

Another way to introduce the concept of the dyadic symbol is to consider the vector equation

$$\mathbf{D} = (\mathbf{A} \cdot \mathbf{B})\mathbf{C} \tag{11.21}$$

The symbol $\mathbf{D}$ represents a vector quantity of magnitude $|\mathbf{A}|\,|\mathbf{B}| \cos (\mathbf{A}, \mathbf{B})\,|\mathbf{C}|$ oriented parallel to $\mathbf{C}$. Now in strict vector notation, a product such as $\mathbf{BC}$, with no dot or cross between the two vectors, is not defined. The dot product in Eq. (11.21) must first be performed to form a scalar, otherwise it is meaningless. An expression in strict vector notation such as

$$\mathbf{D} = \mathbf{A} \cdot \mathbf{BC} \tag{11.22}$$

implies Eq. (11.21). In dyadic notation, a product such as $\mathbf{BC}$ is permitted, and the dyadic symbol $\mathscr{S}$ is used to represent the dyadic (or second-order tensor) quantity $\mathbf{BC}$, or

$$\mathscr{S} = \mathbf{BC} \tag{11.23}$$

Thus Eqs. (11.22) and (11.21) are equivalent to

$$\mathbf{D} = \mathbf{A} \cdot \mathscr{S} \tag{11.24}$$

in dyadic notation. In Cartesian coordinates, if we put

$$\begin{aligned} \mathbf{B} &= B_x \mathbf{i} + B_y \mathbf{j} + B_z \mathbf{k} \\ \mathbf{C} &= C_x \mathbf{i} + C_y \mathbf{j} + C_z \mathbf{k} \end{aligned} \tag{11.25}$$

then

$$\begin{aligned} \mathscr{S} = \mathbf{BC} = \mathbf{ii}B_x C_x &+ \mathbf{ij}B_x C_y + \mathbf{ik}B_x C_z \\ &+ \mathbf{ji}B_y C_x + \mathbf{jj}B_y C_y + \mathbf{jk} B_y C_z \\ &+ \mathbf{ki}B_z C_x + \mathbf{kj}B_z C_y + \mathbf{kk}B_z C_z \end{aligned} \tag{11.26}$$

which is simply the symbolic (dyadic) representation of the tensor equation

$$S_{ij} = B_i C_j$$

Notice that there are no dots or crosses between the unit vectors in Eq. (11.26). It must also be emphasized that the order of the unit vectors in each term must be preserved, and that

$$\mathbf{ij} \neq \mathbf{ji}$$

However, the order of the scalar components is immaterial, e.g.,

$$B_x C_y \mathbf{ij} = \mathbf{ij} B_x C_y = \mathbf{i} B_x C_y \mathbf{j}$$

In general, any dyadic $\mathscr{S}$ (the dyadic symbol for a second-order tensor) may be written as

$$\begin{aligned} \mathscr{S} = \mathbf{ii}S_{xx} &+ \mathbf{ij}S_{xy} + \mathbf{ik}S_{xz} \\ &+ \mathbf{ji}S_{yx} + \mathbf{jj}S_{yy} + \mathbf{jk}S_{yz} \\ &+ \mathbf{ki}S_{zx} + \mathbf{kj}S_{zy} + \mathbf{kk}S_{zz} \end{aligned} \tag{11.27}$$

For each dyadic $\mathscr{S}$ which has components as in Eq. (11.27), we define a conjugate dyadic $\mathscr{S}_c$ such that

$$\begin{aligned} \mathscr{S}_c = \mathbf{ii}S_{xx} &+ \mathbf{ji}S_{xy} + \mathbf{ki}S_{xz} \\ &+ \mathbf{ij}S_{yx} + \mathbf{jj}S_{yy} + \mathbf{kj}S_{yz} \\ &+ \mathbf{ik}S_{zx} + \mathbf{jk}S_{zy} + \mathbf{kk}S_{zz} \end{aligned} \tag{11.28}$$

If a dyadic is equal to its conjugate, or

$$\mathscr{S} = \mathscr{S}_c \tag{11.29}$$

it is called *self-conjugate*, or *symmetric*. If $\mathscr{S}$ is self-conjugate, then

$$S_{xy} = S_{yx} \qquad S_{xz} = S_{zx} \qquad S_{yz} = S_{zy} \tag{11.30}$$

Therefore, a self-conjugate dyadic is the symbolic representation of a symmetric tensor of the second order. (See Section 8.3.) If $\mathscr{S}$ is such that

$$\mathscr{S} = -\mathscr{S}_c \tag{11.31a}$$

we must have

$$S_{xx} = S_{yy} = S_{zz} = 0 \tag{11.31b}$$

and

$$S_{xy} = -S_{yx} \qquad S_{xz} = -S_{zx} \qquad S_{yz} = -S_{zy} \tag{11.31c}$$

In this case, the dyadic is said to be *antisymmetric*.

Dyadics may also be expressed in terms of three vectors, i.e.,

$$\mathscr{S} = \mathbf{i}\mathbf{S}_x + \mathbf{j}\mathbf{S}_y + \mathbf{k}\mathbf{S}_z \tag{11.32}$$

where

$$\begin{aligned} \mathbf{S}_x &= S_{xx}\,\mathbf{i} + S_{xy}\mathbf{j} + S_{xz}\,\mathbf{k} \\ \mathbf{S}_y &= S_{yx}\,\mathbf{i} + S_{yy}\mathbf{j} + S_{yz}\,\mathbf{k} \\ \mathbf{S}_z &= S_{zx}\,\mathbf{i} + S_{zy}\mathbf{j} + S_{zz}\,\mathbf{k} \end{aligned} \tag{11.33}$$

while the conjugate of $\mathscr{S}$ takes the form

$$\mathscr{S}_c = \mathbf{S}_x\mathbf{i} + \mathbf{S}_y\mathbf{j} + \mathbf{S}_z\mathbf{k} \tag{11.34}$$

We shall now define a few operations involving dyadics. Most of these operations are logical extensions of corresponding vector operations. The general rule is that vector products (dot and cross) between a vector and a dyadic are always made between the vector and the adjacent vector in the dyadic.

$$\mathbf{A}\cdot\mathscr{S} = A_x\mathbf{S}_x + A_y\mathbf{S}_y + A_z\mathbf{S}_z \tag{11.35}$$

$$\mathbf{A}\cdot\mathscr{S}_c = (\mathbf{A}\cdot\mathbf{S}_x)\mathbf{i} + (\mathbf{A}\cdot\mathbf{S}_y)\mathbf{j} + (\mathbf{A}\cdot\mathbf{S}_z)\mathbf{k} = \mathscr{S}\cdot\mathbf{A} \tag{11.36}$$

$$\mathbf{A}\times\mathscr{S} = \begin{vmatrix} \mathbf{i} & \mathbf{j} & \mathbf{k} \\ A_x & A_y & A_z \\ \mathbf{S}_x & \mathbf{S}_y & \mathbf{S}_z \end{vmatrix} \tag{11.37}$$

$$\begin{aligned} \operatorname{div}\mathscr{S} = \nabla\cdot\mathscr{S} &= \left(\mathbf{i}\frac{\partial}{\partial x} + \mathbf{j}\frac{\partial}{\partial y} + \mathbf{k}\frac{\partial}{\partial z}\right)\cdot(\mathbf{i}\mathbf{S}_x + \mathbf{j}\mathbf{S}_y + \mathbf{k}\mathbf{S}_z) \\ &= \frac{\partial}{\partial x}\mathbf{S}_x + \frac{\partial}{\partial y}\mathbf{S}_y + \frac{\partial}{\partial z}\mathbf{S}_z \end{aligned} \tag{11.38}$$

$$\operatorname{div} \mathscr{S}_c = \nabla \cdot \mathscr{S}_c = (\nabla \cdot \mathbf{S}_x)\mathbf{i} + (\nabla \cdot \mathbf{S}_y)\mathbf{j} + (\nabla \cdot \mathbf{S}_z)\mathbf{k} \tag{11.39}$$

$$\operatorname{curl} \mathscr{S} = \nabla \times \mathscr{S} = \begin{vmatrix} \mathbf{i} & \mathbf{j} & \mathbf{k} \\ \dfrac{\partial}{\partial x} & \dfrac{\partial}{\partial y} & \dfrac{\partial}{\partial z} \\ \mathbf{S}_x & \mathbf{S}_y & \mathbf{S}_z \end{vmatrix} = \mathbf{i} \times \frac{\partial \mathscr{S}}{\partial x} + \mathbf{j} \times \frac{\partial \mathscr{S}}{\partial y} + \mathbf{k} \times \frac{\partial \mathscr{S}}{\partial z} \tag{11.40}$$

In addition, some special dyadic operators will be used; these will be defined when first encountered. As in vector notation, parentheses will be applied when possible misinterpretation of the meaning of a dyadic expression can result.

In Table 11.1, the scalar, vector, and dyadic quantities, as well as some of the operations we shall use, are presented in both Cartesian tensor notation and vector-dyadic notation. The appropriate places in the text where each operator is first mentioned in each notation are also indicated for easy reference.

11.4. Vector Representation of Stress on a Plane

We shall first discuss the application of the symbolic method in the definition of a stress vector. The stress acting on a *given plane* is a vector; however, the law of vector addition cannot be used for stresses which

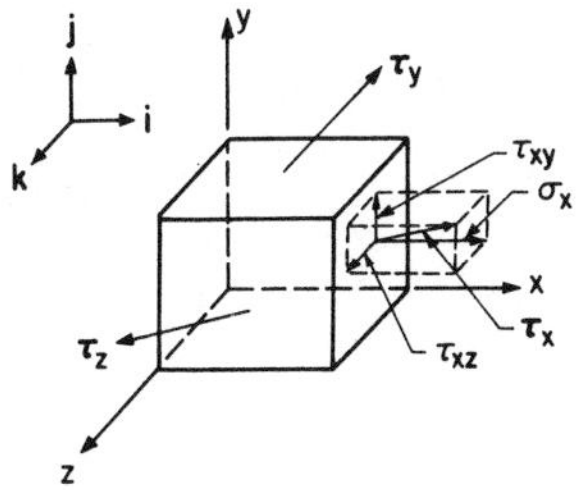

Fig. 11.3 Stress Vectors

act on different planes. Consider, for example, the components of stress σ_x, τ_{xy}, and τ_{xz} acting on the x plane. These can be combined into a single stress vector $\boldsymbol{\tau}_x$ as shown in Fig. 11.3. Thus the stress vector on the x plane is given by

$$\boldsymbol{\tau}_x = \sigma_x \mathbf{i} + \tau_{xy} \mathbf{j} + \tau_{xz} \mathbf{k} \tag{11.41a}$$

Table 11.1

EQUIVALENCE OF QUANTITIES AND OPERATIONS IN TENSOR AND VECTOR-DYADIC NOTATIONS

Quantities and Operations	Cartesian Tensor Notation	Vector and Dyadic Notation	First Defined
Scalar	ϕ	ϕ	Sec. 8.3 Sec. 11.1
First-order tensor, or, vector	u_i	$\mathbf{u}$	Sec. 8.3 Sec. 11.2
Second-order tensor, or, dyadic	S_{ij}	$\mathscr{S}$	Sec. 8.3 Sec. 11.3
Conjugate of a dyadic	S_{ji} for S_{ij}	$\mathscr{S}_c$ for $\mathscr{S}$	Sec. 8.3 Eq. (11.28)
Scalar product	$u_i v_i$	$\mathbf{u} \cdot \mathbf{v}$	Eq. (8.43) Eq. (11.10)
Vector product	$\epsilon_{ijk} u_i v_j$	$\mathbf{u} \times \mathbf{v}$	Eq. (8.60) Eq. (11.12)
Divergence	$u_{i,i}$	$\nabla \cdot \mathbf{u}$	Eq. (8.44) Eq. (11.18)
Gradient	$\phi_{,i}$	$\nabla\phi$	Sec. 8.4 Eq. (11.17)
Curl	$\epsilon_{ijk} u_{k,j}$	$\nabla \times \mathbf{u}$	Sec. 8.7 Eq. (11.19)
Gradient of a vector	$u_{i,j}$	$\nabla\mathbf{u}$	Sec. 8.4 Eq. (11.74)
Conjugate gradient of a vector	$u_{j,i}$ for u_i	$\mathbf{u}\nabla$	— Eq. (11.75)
Divergence of a dyadic	$S_{ij,i}$	$\nabla \cdot \mathscr{S}$	— Eq. (11.38)
Curl of a dyadic	$w_{i\ell} = \epsilon_{ijk} S_{k\ell,j}$	$\nabla \times \mathscr{S}$	— Eq. (11.40)
Conjugate curl	$w_{\ell i} = S_{\ell k,j}\, \epsilon_{ikj}$	$\mathscr{S} \times \nabla$	— Eq. (11.85)
Kronecker delta, or, unity dyadic	δ_{ij}	$\mathscr{L}$	Eq. (8.29) Eq. (11.82)
Laplace	$\phi_{,ii}$	$\nabla^2\phi$	Sec. 10.4 Eq. (11.20)
Biharmonic	$\phi_{,iijj}$	$\nabla^4\phi$ or $\nabla^2\nabla^2\phi$	
Tensor contraction, or, trace of a dyadic	S_{ii}	$\lvert\mathscr{S}\rvert$	Eq. (8.41) Eq. (11.93)

Similarly,

$$\boldsymbol{\tau}_y = \tau_{yx}\mathbf{i} + \sigma_y\mathbf{j} + \tau_{yz}\mathbf{k} \tag{11.41b}$$

$$\boldsymbol{\tau}_z = \tau_{zx}\mathbf{i} + \tau_{zy}\mathbf{j} + \sigma_z\mathbf{k} \tag{11.41c}$$

The relations between the stress vector $\boldsymbol{\tau}_x$ and its components are as follows:

$$\begin{aligned} \sigma_x &= |\boldsymbol{\tau}_x| \cos(\boldsymbol{\tau}_x, \mathbf{i}) = \boldsymbol{\tau}_x \cdot \mathbf{i} \\ \tau_{xy} &= |\boldsymbol{\tau}_x| \cos(\boldsymbol{\tau}_x, \mathbf{j}) = \boldsymbol{\tau}_x \cdot \mathbf{j} \\ \tau_{xz} &= |\boldsymbol{\tau}_x| \cos(\boldsymbol{\tau}_x, \mathbf{k}) = \boldsymbol{\tau}_x \cdot \mathbf{k} \end{aligned} \tag{11.42}$$

and on the y and z planes, we have similar relations which may be obtained by cyclic permutations of Eqs. (11.42).

11.5. Equations of Transformation of Stress

Referring to Fig. 11.4, we have given a uniform state of stress defined by the stress vectors $\boldsymbol{\tau}_x$, $\boldsymbol{\tau}_y$, and $\boldsymbol{\tau}_z$, and we wish to determine the stress vector $\mathbf{p}$ on the oblique plane x'. The vector $\boldsymbol{\mu}_1$, which is the unit

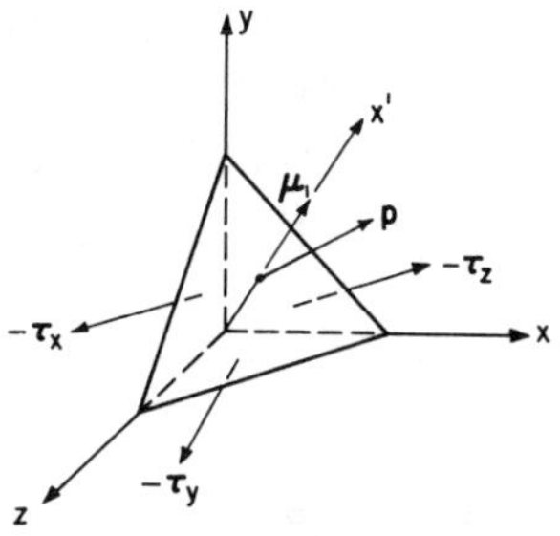

Fig. 11.4 Stress Vector on an Oblique Plane

vector normal to the x' plane, is defined by its direction cosines $\cos(x, x')$, $\cos(y, x')$, and $\cos(z, x')$ with respect to x, y, and z, respectively. We let the areas be represented by A_x, A_y, A_z, and A, where A_x is the area on which $\boldsymbol{\tau}_x$ acts, etc., and A is the area of the oblique plane. Since the tetrahedron is in equilibrium, we have the vector equation

$$A\mathbf{p} = A_x\boldsymbol{\tau}_x + A_y\boldsymbol{\tau}_y + A_z\boldsymbol{\tau}_z$$

which states that $A\mathbf{p}$ is the resultant of $A_x\boldsymbol{\tau}_x$, $A_y\boldsymbol{\tau}_y$, and $A_z\boldsymbol{\tau}_z$. Thus

$$\mathbf{p} = \frac{A_x}{A}\boldsymbol{\tau}_x + \frac{A_y}{A}\boldsymbol{\tau}_y + \frac{A_z}{A}\boldsymbol{\tau}_z \tag{11.43}$$

The area ratios in this equation, from Eqs. (1.23), are the direction cosines of x', or (refer to Table 1.1)

$$\frac{A_x}{A} = \cos(x, x') = a_{11}$$
$$\frac{A_y}{A} = \cos(y, x') = a_{21} \tag{11.44}$$
$$\frac{A_z}{A} = \cos(z, x') = a_{31}$$

Substituting these relations into Eq. (11.43), we obtain

$$\mathbf{p} = a_{11}\boldsymbol{\tau}_x + a_{21}\boldsymbol{\tau}_y + a_{31}\boldsymbol{\tau}_z \tag{11.45}$$

In component form, this becomes

$$p_x = \sigma_x a_{11} + \tau_{yx} a_{21} + \tau_{zx} a_{31}$$
$$p_y = \tau_{xy} a_{11} + \sigma_y a_{21} + \tau_{zy} a_{31} \tag{11.46}$$
$$p_z = \tau_{xz} a_{11} + \tau_{yz} a_{21} + \sigma_z a_{31}$$

where p_x, p_y, and p_z are the components of $\mathbf{p}$ in the x, y, and z directions. We now express the unit vector $\boldsymbol{\mu}_1$ as

$$\boldsymbol{\mu}_1 = a_{11}\mathbf{i} + a_{21}\mathbf{j} + a_{31}\mathbf{k} \tag{11.47}$$

Utilizing Eqs. (11.41) and (11.47), we may write Eqs. (11.46) as follows:

$$p_x = \boldsymbol{\tau}_x \cdot \boldsymbol{\mu}_1 \qquad p_y = \boldsymbol{\tau}_y \cdot \boldsymbol{\mu}_1 \qquad p_z = \boldsymbol{\tau}_z \cdot \boldsymbol{\mu}_1 \tag{11.48}$$

The stress vector $\mathbf{p}$ now becomes

$$\mathbf{p} = p_x\mathbf{i} + p_y\mathbf{j} + p_z\mathbf{k} = (\boldsymbol{\tau}_x \cdot \boldsymbol{\mu}_1)\mathbf{i} + (\boldsymbol{\tau}_y \cdot \boldsymbol{\mu}_1)\mathbf{j} + (\boldsymbol{\tau}_z \cdot \boldsymbol{\mu}_1)\mathbf{k} \tag{11.49}$$

It can be seen from Eqs. (11.48) that the component of stress on x' plane in the x direction is equal to the component of stress on the x plane in x' direction, in accordance with Eq. (9.6).

We now express Eq. (11.49) in the form

$$\mathbf{p} = (\mathbf{i}\boldsymbol{\tau}_x + \mathbf{j}\boldsymbol{\tau}_y + \mathbf{k}\boldsymbol{\tau}_z) \cdot \boldsymbol{\mu}_1 \tag{11.50}$$

or

$$\mathbf{p} = \mathcal{T} \cdot \boldsymbol{\mu}_1 \tag{11.51}$$

where

$$\mathcal{T} = \mathbf{i}\boldsymbol{\tau}_x + \mathbf{j}\boldsymbol{\tau}_y + \mathbf{k}\boldsymbol{\tau}_z \tag{11.52}$$

is the stress dyadic, or the symbolic representation of the stress tensor.

(Refer to Eq. (11.32).) In terms of its stress components, the stress dyadic $\mathscr{T}$ is

$$\begin{aligned}\mathscr{T} &= \mathbf{ii}\sigma_x + \mathbf{ij}\tau_{xy} + \mathbf{ik}\tau_{xz} \\ &\quad + \mathbf{ji}\tau_{yx} + \mathbf{jj}\sigma_y + \mathbf{jk}\tau_{yz} \\ &\quad + \mathbf{ki}\tau_{zx} + \mathbf{kj}\tau_{zy} + \mathbf{kk}\sigma_z\end{aligned} \tag{11.53}$$

and since the stress tensor is symmetric, the stress dyadic is also symmetric, i.e.,

$$\mathscr{T} = \mathscr{T}_c \tag{11.54}$$

The normal stress $\sigma_{x'}$ may now be found, since this is the projection of $\mathbf{p}$ in the direction of x' (or $\boldsymbol{\mu}_1$), so that

$$\sigma_{x'} = \boldsymbol{\mu}_1 \cdot \mathbf{p} = \boldsymbol{\mu}_1 \cdot (\mathscr{T} \cdot \boldsymbol{\mu}_1) = \boldsymbol{\mu}_1 \cdot \mathscr{T} \cdot \boldsymbol{\mu}_1 \tag{11.55}$$

or

$$\begin{aligned}\sigma_{x'} &= \sigma_x a_{11}^2 + \sigma_y a_{21}^2 + \sigma_z a_{31}^2 \\ &\quad + 2\tau_{xy} a_{11} a_{21} + 2\tau_{xz} a_{11} a_{31} + 2\tau_{zy} a_{21} a_{31}\end{aligned} \tag{11.56}$$

which was derived earlier in Chapter 1 as Eq. (1.25) and also as Eq. (9.2).

The shear stress components on the x' plane are

$$\tau_{x'y'} = \boldsymbol{\mu}_2 \cdot \mathbf{p} = \boldsymbol{\mu}_2 \cdot \mathscr{T} \cdot \boldsymbol{\mu}_1 \tag{11.57}$$

$$\tau_{x'z'} = \boldsymbol{\mu}_3 \cdot \mathbf{p} = \boldsymbol{\mu}_3 \cdot \mathscr{T} \cdot \boldsymbol{\mu}_1 \tag{11.58}$$

where $\boldsymbol{\mu}_2$ and $\boldsymbol{\mu}_3$ are unit vectors in the y' and z' directions, respectively. After substituting the value of $\mathscr{T}$ from Eq. (11.53) and expressing $\boldsymbol{\mu}_1$ and $\boldsymbol{\mu}_2$ in terms of the direction cosines, the expression for $\tau_{x'y'}$ becomes

$$\begin{aligned}\tau_{x'y'} &= \sigma_x a_{11} a_{12} + \sigma_y a_{21} a_{22} + \sigma_z a_{31} a_{32} \\ &\quad + \tau_{xy}[a_{11} a_{22} + a_{21} a_{12}] \\ &\quad + \tau_{yz}[a_{21} a_{32} + a_{31} a_{22}] \\ &\quad + \tau_{zx}[a_{31} a_{12} + a_{11} a_{32}]\end{aligned} \tag{11.59}$$

which was also derived in Chapter 1 and Chapter 9.

If we choose the x, y, and z directions to coincide with the principal directions, the stress dyadic becomes

$$\mathscr{T} = \mathbf{i}\boldsymbol{\tau}_1 + \mathbf{j}\boldsymbol{\tau}_2 + \mathbf{k}\boldsymbol{\tau}_3 = \sigma_1 \mathbf{ii} + \sigma_2 \mathbf{jj} + \sigma_3 \mathbf{kk} \tag{11.60}$$

where $\boldsymbol{\tau}_1$, $\boldsymbol{\tau}_2$, and $\boldsymbol{\tau}_3$ are the stress vectors on the principal planes, and σ_1, σ_2, and σ_3 are the principal stresses.

11.6. Equations of Equilibrium

We now consider a continuous variation of stress and the resulting equation of equilibrium in the x direction,

$$\frac{\partial \sigma_x}{\partial x} + \frac{\partial \tau_{xy}}{\partial y} + \frac{\partial \tau_{xz}}{\partial z} + F_x = 0$$

where F_x is the x component of the body force intensity. The divergence of $\boldsymbol{\tau}_x$ is given by

$$\operatorname{div} \boldsymbol{\tau}_x = \nabla \cdot \boldsymbol{\tau}_x = \frac{\partial \sigma_x}{\partial x} + \frac{\partial \tau_{xy}}{\partial y} + \frac{\partial \tau_{xz}}{\partial z} \tag{11.61}$$

so that this equation of equilibrium may be written

$$\operatorname{div} \boldsymbol{\tau}_x + F_x = 0 \tag{11.62}$$

Similarly

$$\operatorname{div} \boldsymbol{\tau}_y + F_y = 0 \tag{11.63}$$

and

$$\operatorname{div} \boldsymbol{\tau}_z + F_z = 0 \tag{11.64}$$

Multiplying Eq. (11.62) by $\mathbf{i}$, Eq. (11.63) by $\mathbf{j}$, and Eq. (11.64) by $\mathbf{k}$ and adding, we get

$$\mathbf{i} \operatorname{div} \boldsymbol{\tau}_x + \mathbf{j} \operatorname{div} \boldsymbol{\tau}_y + \mathbf{k} \operatorname{div} \boldsymbol{\tau}_z + \mathbf{F} = 0 \tag{11.65}$$

where $\mathbf{F} = F_x \mathbf{i} + F_y \mathbf{j} + F_z \mathbf{k}$. Thus, in dyadic notation, we have

$$\operatorname{div}(\boldsymbol{\tau}_x \mathbf{i} + \boldsymbol{\tau}_y \mathbf{j} + \boldsymbol{\tau}_z \mathbf{k}) + \mathbf{F} = 0 \tag{11.66}$$

since $\operatorname{div}(\boldsymbol{\tau}_x \mathbf{i}) = \mathbf{i} \operatorname{div}(\boldsymbol{\tau}_x)$, etc. And from Eqs. (11.52) and (11.54) we see that Eq. (11.66) may be more concisely written as

$$\operatorname{div} \mathscr{T} + \mathbf{F} = 0 \tag{11.67}$$

11.7. Displacement and Strain

The displacement vector[1] $\mathbf{u}$ is defined by

$$\mathbf{u} = u\mathbf{i} + v\mathbf{j} + w\mathbf{k} \tag{11.68}$$

The divergence of $\mathbf{u}$ gives the dilatation ϵ, or

$$\begin{aligned} \nabla \cdot \mathbf{u} = \operatorname{div} \mathbf{u} &= \frac{\partial u}{\partial x} + \frac{\partial v}{\partial y} + \frac{\partial w}{\partial z} \\ &= \epsilon_x + \epsilon_y + \epsilon_z = \epsilon \end{aligned} \tag{11.69}$$

[1] To be consistent with the vector notation used in Chapter 9, we are using $\mathbf{u}$ to represent the displacement vector. Our notation for the magnitude of $\mathbf{u}$ is $|\mathbf{u}|$, so that this should not be confused with the x component of displacement, u.

We define the strain vectors $\mathbf{e}_x$, $\mathbf{e}_y$, and $\mathbf{e}_z$ by

$$\mathbf{e}_x = \epsilon_x \mathbf{i} + \tfrac{1}{2}\gamma_{xy}\mathbf{j} + \tfrac{1}{2}\gamma_{xz}\mathbf{k} \qquad (x, y, z; \mathbf{i}, \mathbf{j}, \mathbf{k}) \tag{11.70}$$

In terms of displacements, the vector Eq. (11.70) may be written as

$$\mathbf{e}_x = \frac{\partial u}{\partial x}\mathbf{i} + \frac{1}{2}\left(\frac{\partial v}{\partial x} + \frac{\partial u}{\partial y}\right)\mathbf{j} + \frac{1}{2}\left(\frac{\partial w}{\partial x} + \frac{\partial u}{\partial z}\right)\mathbf{k}$$

or

$$\mathbf{e}_x = \frac{1}{2}\left(\nabla u + \frac{\partial \mathbf{u}}{\partial x}\right) \qquad (x, y, z; u, v, w) \tag{11.71}$$

These three vector equations are equivalent to the six strain-displacement relations, Eqs. (2.1). If we introduce a strain dyadic $\mathscr{E}$ where

$$\mathscr{E} = \mathbf{i}\mathbf{e}_x + \mathbf{j}\mathbf{e}_y + \mathbf{k}\mathbf{e}_z \tag{11.72}$$

then Eqs. (11.71) may be written as one dyadic equation

$$\mathscr{E} = \tfrac{1}{2}(\nabla\mathbf{u} + \mathbf{u}\nabla) \tag{11.73}$$

where $\mathbf{u}\nabla$ is defined as the conjugate of $\nabla\mathbf{u}$, or, explicitly,

$$\nabla\mathbf{u} = \mathbf{i}\frac{\partial}{\partial x}(\mathbf{u}) + \mathbf{j}\frac{\partial}{\partial y}(\mathbf{u}) + \mathbf{k}\frac{\partial}{\partial z}(\mathbf{u}) \tag{11.74}$$

and

$$\mathbf{u}\nabla = \frac{\partial}{\partial x}(\mathbf{u})\mathbf{i} + \frac{\partial}{\partial y}(\mathbf{u})\mathbf{j} + \frac{\partial}{\partial z}(\mathbf{u})\mathbf{k} \tag{11.75}$$

i.e., the operator $\nabla = \mathbf{i}\frac{\partial}{\partial x} + \mathbf{j}\frac{\partial}{\partial y} + \mathbf{k}\frac{\partial}{\partial z}$ in $\mathbf{u}\nabla$ is applied from the right.

11.8. Generalized Hooke's Law and Navier's Equation

By using generalized Hooke's law and Eq. (11.69), we see that

$$\begin{aligned} \sigma_x &= 2G\frac{\partial u}{\partial x} + \lambda \operatorname{div} \mathbf{u} \\ \tau_{xy} &= G\left(\frac{\partial v}{\partial x} + \frac{\partial u}{\partial y}\right) \\ \tau_{xz} &= G\left(\frac{\partial w}{\partial x} + \frac{\partial u}{\partial z}\right) \end{aligned} \tag{11.76}$$

Multiplying the first equation by $\mathbf{i}$, the second by $\mathbf{j}$, and the third by $\mathbf{k}$ and adding, we obtain

$$\boldsymbol{\tau}_x = G\left(\nabla u + \frac{\partial \mathbf{u}}{\partial x}\right) + (\lambda \operatorname{div} \mathbf{u})\mathbf{i} \tag{11.77}$$

Similarly,

$$\boldsymbol{\tau}_y = G\left(\nabla v + \frac{\partial \mathbf{u}}{\partial y}\right) + (\lambda \operatorname{div} \mathbf{u})\mathbf{j} \tag{11.78}$$

and

$$\boldsymbol{\tau}_z = G\left(\nabla w + \frac{\partial \mathbf{u}}{\partial z}\right) + (\lambda \operatorname{div} \mathbf{u})\mathbf{k} \tag{11.79}$$

or, in slightly different form,

$$\boldsymbol{\tau}_x = 2G(\epsilon_x \mathbf{i} + \tfrac{1}{2}\gamma_{xy}\mathbf{j} + \tfrac{1}{2}\gamma_{xz}\mathbf{k}) + \lambda(\operatorname{div} \mathbf{u})\mathbf{i}$$

or

$$\boldsymbol{\tau}_x = 2G\mathbf{e}_x + \lambda(\operatorname{div} \mathbf{u})\mathbf{i} \quad (x, y, z; \mathbf{i}, \mathbf{j}, \mathbf{k}) \tag{11.80}$$

Equations (11.80) can be combined into the single dyadic equation

$$\mathscr{T} = 2G\mathscr{E} + \mathscr{L}\lambda\epsilon \tag{11.81}$$

where

$$\mathscr{L} = \mathbf{ii} + \mathbf{jj} + \mathbf{kk} \tag{11.82}$$

is called the *unity dyadic*. Equations (11.81) are identical to Eqs. (3.18).

By substituting Eqs. (11.73) and (11.69) into Eq. (11.81), we obtain the dyadic form of the stress-displacement relations,

$$\mathscr{T} = G(\nabla\mathbf{u} + \mathbf{u}\nabla) + \mathscr{L}\lambda\nabla\cdot\mathbf{u} \tag{11.83}$$

Substituting this expression in Eq. (11.67), we get

$$G\nabla^2\mathbf{u} + G\nabla\cdot(\mathbf{u}\nabla) + \nabla\cdot(\mathscr{L}\lambda\nabla\cdot\mathbf{u}) + \mathbf{F} = 0$$

This equation may be rearranged as

$$G\nabla^2\mathbf{u} + (\lambda + G)\nabla\nabla\cdot\mathbf{u} + \mathbf{F} = 0 \tag{11.84}$$

since

$$\nabla\cdot(\mathbf{u}\nabla) = (\nabla\cdot\mathbf{u})\nabla = \nabla(\nabla\cdot\mathbf{u})$$

and

$$\nabla\cdot(\mathscr{L}\nabla\cdot\mathbf{u}) = \nabla(\nabla\cdot\mathbf{u})$$

This represents the three scalar equations of equilibrium in terms of displacement (4.18), or Navier's equations. Other forms of Eq. (11.84) are given in Problem 11-1.

11.9. Equations of Compatibility

In this section we shall develop the compatibility equations in terms of strain in dyadic notation. In order to accomplish this, we need the definitions of the operations of curl and conjugate curl as applied to dyadic quantities. The curl of a dyadic, from Eq. (11.40), is given by

$$\nabla \times \mathscr{F} = \begin{vmatrix} \mathbf{i} & \mathbf{j} & \mathbf{k} \\ \dfrac{\partial}{\partial x} & \dfrac{\partial}{\partial y} & \dfrac{\partial}{\partial z} \\ \mathbf{F}_x & \mathbf{F}_y & \mathbf{F}_z \end{vmatrix} = \mathbf{i} \times \frac{\partial \mathscr{F}}{\partial x} + \mathbf{j} \times \frac{\partial \mathscr{F}}{\partial y} + \mathbf{k} \times \frac{\partial \mathscr{F}}{\partial z}$$

where

$$\mathscr{F} = \mathbf{i}\,\mathbf{F}_x + \mathbf{j}\mathbf{F}_y + \mathbf{k}\mathbf{F}_z$$

is any dyadic. The conjugate curl is defined as

$$\mathscr{F} \times \nabla = \frac{\partial \mathscr{F}}{\partial x} \times \mathbf{i} + \frac{\partial \mathscr{F}}{\partial y} \times \mathbf{j} + \frac{\partial \mathscr{F}}{\partial z} \times \mathbf{k} \tag{11.85}$$

Using these definitions, we can show that

$$\nabla \times \nabla \mathscr{F} + \mathscr{F} \nabla \times \nabla = 0 \tag{11.86}$$

We next observe that Eqs. (10.3) may be expressed in dyadic notation (see also Eq. (11.73)) as

$$\mathscr{E} = \tfrac{1}{2}(\nabla \mathbf{u} + \mathbf{u} \nabla) \tag{11.73}$$

$$\Omega = \tfrac{1}{2}(\nabla \mathbf{u} - \mathbf{u} \nabla) \tag{11.87}$$

where Ω is the rotation dyadic. Thus, adding Eqs. (11.73) and (11.87), we find that

$$\mathscr{E} + \Omega = \nabla \mathbf{u} \tag{11.88}$$

Taking the curl of this expression and using Eq. (11.86), we get

$$\nabla \times \mathscr{E} + \nabla \times \Omega = 0 \tag{11.89}$$

But from Eq. (11.87) we see that

$$\nabla \times \Omega = -\tfrac{1}{2} \nabla \times (\mathbf{u} \nabla) = -\tfrac{1}{2}(\nabla \times \mathbf{u}) \nabla \tag{11.90}$$

Thus, applying the conjugate curl to Eq. (11.89), using Eq. (11.90), we have

$$\nabla \times \mathscr{E} \times \nabla = 0 \tag{11.91}$$

which, on expansion, gives (identically) the compatibility equations (2.2). Another form of Eqs. (11.91) is given in Problem 11-3.

The compatibility equations in terms of stress may be written as a single dyadic equation, by using the result of Problem 11-2, in the form

$$\nabla^2 \mathscr{T} + \left(\frac{1}{1+\nu}\right) \nabla\nabla\Theta + \left(\frac{\nu}{1-\nu}\right) \mathscr{L}\nabla \cdot \mathbf{F} + \mathbf{F}\nabla + \nabla\mathbf{F} = 0 \qquad (11.92)$$

11.10. Strain Energy

In order to express the strain energy density U_0 in dyadic notation, it is convenient to define a new symbol. The trace $|\mathscr{A}|$ of a dyadic $\mathscr{A}$ is

Table 11.2

GOVERNING EQUATIONS IN DYADIC NOTATION

$\operatorname{div} \mathscr{T} + \mathbf{F} = 0$ (11.67)

$\mathscr{E} = (1/2)(\nabla\mathbf{u} + \mathbf{u}\nabla)$ (11.73)

$\mathscr{T} = 2G\mathscr{E} + \mathscr{L}\lambda\epsilon$ (11.81)

$\mathscr{T}, \mathscr{E}, \mathbf{u}$

$\operatorname{div} \mathscr{T} + \mathbf{F} = 0$

$\nabla \times \mathscr{E} \times \nabla = 0$ (11.91)

$\mathscr{E} = [(1+\nu)/E]\mathscr{T} - (\nu/E)\mathscr{L}\Theta$

$\mathscr{T}, \mathscr{E}$

$\operatorname{div} \mathscr{T} + \mathbf{F} = 0$

$\mathscr{T} = G(\nabla\mathbf{u} + \mathbf{u}\nabla) + \mathscr{L}\lambda\nabla\cdot\mathbf{u}$ (11.83)

$\mathscr{T}, \mathbf{u}$

$\operatorname{div} \mathscr{T} + \mathbf{F} = 0$

$\nabla^2\mathscr{T} + [1/(1+\nu)]\nabla\nabla\Theta + \mathscr{L}[\nu/(1-\nu)]\nabla\cdot\mathbf{F} + \mathbf{F}\nabla + \nabla\mathbf{F} = 0$ (11.92)

$\mathscr{T}$

$G\nabla^2\mathbf{u} + (\lambda + G)\nabla\nabla\cdot\mathbf{u} + \mathbf{F} = 0$ (11.84)

$\mathbf{u}$

Stress Boundary Condition $\mathbf{T}^{\mu} = \mathscr{T} \cdot \boldsymbol{\mu}$

Displacement Boundary Condition $\mathbf{u} = \mathbf{u}_b$

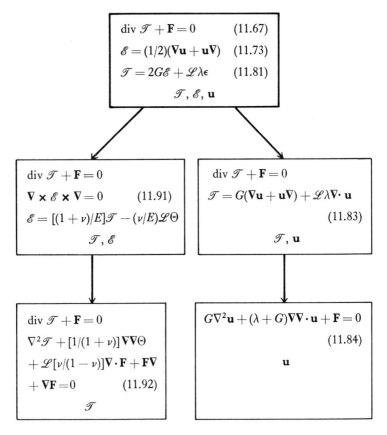

a scalar which is defined by

$$|\mathscr{A}| = A_{xx} + A_{yy} + A_{zz} \tag{11.93}$$

In dyadic notation the strain energy density expressions in Eqs. (7.6), (7.7), and (7.8) are

$$U_0 = \tfrac{1}{2}|\mathscr{T} \cdot \mathscr{E}| \tag{11.94}$$

$$U_0 = -\frac{\nu}{2E}|\mathscr{T}|^2 + \frac{1}{4G}|\mathscr{T} \cdot \mathscr{T}| \tag{11.95}$$

$$U_0 = \frac{\lambda}{2}|\mathscr{E}|^2 + G|\mathscr{E} \cdot \mathscr{E}| \tag{11.96}$$

respectively. These three equations are also equivalent to Eqs. (10.41), (10.42), and (10.43), respectively.

11.11. Governing Equations of Elasticity

The governing equations of elasticity are summarized in Table 11.2. The 15 governing scalar equations expressed in dyadic notation are depicted in the center box, along with the dependent variables in this formulation. The development of the stress formulation is illustrated in the left-hand boxes, while that of the displacement formulation is shown in the right-hand boxes. The dependent variables for each system of equations are also placed in the appropriate boxes.

Referring to Eq. (11.51), we can write the stress boundary conditions in the form

$$\mathbf{T}^{\mu} = \mathscr{T} \cdot \boldsymbol{\mu} \tag{11.97}$$

where $\mathbf{T}^{\mu}$ is the prescribed surface force vector, $\mathscr{T}$ is the stress dyadic evaluated on the boundary, and $\boldsymbol{\mu}$ is the outward unit normal vector on the boundary.

PROBLEMS

11-1 Show that the equilibrium equations in terms of displacement, Eq. (11.84), or Navier's equations, may be written in the following vector forms,

$$(\lambda + 2G) \operatorname{grad} \operatorname{div} \mathbf{u} - G \operatorname{curl} \operatorname{curl} \mathbf{u} + \mathbf{F} = 0$$

$$\frac{2(1-\nu)G}{1-2\nu} \operatorname{grad} \epsilon - 2G \operatorname{curl} \boldsymbol{\omega} + \mathbf{F} = 0$$

where

$$\boldsymbol{\omega} = \tfrac{1}{2} \text{ curl } \mathbf{u}$$

11-2 Show that the Beltrami-Michell compatibility equations in terms of stress, Eqs. (4.25), in vector form are

$$\nabla^2 \boldsymbol{\tau}_x + \frac{1}{1+\nu} \text{grad} \left(\frac{\partial \Theta}{\partial x} \right) + \mathbf{i} \frac{\nu}{1-\nu} \text{div } \mathbf{F} + \text{grad } F_x + \frac{\partial}{\partial x} \mathbf{F} = 0 \quad (x, y, z)$$

where

$$\Theta = \sigma_x + \sigma_y + \sigma_z$$

11-3 Show by expansion that Eq. (11.91) is equivalent to the expression

$$\nabla^2 \mathscr{E} + \nabla\nabla\epsilon - (\nabla \cdot \mathscr{E}\nabla + \nabla\nabla \cdot \mathscr{E}) = 0$$

The latter, on expansion, gives compatibility equations which are linear combinations of those obtained from (11.91).

11-4 Show that strain energy density U_0 expressions, Eqs. (7.6), (7.7), and (7.8), in vector form are

$$U_0 = \tfrac{1}{2}(\boldsymbol{\tau}_x \cdot \mathbf{e}_x + \boldsymbol{\tau}_y \cdot \mathbf{e}_y + \boldsymbol{\tau}_z \cdot \mathbf{e}_z)$$

$$U_0 = -\frac{1}{2}\frac{\nu}{E}\Theta^2 + \frac{1}{4G}(\boldsymbol{\tau}_x \cdot \boldsymbol{\tau}_x + \boldsymbol{\tau}_y \cdot \boldsymbol{\tau}_y + \boldsymbol{\tau}_z \cdot \boldsymbol{\tau}_z)$$

$$U_0 = \tfrac{1}{2}\lambda\epsilon^2 + G(\mathbf{e}_x \cdot \mathbf{e}_x + \mathbf{e}_y \cdot \mathbf{e}_y + \mathbf{e}_z \cdot \mathbf{e}_z)$$

where

$$\epsilon = \epsilon_x + \epsilon_y + \epsilon_z$$

$$\Theta = \sigma_x + \sigma_y + \sigma_z$$

11-5 Show that

$$\nabla(\boldsymbol{\Phi} \cdot \boldsymbol{\rho}) = \boldsymbol{\Phi} + (\nabla\boldsymbol{\Phi}) \cdot \boldsymbol{\rho}$$

where

$$\boldsymbol{\rho} = x\mathbf{i} + y\mathbf{j} + z\mathbf{k}$$

11-6 Using Table 11.1, show that the governing equations of Tables 10.1 and 11.2 are equivalent.

12

Orthogonal Curvilinear Coordinates

12.1. Introduction

As discussed in Section 5.4, in treating elasticity problems it is often necessary to write the governing equations in curvilinear coordinates. This is mainly because the shape of the boundary surface of certain bodies can be expressed most conveniently in certain curvilinear coordinates, i.e., one coordinate is constant on the boundary. As a result, the boundary conditions can be expressed in simple forms. If the body under consideration is bounded by a sphere, for example, the boundary is simply $r = a$ in spherical coordinates, where a is the radius of the sphere. The most general method of deriving equations in curvilinear coordinates is by using the general tensor calculus, which can handle nonorthogonal coordinate systems of any dimensions. In this text, however, we shall restrict our discussion to three-dimensional *orthogonal* curvilinear coordinates. By this restriction, we introduce only one concept which is not present in the discussion of Cartesian systems, i.e., the curvilinear nature of the coordinates. After absorbing the material in this chapter, it is hoped that the student will be well prepared to study the general tensor calculus.

12.2. Scale Factors

Consider a set of three independent functions of the Cartesian variables x_1, x_2, x_3,

$$\alpha_m = \alpha_m(x_1, x_2, x_3) \qquad (m = 1, 2, 3) \tag{12.1}$$

We shall suppose that these may be solved for x_1, x_2, x_3 in terms of α_1, α_2, α_3, or

$$x_n = x_n(\alpha_1, \alpha_2, \alpha_3) \qquad (n = 1, 2, 3) \tag{12.2}$$

The functions in Eqs. (12.1) and (12.2) are assumed to be single-valued and to have continuous derivatives. Each of the equations

$$\alpha_m(x_1, x_2, x_3) = \text{const.} \qquad (m = 1, 2, 3) \tag{12.3}$$

represents a surface (Fig. 12.1). If the constant in the equation is varied, we get a family of surfaces; all together, Eqs. (12.3) represent three families of surfaces. The lines of intersection of these surfaces form three families of curved lines, which will be used as the coordinate lines in our curvilinear coordinate system. That is, any point in space can be defined by specifying the functions α_m at the point. The local coordinate directions are tangent to the three coordinate lines intersecting at the point, the positive directions of the axes being in the directions of increasing α_m. If the three local coordinate axes are mutually perpendicular at each point, the curvilinear coordinate system is said to be

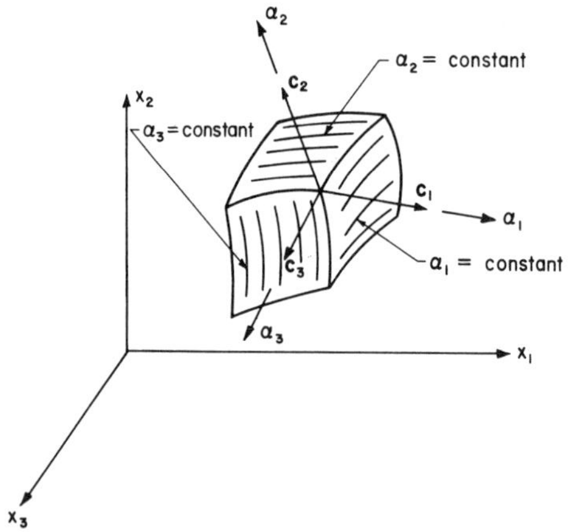

Fig. 12.1 Curvilinear Coordinates

orthogonal. In the case of cylindrical coordinates, $\alpha_1 = r$, $\alpha_2 = \theta$, $\alpha_3 = z$ (see Section 12.7). Along the line of intersection of α_2 = const. and α_3 = const., only α_1 can vary; therefore this line is designated as the α_1 line, as shown in Fig. 12.1. We shall consider only functions $\alpha_m(x_1, x_2, x_3)$ for which the α_m coordinates are orthogonal. In order that the coordinates be independent the functional determinant $\mathcal{J}$ (the Jacobian),

$$\mathcal{J}\left(\frac{x_1, x_2, x_3}{\alpha_1, \alpha_2, \alpha_3}\right) = \begin{vmatrix} \frac{\partial x_1}{\partial \alpha_1} & \frac{\partial x_2}{\partial \alpha_1} & \frac{\partial x_3}{\partial \alpha_1} \\ \frac{\partial x_1}{\partial \alpha_2} & \frac{\partial x_2}{\partial \alpha_2} & \frac{\partial x_3}{\partial \alpha_2} \\ \frac{\partial x_1}{\partial \alpha_3} & \frac{\partial x_2}{\partial \alpha_3} & \frac{\partial x_3}{\partial \alpha_3} \end{vmatrix} \tag{12.4}$$

must not vanish (also see Problem 12-1).

We shall use the notation a_{mn} to represent the direction cosines between the α_1, α_2, α_3 and x_1, x_2, x_3 coordinate directions, as shown in Table 12.1. Since the α coordinate system is orthogonal at every point, the

Table 12.1

TABLE OF DIRECTION COSINES BETWEEN THE CURVILINEAR COORDINATES α_1, α_2, α_3 AND THE CARTESIAN COORDINATES x_1, x_2, x_3

	α_1	α_2	α_3
x_1	a_{11}	a_{12}	a_{13}
x_2	a_{21}	a_{22}	a_{23}
x_3	a_{31}	a_{32}	a_{33}

relations between the direction cosines, Eqs. (8.11), are valid. These relations are rewritten below for easy reference,[1]

$$\begin{aligned} \sum_{\ell} a_{\ell m} a_{\ell n} &= 1 \quad \text{if} \quad m = n \\ &= 0 \quad \text{if} \quad m \neq n \\ \sum_{\ell} a_{m\ell} a_{n\ell} &= 1 \quad \text{if} \quad m = n \\ &= 0 \quad \text{if} \quad m \neq n \end{aligned} \tag{12.5}$$

[1] To avoid possible misinterpretation, all summation in this section will be indicated by a summation sign. Repeated indices here will not imply summation. All summations are over the range, 1, 2, and 3, since only a three-dimensional space is being considered.

Let us place three unit vectors $\mathbf{c}_1$, $\mathbf{c}_2$, and $\mathbf{c}_3$, tangent to the coordinate lines α_1, α_2, and α_3, respectively, at the point under consideration, as shown in Fig. 12.1. We shall consider only right-handed coordinate systems, such that $\mathbf{c}_m \times \mathbf{c}_n = \mathbf{c}_p$, where the numerals taken by m, n, and p are unequal in cyclic order of 123. These vectors are of unit length just as the unit vectors **i**, **j**, and **k** used in a Cartesian coordinate system. If the units of length are inches, then these vectors are all one inch in length. There is a major difference between $\mathbf{c}_1$, $\mathbf{c}_2$, $\mathbf{c}_3$ and **i**, **j**, **k** in that the former change their *directions* from point to point in space, whereas the latter remain fixed in direction. The rates of change of these vectors with respect to increments in coordinates α_m will be derived in the next section.

Another major difference between the curvilinear coordinates and the Cartesian coordinates is that x_1, x_2, and x_3 are all measured in length (say, inches) while α_1, α_2, and α_3 may be of any dimension, not necessarily length. For instance, in cylindrical coordinates, $\alpha_1 = r$, $\alpha_2 = \theta$, and $\alpha_3 = x_3$, where θ is an angle, not a length. The direction cosines between the axes of two *Cartesian* coordinate systems are equal to the partial derivatives of the coordinates in one system with respect to those in the other system, as shown by Eqs. (8.25), or

$$\frac{\partial x_i}{\partial x'_j} = \frac{\partial x'_j}{\partial x_i} = a_{ij}$$

Since the curvilinear coordinates are not always linear dimensions, their derivatives with respect to coordinates in another system are not the direction cosines. In order to determine the relations between the direction cosines and the functional derivatives, we must find expressions for the lengths along the coordinate lines. Therefore we let $h_n \, d\alpha_n$ be the length element (inches) corresponding to a change of $d\alpha_n$. The quantity h_n is called a *scale factor for the coordinate* α_n; in general, h_n changes from point to point in space. The direction cosine a_{11} is equal to the ratio of the change in length along α_1 to an infinitesimal change in length along x_1, or

$$a_{11} = h_1 \frac{\partial \alpha_1}{\partial x_1}$$

In a similar manner, the ratio of the change in length along x_1 to an infinitesimal change in length along α_1 is also equal to a_{11}, or

$$a_{11} = \frac{1}{h_1} \frac{\partial x_1}{\partial \alpha_1}$$

Or, in general,

$$a_{mn} = \frac{1}{h_n}\frac{\partial x_m}{\partial \alpha_n} = h_n \frac{\partial \alpha_n}{\partial x_m} \tag{12.6}$$

Notice that the repetition of n does not imply any summation process. Similarly, the direction cosines between two curvilinear coordinate systems α_m and α'_n are

$$a_{mn} = \frac{h_m\,\partial \alpha_m}{h'_n \partial \alpha'_n} = \frac{h'_n \partial \alpha'_n}{h_m\,\partial \alpha_m}$$

where a_{mn} stands for the cosine of the angle between the α_m and α'_n directions at a point.

From Eqs. (12.5) and (12.6), we see that

$$\sum_\ell \frac{\partial x_\ell}{\partial \alpha_m}\frac{\partial x_\ell}{\partial \alpha_n} = \sum_\ell \frac{\partial \alpha_m}{\partial x_\ell}\frac{\partial \alpha_n}{\partial x_\ell} = \sum_\ell \frac{\partial x_\ell}{\partial \alpha_m}\frac{\partial \alpha_n}{\partial x_\ell} = 0 \qquad \text{if} \quad m \neq n \tag{12.7}$$

since $h_i \neq 0$. We also have

$$\sum_\ell \left(\frac{\partial x_\ell}{\partial \alpha_n}\right)^2 = h_n^{\,2}$$
$$\sum_\ell \left(\frac{\partial \alpha_n}{\partial x_\ell}\right)^2 = \frac{1}{h_n^{\,2}} \tag{12.8}$$

The coefficients g_{mn} defined by

$$g_{mn} = \sum_\ell \frac{\partial x_\ell}{\partial \alpha_m}\frac{\partial x_\ell}{\partial \alpha_n}$$

are called the *metric coefficients*. If the curvilinear coordinate system is orthogonal, we see from Eqs. (12.7) and (12.8) that

$$g_{mn} = 0 \qquad (m \neq n)$$
$$g_{mn} = h_m^{\,2} \qquad (m = n)$$

In unabridged form, Eq. (12.8) is

$$\left(\frac{\partial x_1}{\partial \alpha_m}\right)^2 + \left(\frac{\partial x_2}{\partial \alpha_m}\right)^2 + \left(\frac{\partial x_3}{\partial \alpha_m}\right)^2 = h_m^{\,2}$$
$$\left(\frac{\partial \alpha_m}{\partial x_1}\right)^2 + \left(\frac{\partial \alpha_m}{\partial x_2}\right)^2 + \left(\frac{\partial \alpha_m}{\partial x_3}\right)^2 = \frac{1}{h_m^{\,2}}$$

Let $\mathbf{s} = \mathbf{i}x_1 + \mathbf{j}x_2 + \mathbf{k}x_3$ be the position vector. The infinitesimal increment of $\mathbf{s}$ is then $d\mathbf{s} = \mathbf{i}dx_1 + \mathbf{j}\,dx_2 + \mathbf{k}\,dx_3$. In the α_n coordinates, we have

$$d\mathbf{s} = \frac{\partial \mathbf{s}}{\partial \alpha_1} d\alpha_1 + \frac{\partial \mathbf{s}}{\partial \alpha_2} d\alpha_2 + \frac{\partial \mathbf{s}}{\partial \alpha_3} d\alpha_3 = h_1\, d\alpha_1\, \mathbf{c}_1 + h_2\, d\alpha_2\, \mathbf{c}_2 + h_3\, d\alpha_3\, \mathbf{c}_3 \tag{12.9}$$

where the last expression follows from a projection of $d\mathbf{s}$ in the α_n directions. The magnitude of $d\mathbf{s}$ is given by

$$(ds)^2 = \sum_m (dx_m)^2 = \sum_n (h_n\, d\alpha_n)^2 \tag{12.10}$$

12.3. Derivatives of the Unit Vectors

Since the unit vectors $\mathbf{c}_1$, $\mathbf{c}_2$, and $\mathbf{c}_3$ vary in direction from point to point, it is most important to know the expressions for the rates of change of these vectors due to changes of α_n. Let us consider the derivative $\partial \mathbf{c}_1/\partial \alpha_1$. The components, in Cartesian coordinates, of the unit vector $\mathbf{c}_1$ are the direction cosines a_{m1}; therefore, from Eq. (12.6),

$$\frac{\partial \mathbf{c}_1}{\partial \alpha_1} = \frac{\partial}{\partial \alpha_1}\left[\frac{\mathbf{i}\partial x_1}{h_1\, \partial \alpha_1} + \frac{\mathbf{j}\partial x_2}{h_1\, \partial \alpha_1} + \frac{\mathbf{k}\partial x_3}{h_1\, \partial \alpha_1}\right] \tag{12.11}$$

$$= h_1\, \mathbf{c}_1 \frac{\partial}{\partial \alpha_1}\left(\frac{1}{h_1}\right) + \frac{1}{h_1}\left[\mathbf{i}\frac{\partial^2 x_1}{\partial {\alpha_1}^2} + \mathbf{j}\frac{\partial^2 x_2}{\partial {\alpha_1}^2} + \mathbf{k}\frac{\partial^2 x_3}{\partial {\alpha_1}^2}\right]$$

Noting that the α_n components of the unit vectors $\mathbf{i}$, $\mathbf{j}$, $\mathbf{k}$ are a_{mn}, and utilizing Eqs. (12.7) and (12.8), Eq. (12.11) can be written as

$$\frac{\partial \mathbf{c}_1}{\partial \alpha_1} = -\frac{\mathbf{c}_1}{h_1}\frac{\partial h_1}{\partial \alpha_1} + \frac{1}{h_1}\left[\frac{\partial^2 x_1}{\partial {\alpha_1}^2}\sum_n \frac{\mathbf{c}_n\, \partial x_1}{h_n\, \partial \alpha_n} + \frac{\partial^2 x_2}{\partial {\alpha_1}^2}\sum_n \frac{\mathbf{c}_n\, \partial x_2}{h_n\, \partial \alpha_n} + \frac{\partial^2 x_3}{\partial {\alpha_1}^2}\sum_n \frac{\mathbf{c}_n\, \partial x_3}{h_n\, \partial \alpha_n}\right]$$

$$= -\frac{\mathbf{c}_1}{h_1}\frac{\partial h_1}{\partial \alpha_1} + \frac{1}{h_1}\left\{\frac{\mathbf{c}_1}{h_1}\left[\frac{\partial^2 x_1}{\partial {\alpha_1}^2}\frac{\partial x_1}{\partial \alpha_1} + \frac{\partial^2 x_2}{\partial {\alpha_1}^2}\frac{\partial x_2}{\partial \alpha_1} + \frac{\partial^2 x_3}{\partial {\alpha_1}^2}\frac{\partial x_3}{\partial \alpha_1}\right]\right.$$

$$\left. + \frac{\mathbf{c}_2}{h_2}\sum_n \frac{\partial^2 x_n}{\partial {\alpha_1}^2}\frac{\partial x_n}{\partial \alpha_2} + \frac{\mathbf{c}_3}{h_3}\sum_n \frac{\partial^2 x_n}{\partial {\alpha_1}^2}\frac{\partial x_n}{\partial \alpha_3}\right\}$$

$$= -\frac{\mathbf{c}_1}{h_1}\frac{\partial h_1}{\partial \alpha_1} + \frac{1}{h_1}\left\{\frac{\mathbf{c}_1}{h_1}\sum_n \frac{1}{2}\frac{\partial}{\partial \alpha_1}\left(\frac{\partial x_n}{\partial \alpha_1}\right)^2 + \frac{\mathbf{c}_2}{h_2}\sum_n \left[\frac{\partial}{\partial \alpha_1}\left(\frac{\partial x_n}{\partial \alpha_1}\frac{\partial x_n}{\partial \alpha_2}\right)\right.\right.$$

$$\left.\left. - \frac{1}{2}\frac{\partial}{\partial \alpha_2}\left(\frac{\partial x_n}{\partial \alpha_1}\right)^2\right] + \frac{\mathbf{c}_3}{h_3}\left[-h_1\frac{\partial h_1}{\partial \alpha_3}\right]\right\}$$

or

$$\frac{\partial \mathbf{c}_1}{\partial \alpha_1} = -\frac{\mathbf{c}_2}{h_2}\frac{\partial h_1}{\partial \alpha_2} - \frac{\mathbf{c}_3}{h_3}\frac{\partial h_1}{\partial \alpha_3} \tag{12.12}$$

By similar procedures, it can be shown (see Problem 12-2) that the following formulas are true:[2]

$$\begin{aligned}
&\frac{\partial \mathbf{c}_1}{\partial \alpha_1} = -\frac{\mathbf{c}_2}{h_2}\frac{\partial h_1}{\partial \alpha_2} - \frac{\mathbf{c}_3}{h_3}\frac{\partial h_1}{\partial \alpha_3} \qquad \frac{\partial \mathbf{c}_1}{\partial \alpha_2} = \frac{\mathbf{c}_2}{h_1}\frac{\partial h_2}{\partial \alpha_1} \qquad \frac{\partial \mathbf{c}_1}{\partial \alpha_3} = \frac{\mathbf{c}_3}{h_1}\frac{\partial h_3}{\partial \alpha_1} \\
&\frac{\partial \mathbf{c}_2}{\partial \alpha_2} = -\frac{\mathbf{c}_3}{h_3}\frac{\partial h_2}{\partial \alpha_3} - \frac{\mathbf{c}_1}{h_1}\frac{\partial h_2}{\partial \alpha_1} \qquad \frac{\partial \mathbf{c}_2}{\partial \alpha_3} = \frac{\mathbf{c}_3}{h_2}\frac{\partial h_3}{\partial \alpha_2} \qquad \frac{\partial \mathbf{c}_2}{\partial \alpha_1} = \frac{\mathbf{c}_1}{h_2}\frac{\partial h_1}{\partial \alpha_2} \qquad (12.13) \\
&\frac{\partial \mathbf{c}_3}{\partial \alpha_3} = -\frac{\mathbf{c}_1}{h_1}\frac{\partial h_3}{\partial \alpha_1} - \frac{\mathbf{c}_2}{h_2}\frac{\partial h_3}{\partial \alpha_2} \qquad \frac{\partial \mathbf{c}_3}{\partial \alpha_1} = \frac{\mathbf{c}_1}{h_3}\frac{\partial h_1}{\partial \alpha_3} \qquad \frac{\partial \mathbf{c}_3}{\partial \alpha_2} = \frac{\mathbf{c}_2}{h_3}\frac{\partial h_2}{\partial \alpha_3}
\end{aligned}$$

The significance of Eqs. (12.13) can be demonstrated by graphic means as shown in Fig. 12.2. The meaning of $\partial \mathbf{c}_1/\partial \alpha_1$ is demonstrated in

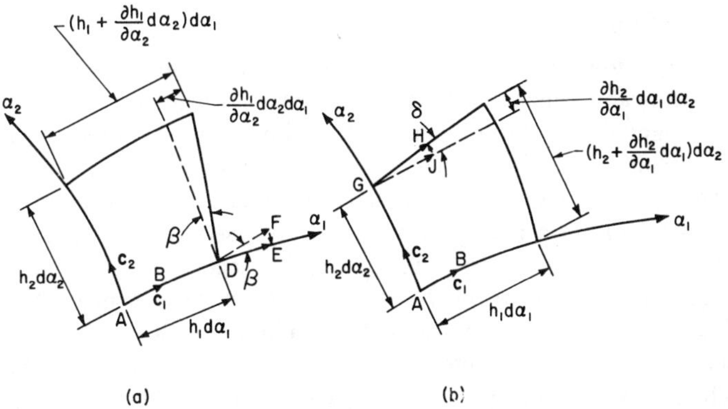

Fig. 12.2 Rates of Change of Unit Vectors Along Curvilinear Coordinates

Fig. 12.2(a), where $\mathbf{c}_1$ at A is represented by AB. At point D, a distance $h_1\, d\alpha_1$ away from A along the α_1 coordinate line, the unit vector in the α_1 direction is shown as DE. The change of $\mathbf{c}_1$ from A to D is the vector FE, with a magnitude of $\beta = (1/h_2)(\partial h_1/\partial \alpha_2)d\alpha_1$ and a direction opposite to that of $\mathbf{c}_2$ if $\partial h_1/\partial \alpha_2$ is positive. Therefore, we have $\partial \mathbf{c}_1/\partial \alpha_1 = -(\mathbf{c}_2/h_2)(\partial h_1/\partial \alpha_2)$ plus a similar component in the α_3 direction. Figure 12.2(b) gives a graphic representation of the formula for $\partial \mathbf{c}_1/\partial \alpha_2$.

[2] These equations may be expressed in a more concise form by using Christoffel symbols. If we define

$$\left\{ {i \atop i\ \ i} \right\} = \frac{1}{h_i}\frac{\partial h_i}{\partial \alpha_i} \qquad \left\{ {i \atop i\ \ j} \right\} = \left\{ {i \atop j\ \ i} \right\} = \frac{1}{h_i}\frac{\partial h_i}{\partial \alpha_j}$$

$$\left\{ {j \atop i\ \ i} \right\} = -\frac{h_i}{(h_j)^2}\frac{\partial h_i}{\partial \alpha_j} \qquad \left\{ {i \atop j\ \ k} \right\} = 0, \quad \text{if } i \neq j \neq k,$$

then Eqs. (12.13) are equivalent to

$$\frac{\partial}{\partial \alpha_j}(h_i \mathbf{c}_i) = \sum_n h_n \mathbf{c}_n \left\{ {n \atop i\ \ j} \right\}$$

The unit vector $\mathbf{c}_1$ is in the direction of increasing α_1, thus it is normal to the $\alpha_1 = $ const. surface. The curvature of this surface in the α_2 direction is measured by the change of direction of $\mathbf{c}_1$ in the α_2 direction for a displacement $h_2\, d\alpha_2$. From the geometry of Fig. 12.2(b), we see that

$$\frac{1}{R_2} = \mathbf{c}_2 \cdot \left(\frac{\partial \mathbf{c}_1}{h_2\, \partial \alpha_2}\right) = \frac{1}{h_1 h_2} \frac{\partial h_2}{\partial \alpha_1} \tag{12.14}$$

where R_2 is the radius of curvature of the α_2 coordinate line. We also observe that the area of an element lying in a plane normal to the α_3 direction is given by $h_1 h_2\, d\alpha_1\, d\alpha_2$ (Fig. 12.2), and the volume of a three-dimensional element is $h_1 h_2 h_3\, d\alpha_1\, d\alpha_2\, d\alpha_3$. (See Problem 12-3.)

12.4. Vector Operators

All equations in Cartesian coordinates which do not involve space derivatives and which pertain to properties *at a point* are also applicable to orthogonal curvilinear coordinates. For instance, Eqs. (8.11) and (8.25), which were first derived for Cartesian coordinates, can be used for the α_n coordinates. If space derivatives are involved, however, the Cartesian equations and operators cannot be used for curvilinear coordinates by merely changing x_n to α_n. In this section we shall derive the expressions for some of the vector operators in the α_n coordinates.

Let us consider the operator ∇, the gradient operator. When applied to a scalar ϕ, it gives a vector $\nabla\phi$, the components of which are the rates of change of ϕ with respect to the space variables. Along the direction of this vector, ϕ has the greatest rate of change, which is equal to the magnitude of $\nabla\phi$. Let the components of $\nabla\phi$ in the α_n directions be f_n, which are to be determined; then

$$\nabla\phi = f_1 \mathbf{c}_1 + f_2 \mathbf{c}_2 + f_3 \mathbf{c}_3 \tag{12.15}$$

The increment of ϕ due to a change of position $d\mathbf{s}$ is, according to Eqs. (12.9) and (12.15),

$$d\phi = \nabla\phi \cdot d\mathbf{s} = h_1 f_1\, d\alpha_1 + h_2 f_2\, d\alpha_2 + h_3 f_3\, d\alpha_3 \tag{12.16}$$

But

$$d\phi = \frac{\partial \phi}{\partial \alpha_1} d\alpha_1 + \frac{\partial \phi}{\partial \alpha_2} d\alpha_2 + \frac{\partial \phi}{\partial \alpha_3} d\alpha_3 \tag{12.17}$$

Combining Eqs. (12.16) and (12.17), we obtain, since the $d\alpha_n$ are arbitrary,

$$f_1 = \frac{1}{h_1}\frac{\partial \phi}{\partial \alpha_1} \qquad f_2 = \frac{1}{h_2}\frac{\partial \phi}{\partial \alpha_2} \qquad f_3 = \frac{1}{h_3}\frac{\partial \phi}{\partial \alpha_3}$$

or

$$\operatorname{grad}\phi = \nabla\phi = \frac{1}{h_1}\frac{\partial\phi}{\partial\alpha_1}\mathbf{c}_1 + \frac{1}{h_2}\frac{\partial\phi}{\partial\alpha_2}\mathbf{c}_2 + \frac{1}{h_3}\frac{\partial\phi}{\partial\alpha_3}\mathbf{c}_3$$
$$= \sum_n \mathbf{c}_n \frac{1}{h_n}\frac{\partial\phi}{\partial\alpha_n} \tag{12.18}$$

Since $\nabla\phi$ is a vector, its components in α_n coordinates, $\frac{1}{h_n}\frac{\partial\phi}{\partial\alpha_n}$, are related to its components in another coordinate system by equations of the form of Eqs. (8.12).

We shall next derive the expression for the divergence in orthogonal curvilinear coordinates. The divergence of a vector **A** is, according to the standard definition,

$$\begin{aligned}\operatorname{div}\mathbf{A} &= \nabla\cdot\mathbf{A} = \nabla\cdot(A_1\mathbf{c}_1 + A_2\mathbf{c}_2 + A_3\mathbf{c}_3)\\ &= \nabla\cdot(A_1\mathbf{c}_1) + \nabla\cdot(A_2\mathbf{c}_2) + \nabla\cdot(A_3\mathbf{c}_3)\end{aligned} \tag{12.19}$$

From Eq. (12.18), we see that

$$\nabla\alpha_1 = \frac{1}{h_1}\mathbf{c}_1 \tag{12.20}$$

since α_m are independent coordinates. Similarly,

$$\nabla\alpha_2 = \frac{1}{h_2}\mathbf{c}_2 \qquad \nabla\alpha_3 = \frac{1}{h_3}\mathbf{c}_3$$

or

$$\nabla\alpha_m = \frac{1}{h_m}\mathbf{c}_m \tag{12.21}$$

From the definition of a vector product, Eqs. (11.12) or (11.15), we have

$$\nabla\alpha_1 \times \nabla\alpha_2 = \frac{\mathbf{c}_1}{h_1} \times \frac{\mathbf{c}_2}{h_2} = \frac{1}{h_1 h_2}\mathbf{c}_3$$

or, in general,

$$\begin{aligned}\mathbf{c}_1 &= h_2 h_3\,\nabla\alpha_2 \times \nabla\alpha_3\\ \mathbf{c}_2 &= h_3 h_1\,\nabla\alpha_3 \times \nabla\alpha_1\\ \mathbf{c}_3 &= h_1 h_2\,\nabla\alpha_1 \times \nabla\alpha_2\end{aligned} \tag{12.22}$$

Substituting (12.22) into the first term on the right-hand side of Eq.

(12.19), we obtain (see Problem 12-5)

$$\begin{aligned}\nabla\cdot(A_1\,\mathbf{c}_1) &= \nabla\cdot(A_1\,h_2\,h_3\,\nabla\alpha_2\times\nabla\alpha_3)\\ &= \nabla(A_1\,h_2\,h_3)\cdot(\nabla\alpha_2\times\nabla\alpha_3)+A_1\,h_2\,h_3\,\nabla\cdot(\nabla\alpha_2\times\nabla\alpha_3)\\ &= \left[\frac{\mathbf{c}_1}{h_1}\frac{\partial}{\partial\alpha_1}(A_1\,h_2\,h_3)+\frac{\mathbf{c}_2}{h_2}\frac{\partial}{\partial\alpha_2}(A_1\,h_2\,h_3)\right.\\ &\quad\left.+\frac{\mathbf{c}_3}{h_3}\frac{\partial}{\partial\alpha_3}(A_1\,h_2\,h_3)\right]\cdot\left(\frac{\mathbf{c}_1}{h_2\,h_3}\right)+0\\ &= \frac{1}{h_1\,h_2\,h_3}\frac{\partial}{\partial\alpha_1}(A_1\,h_2\,h_3) \qquad (12.23)\end{aligned}$$

Similarly, the other two terms of Eq. (12.19) are determined and we get the expression for the divergence,

$$\begin{aligned}\operatorname{div}\mathbf{A}=\nabla\cdot\mathbf{A} &= \frac{1}{h_1\,h_2\,h_3}\left[\frac{\partial}{\partial\alpha_1}(A_1\,h_2\,h_3)+\frac{\partial}{\partial\alpha_2}(A_2\,h_3\,h_1)\right.\\ &\quad\left.+\frac{\partial}{\partial\alpha_3}(A_3\,h_1\,h_2)\right]\\ &= \frac{1}{h_1\,h_2\,h_3}\sum_n\frac{\partial}{\partial\alpha_n}\left(h_1\,h_2\,h_3\,\frac{A_n}{h_n}\right) \qquad (12.24)\end{aligned}$$

This expression can also be derived by considering the physical meaning of $\nabla\cdot\mathbf{A}$ as applied to an infinitesimal element. (See Problem 12-6.)

For the curl operator, or $\nabla\times$, we have

$$\begin{aligned}\nabla\times\mathbf{A} &= \nabla\times(\mathbf{c}_1\,A_1+\mathbf{c}_2\,A_2+\mathbf{c}_3\,A_3)\\ &= \nabla\times(A_1\,h_1\,\nabla\alpha_1)+\nabla\times(A_2\,h_2\,\nabla\alpha_2)+\nabla\times(A_3\,h_3\,\nabla\alpha_3) \qquad (12.25)\end{aligned}$$

where Eq. (12.21) has been used. The first term on the right-hand side of Eq. (12.25) can be expressed as (see Problem 12-7)

$$\nabla\times(A_1\,h_1\,\nabla\alpha_1)=\frac{\mathbf{c}_2}{h_3\,h_1}\frac{\partial}{\partial\alpha_3}(A_1\,h_1)-\frac{\mathbf{c}_3}{h_2\,h_1}\frac{\partial}{\partial\alpha_2}(A_1\,h_1) \qquad (12.26)$$

Therefore,

$$\begin{aligned}\operatorname{curl} \mathbf{A} = \nabla \times \mathbf{A} &= \frac{\mathbf{c}_1}{h_2 h_3}\left[\frac{\partial(A_3 h_3)}{\partial \alpha_2} - \frac{\partial(A_2 h_2)}{\partial \alpha_3}\right] \\ &\quad + \frac{\mathbf{c}_2}{h_3 h_1}\left[\frac{\partial(A_1 h_1)}{\partial \alpha_3} - \frac{\partial(A_3 h_3)}{\partial \alpha_1}\right] \\ &\quad + \frac{\mathbf{c}_3}{h_1 h_2}\left[\frac{\partial(A_2 h_2)}{\partial \alpha_1} - \frac{\partial(A_1 h_1)}{\partial \alpha_2}\right] \\ &= \frac{1}{h_1 h_2 h_3}\begin{vmatrix} h_1 \mathbf{c}_1 & h_2 \mathbf{c}_2 & h_3 \mathbf{c}_3 \\ \dfrac{\partial}{\partial \alpha_1} & \dfrac{\partial}{\partial \alpha_2} & \dfrac{\partial}{\partial \alpha_3} \\ h_1 A_1 & h_2 A_2 & h_3 A_3 \end{vmatrix} \end{aligned} \tag{12.27}$$

The Laplace operator can be derived easily from Eqs. (12.18) and (12.24), with $\mathbf{A} = \nabla\phi$,

$$\nabla^2\phi = \nabla \cdot \nabla\phi = \frac{1}{h_1 h_2 h_3} \sum_n \frac{\partial}{\partial \alpha_n}\left[\frac{h_1 h_2 h_3}{h_n^{\,2}} \frac{\partial \phi}{\partial \alpha_n}\right] \tag{12.28}$$

The Laplace operator can also operate on a vector, which results in another vector. From the vector formula

$$\nabla^2 \mathbf{A} = \operatorname{grad}(\operatorname{div} \mathbf{A}) - \operatorname{curl}(\operatorname{curl} \mathbf{A}) \tag{12.29}$$

and the expressions for grad, div, and curl, we can write $\nabla^2\mathbf{A}$ in terms of its components in α_n coordinates. The resulting general expression for $\nabla^2\mathbf{A}$, however, is very lengthy and will not be given here. For most of the systems of interest to us, such as cylindrical or spherical coordinates, the form of $\nabla^2\mathbf{A}$ is comparatively simple and will be given in Section 12.7.

Another way of deriving $\nabla^2\mathbf{A}$ is to write $\mathbf{A} = \mathbf{c}_1 A_1 + \mathbf{c}_2 A_2 + \mathbf{c}_3 A_3$, and utilize the formula

$$\nabla^2(\phi\psi) = \phi\nabla^2\psi + 2\nabla\phi \cdot \nabla\psi + \psi\nabla^2\phi$$

then

$$\begin{aligned}\nabla^2\mathbf{A} &= \nabla^2(\mathbf{c}_1 A_1 + \mathbf{c}_2 A_2 + \mathbf{c}_3 A_3) \\ &= \mathbf{c}_1 \nabla^2 A_1 + \mathbf{c}_2 \nabla^2 A_2 + \mathbf{c}_3 \nabla^2 A_3 + A_1 \nabla^2 \mathbf{c}_1 \\ &\quad + A_2 \nabla^2 \mathbf{c}_2 + A_3 \nabla^2 \mathbf{c}_3 + 2\nabla A_1 \cdot \nabla \mathbf{c}_1 \\ &\quad + 2\nabla A_2 \cdot \nabla \mathbf{c}_2 + 2\nabla A_3 \cdot \nabla \mathbf{c}_3 \end{aligned} \tag{12.30}$$

With the help of Eqs. (12.13) for the derivatives of $\mathbf{c}_n$, and Eq. (12.35) the general expression can also be obtained.

12.5. Dyadic Notation and Dyadic Operators

As discussed in Chapter 11, the symbolic notation for second-order tensors is called dyadic notation, where a second-order tensor is represented by a dyadic. Since it is symbolic in nature, all equations written in dyadic notation must be valid in any coordinate system. In Chapter 11, we dealt with the Cartesian components of vectors and dyadics. In this section, we shall extend the discussion of Chapter 11 and see how the operators and equations in dyadic notation can be applied to curvilinear coordinates.

A dyadic has nine components in any curvilinear coordinate system, or

$$\begin{aligned}\mathcal{S} &= \mathbf{c}_1 S_{11} \mathbf{c}_1 + \mathbf{c}_1 S_{12} \mathbf{c}_2 + \mathbf{c}_1 S_{13} \mathbf{c}_3 + \mathbf{c}_2 S_{21} \mathbf{c}_1 + \mathbf{c}_2 S_{22} \mathbf{c}_2 \\ &\qquad + \mathbf{c}_2 S_{23} \mathbf{c}_3 + \mathbf{c}_3 S_{31} \mathbf{c}_1 + \mathbf{c}_3 S_{32} \mathbf{c}_2 + \mathbf{c}_3 S_{33} \mathbf{c}_3 \\ &= \sum_{m,n} \mathbf{c}_m S_{mn} \mathbf{c}_n \\ &= \mathbf{c}_1 \mathbf{S}_1 + \mathbf{c}_2 \mathbf{S}_2 + \mathbf{c}_3 \mathbf{S}_3 \\ &= \sum_m \mathbf{c}_m \mathbf{S}_m \end{aligned} \tag{12.31}$$

where $\mathbf{S}_m = \sum_j S_{mj} \mathbf{c}_j$, as in Chapter 11, and $\mathbf{c}_1$, $\mathbf{c}_2$, $\mathbf{c}_3$ are the unit vectors along the α_1, α_2, α_3 coordinate directions, respectively. We are denoting multiple summations by a single $\sum$. The number of summations required is then equal to the number of indices underneath the sign $\sum$. For example, the notation $\sum_{m,n}$ represents a double sum over the range (1, 2, 3) of m and n. The conjugate $\mathcal{S}_c$ of a dyadic $\mathcal{S}$ is defined as

$$\mathcal{S}_c = \mathbf{S}_1 \mathbf{c}_1 + \mathbf{S}_2 \mathbf{c}_2 + \mathbf{S}_3 \mathbf{c}_3 = \sum_m \mathbf{S}_m \mathbf{c}_m$$

Any vector product (dot or cross) between a vector and a dyadic is always made between the vector and the adjacent vector in the dyadic; for instance, the dot product between a dyadic and a vector is defined as follows

$$\begin{aligned}\mathcal{A} \cdot \mathbf{B} &= (\mathbf{c}_1 \mathbf{A}_1 + \mathbf{c}_2 \mathbf{A}_2 + \mathbf{c}_3 \mathbf{A}_3) \cdot (\mathbf{c}_1 B_1 + \mathbf{c}_2 B_2 + \mathbf{c}_3 B_3) \\ &= \mathbf{c}_1 (A_{11} B_1 + A_{12} B_2 + A_{13} B_3) + \mathbf{c}_2 (A_{21} B_1 + A_{22} B_2 \\ &\quad + A_{23} B_3) + \mathbf{c}_3 (A_{31} B_1 + A_{32} B_2 + A_{33} B_3) \\ &= \sum_{m,n} \mathbf{c}_m A_{mn} B_n \end{aligned} \tag{12.32}$$

Also (see Problem 12-9),

$$\mathbf{B} \cdot \mathcal{A} = \sum_{m,n} B_m A_{mn} \mathbf{c}_n \neq \mathcal{A} \cdot \mathbf{B} \tag{12.33}$$

The dot product between two dyadics gives another dyadic,

$$\mathscr{A} \cdot \mathscr{B} = \sum_{m,n} \mathbf{c}_m [\sum_i A_{mi} B_{in}] \mathbf{c}_n \tag{12.34}$$

The definitions of vector differential operations applied to dyadics are simple extensions of the corresponding operations on vectors. In particular,

$$\text{grad } \mathbf{A} = \nabla \mathbf{A} = \sum_m \frac{\mathbf{c}_m}{h_m} \frac{\partial \mathbf{A}}{\partial \alpha_m} \tag{12.35}$$

and

$$(\nabla \mathbf{A})_c = \mathbf{A} \nabla = \sum_m \frac{1}{h_m} \frac{\partial \mathbf{A}}{\partial \alpha_m} \mathbf{c}_m \tag{12.36}$$

It is too cumbersome to write out all the components for grad **A**, but again, in most coordinates of interest to us, once h_m and α_m are given, Eq. (12.35) can be evaluated easily. It must be kept in mind that

$$\frac{\partial(\mathbf{c}_m A_m)}{\partial \alpha_n} = \mathbf{c}_m \frac{\partial A_m}{\partial \alpha_n} + A_m \frac{\partial \mathbf{c}_m}{\partial \alpha_n} \tag{12.37}$$

The divergence, when applied to a dyadic, gives

$$\begin{aligned} \text{div } \mathscr{S} &= \nabla \cdot \mathscr{S} = \nabla \cdot (\mathbf{c}_1 \mathbf{S}_1 + \mathbf{c}_2 \mathbf{S}_2 + \mathbf{c}_3 \mathbf{S}_3) \\ &= \frac{1}{h_1 h_2 h_3} \sum_n \frac{\partial}{\partial \alpha_n} \left(h_1 h_2 h_3 \frac{\mathbf{S}_n}{h_n} \right) \end{aligned} \tag{12.38}$$

and

$$\text{div } \mathscr{S}_c = \nabla \cdot \mathscr{S}_c = \nabla \cdot (\mathbf{S}_1 \mathbf{c}_1) + \nabla \cdot (\mathbf{S}_2 \mathbf{c}_2) + \nabla \cdot (\mathbf{S}_3 \mathbf{c}_3) \tag{12.39}$$

We shall also need the unity dyadic $\mathscr{L}$, defined as

$$\mathscr{L} = \mathbf{c}_1 \mathbf{c}_1 + \mathbf{c}_2 \mathbf{c}_2 + \mathbf{c}_3 \mathbf{c}_3 = \sum_n \mathbf{c}_n \mathbf{c}_n \tag{12.40}$$

The curl of a dyadic $\mathscr{A}$ may be obtained from Eq. (12.27) by replacing A_1, A_2, A_3 by $\mathbf{A}_1$, $\mathbf{A}_2$, $\mathbf{A}_3$, respectively.

12.6. Governing Equations of Elasticity in Dyadic Notation

In Chapter 5 we mentioned that there are four approaches which can be used to obtain the governing equations in a particular curvilinear coordinate system. These are:

1. Transformation from Cartesian equations,
2. Deriving each equation in the particular coordinate system,
3. Symbolic (vector and dyadic) approach, and
4. General tensor approach.

In Chapter 5, the elasticity equations in polar coordinates were derived by the first two approaches, i.e., transformation from the Cartesian equations and derivation of each equation from an infinitesimal polar element. In Chapter 11, the equations in dyadic notation were derived by using the corresponding equations written in terms of Cartesian coordinates. These dyadic (and vector) equations are applicable to any curvilinear orthogonal coordinate system, although they were first derived using Cartesian coordinates. For instance, the three Navier equations were first derived in Chapter 4 using Cartesian coordinates as Eqs. (4.18). If the first of (4.18) is multiplied by **i**, the second by **j**, and the third by **k**, we get the vector form of Navier's equations (11.85). A word of caution must be given here. If the equation involved is a scalar or vector equation in terms of scalar and vector quantities, *vector notation* is sufficient. However, if the equation involves second-order tensors, then the vector form of the Cartesian equations is not applicable in curvilinear coordinate systems. When second-order tensors are involved, the equation must be expressed in *dyadic notation* before applying it to curvilinear coordinate systems. For instance, the equilibrium equation in the x direction, (11.62), is

$$\operatorname{div} \boldsymbol{\tau}_x + F_x = 0$$

We cannot use this equation in a curvilinear coordinate system because stress is a second-order tensor, i.e., the equation necessarily makes reference to a particular coordinate axis. However, the equation (11.67)

$$\operatorname{div} \mathscr{T} + \mathbf{F} = 0$$

is true in any coordinate system because it is written in dyadic notation.

The symbolic equations (the third approach listed above) can be obtained either through the Cartesian equations, as in Chapter 11, or in a direct manner. For example, the equilibrium equations may be derived directly by following the general procedure of Section 9.3. Equilibrium requires that the vector sum of all forces on any portion of the body vanish, or

$$\int_V \mathbf{F}\, dV + \int_S \mathscr{T} \cdot \boldsymbol{\nu}\, dS = 0 \tag{12.41}$$

Applying the Gauss theorem, Eq. (8.63), we have

$$\int_V (\mathbf{F} + \operatorname{div} \mathscr{T}) dV = 0$$

but V is arbitrary, so that

$$\operatorname{div} \mathscr{T} + \mathbf{F} = 0 \tag{12.42}$$

As shown in Chapter 11, the strain-displacement relations as derived by using Cartesian coordinates, when written in dyadic notation are

$$\mathscr{E} = \tfrac{1}{2}(\nabla\mathbf{u} + \mathbf{u}\nabla) \tag{12.43}$$

This dyadic equation is true in any coordinate system and the geometrical meaning of the components of $\mathscr{E}$ are still the same as before, i.e., normal strain gives the elongation per unit length and shear strain gives one-half the decrease of the angle between two orthogonal lines.

By using orthogonal curvilinear coordinates in the symbolic approach, the strain-displacement relations may also be derived as follows. Consider a particle located at point $P_0(\alpha_1, \alpha_2, \alpha_3)$ in space and a neighboring particle at point $P(\alpha_1 + d\alpha_1, \alpha_2 + d\alpha_2, \alpha_3 + d\alpha_3)$. The space vector connecting these two points is denoted by $d\mathbf{s}$. The particle at P_0, after going through a displacement $\mathbf{u}(\alpha_1, \alpha_2, \alpha_3)$, reaches a position P'_0, while the particle at P is displaced to P' through a displacement $\mathbf{u}(\alpha_1 + d\alpha_1, \alpha_2 + d\alpha_2, \alpha_3 + d\alpha_3)$, as shown in Figure 12.3.

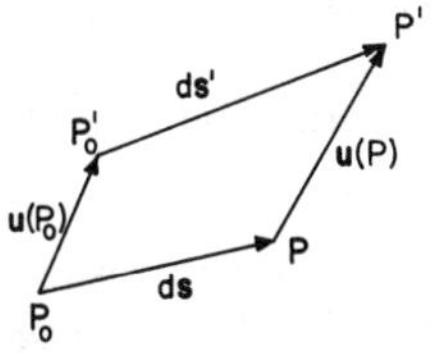

Fig. 12.3 Displacement of Two Neighboring Particles

From vector analysis, we see that

$$d\mathbf{s}' = d\mathbf{s} + \mathbf{u}(P) - \mathbf{u}(P_0)$$

where the notation $\mathbf{u}(P_0)$ means that $\mathbf{u}$ is to be evaluated at $P_0(\alpha_1, \alpha_2, \alpha_3)$. But,

$$\begin{aligned}\mathbf{u}(P) &= \mathbf{u}(P_0) + d\mathbf{u} = \mathbf{u}(P_0) + \frac{\partial \mathbf{u}}{\partial \alpha_1} d\alpha_1 + \frac{\partial \mathbf{u}}{\partial \alpha_2} d\alpha_2 + \frac{\partial \mathbf{u}}{\partial \alpha_3} d\alpha_3 \\ &= \mathbf{u}(P_0) + d\mathbf{s} \cdot \nabla\mathbf{u}\end{aligned}$$

where Eqs. (12.9) and (12.35) have been utilized. It follows that

$$d\mathbf{s}' = d\mathbf{s} + d\mathbf{s} \cdot \nabla\mathbf{u} \tag{12.44}$$

The quantity $\nabla\mathbf{u}$ is a dyadic representing the displacement of P relative to P_0. Decomposing this dyadic into its symmetric and antisymmetric parts, we have

$$\begin{aligned}\nabla\mathbf{u} &= \tfrac{1}{2}(\nabla\mathbf{u} + \mathbf{u}\nabla) + \tfrac{1}{2}(\nabla\mathbf{u} - \mathbf{u}\nabla) \\ &= \mathscr{E} + \Omega\end{aligned} \tag{12.45}$$

where $\mathscr{E}$, the strain dyadic, is as defined by Eq. (12.43), and

$$\Omega = \tfrac{1}{2}(\nabla\mathbf{u} - \mathbf{u}\nabla) = -\tfrac{1}{2}(\text{curl}\,\mathbf{u}) \times \mathscr{L} \tag{12.46}$$

is the rotation dyadic.

The Hooke's law equations pertain to properties at a point and do not involve space derivatives; consequently, Eqs. (10.23) and (10.25), which were derived for Cartesian coordinates, are also applicable to other orthogonal coordinate systems. In dyadic form, these are

$$\mathscr{T} = 2G\mathscr{E} + \mathscr{L}\lambda\epsilon \tag{12.47}$$

or

$$\mathscr{E} = \frac{1+\nu}{E}\mathscr{T} - \frac{\nu}{E}\mathscr{L}\Theta \tag{12.48}$$

where $\epsilon = \nabla \cdot \mathbf{u}$ and $\Theta = \sum_i \tau_{ii}$.

Thus, we have derived the 15 basic equations, (12.42), (12.43), and (12.47). The rest of the elasticity equations can be obtained by combining these 15 equations as in Chapter 11.

12.7. Summary of Vector and Dyadic Operators in Cylindrical and Spherical Coordinates

In this section we shall give the expressions for various operators in cylindrical and spherical coordinates. These two coordinate systems

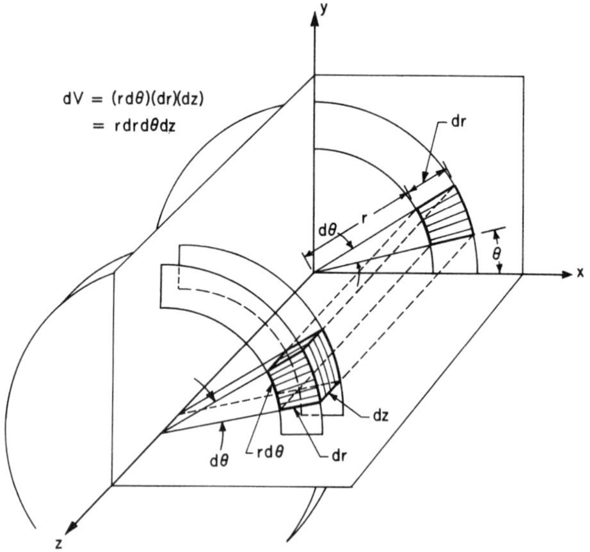

Fig. 12.4 An Element in Cylindrical Coordinates

are among the most important ones in engineering application. An element in cylindrical coordinates is shown in Fig. 12.4 and that in spherical coordinates in Fig. 12.5.

For cylindrical coordinates,

$$\alpha_1 = r \qquad \alpha_2 = \theta \qquad \alpha_3 = z \qquad h_1 = 1 \qquad h_2 = r \qquad h_3 = 1$$

$$\mathbf{c}_1 = \mathbf{c}_r \qquad \mathbf{c}_2 = \mathbf{c}_\theta \qquad \mathbf{c}_3 = \mathbf{c}_z$$

$$\mathbf{A} = \mathbf{c}_r A_r + \mathbf{c}_\theta A_\theta + \mathbf{c}_z A_z$$

Thus,

$$\frac{\partial \mathbf{c}_m}{\partial \alpha_n} = \begin{pmatrix} 0 & \mathbf{c}_\theta & 0 \\ 0 & -\mathbf{c}_r & 0 \\ 0 & 0 & 0 \end{pmatrix}$$

$$\text{grad } \phi = \mathbf{c}_r \frac{\partial \phi}{\partial r} + \mathbf{c}_\theta \frac{\partial \phi}{r \partial \theta} + \mathbf{c}_z \frac{\partial \phi}{\partial z}$$

$$\text{div } \mathbf{A} = \frac{1}{r} \frac{\partial}{\partial r} (r A_r) + \frac{\partial A_\theta}{r \partial \theta} + \frac{\partial A_z}{\partial z}$$

$$\text{curl } \mathbf{A} = \mathbf{c}_r \left(\frac{\partial A_z}{r \partial \theta} - \frac{\partial A_\theta}{\partial z} \right) + \mathbf{c}_\theta \left(\frac{\partial A_r}{\partial z} - \frac{\partial A_z}{\partial r} \right) + \mathbf{c}_z \left[\frac{1}{r} \frac{\partial (A_\theta r)}{\partial r} - \frac{1}{r} \frac{\partial A_r}{\partial \theta} \right]$$

$$\nabla^2 \phi = \frac{1}{r} \left[\frac{\partial}{\partial r} \left(r \frac{\partial \phi}{\partial r} \right) + \frac{\partial}{\partial \theta} \left(\frac{1}{r} \frac{\partial \phi}{\partial \theta} \right) + \frac{\partial}{\partial z} \left(r \frac{\partial \phi}{\partial z} \right) \right]$$

$$= \frac{1}{r} \frac{\partial}{\partial r} \left(r \frac{\partial \phi}{\partial r} \right) + \frac{1}{r^2} \frac{\partial^2 \phi}{\partial \theta^2} + \frac{\partial^2 \phi}{\partial z^2}$$

$$= \frac{\partial^2 \phi}{\partial r^2} + \frac{1}{r} \frac{\partial \phi}{\partial r} + \frac{1}{r^2} \frac{\partial^2 \phi}{\partial \theta^2} + \frac{\partial^2 \phi}{\partial z^2}$$

$$\nabla^2 \mathbf{A} = \mathbf{c}_r \left[\nabla^2 A_r - \frac{A_r}{r^2} - \frac{2}{r^2} \frac{\partial A_\theta}{\partial \theta} \right] + \mathbf{c}_\theta \left[\nabla^2 A_\theta - \frac{A_\theta}{r^2} + \frac{2}{r^2} \frac{\partial A_r}{\partial \theta} \right] + \mathbf{c}_z \nabla^2 A_z$$

$$\nabla \mathbf{A} = \frac{\partial A_r}{\partial r} \mathbf{c}_r \mathbf{c}_r + \frac{1}{r} \left(\frac{\partial A_\theta}{\partial \theta} + A_r \right) \mathbf{c}_\theta \mathbf{c}_\theta + \frac{\partial A_z}{\partial z} \mathbf{c}_z \mathbf{c}_z + \frac{\partial A_\theta}{\partial r} \mathbf{c}_r \mathbf{c}_\theta$$
$$+ \frac{1}{r} \left(\frac{\partial A_r}{\partial \theta} - A_\theta \right) \mathbf{c}_\theta \mathbf{c}_r + \frac{\partial A_z}{\partial r} \mathbf{c}_r \mathbf{c}_z + \frac{\partial A_r}{\partial z} \mathbf{c}_z \mathbf{c}_r$$
$$+ \frac{1}{r} \frac{\partial A_z}{\partial \theta} \mathbf{c}_\theta \mathbf{c}_z + \frac{\partial A_\theta}{\partial z} \mathbf{c}_z \mathbf{c}_\theta$$

$$\mathbf{A\nabla} = \frac{\partial A_r}{\partial r}\mathbf{c}_r\mathbf{c}_r + \frac{1}{r}\left(\frac{\partial A_\theta}{\partial \theta} + A_r\right)\mathbf{c}_\theta\mathbf{c}_\theta + \frac{\partial A_z}{\partial z}\mathbf{c}_z\mathbf{c}_z + \frac{\partial A_\theta}{\partial r}\mathbf{c}_\theta\mathbf{c}_r$$
$$+ \frac{1}{r}\left(\frac{\partial A_r}{\partial \theta} - A_\theta\right)\mathbf{c}_r\mathbf{c}_\theta + \frac{\partial A_z}{\partial r}\mathbf{c}_z\mathbf{c}_r + \frac{1}{r}\frac{\partial A_z}{\partial \theta}\mathbf{c}_z\mathbf{c}_\theta$$
$$+ \frac{\partial A_\theta}{\partial z}\mathbf{c}_\theta\mathbf{c}_z + \frac{\partial A_r}{\partial z}\mathbf{c}_r\mathbf{c}_z$$

$$\operatorname{div}\mathscr{S} = \mathbf{c}_r\left[\frac{\partial S_{rr}}{\partial r} + \frac{1}{r}\frac{\partial S_{r\theta}}{\partial \theta} + \frac{\partial S_{rz}}{\partial z} + \frac{S_{rr} - S_{\theta\theta}}{r}\right]$$
$$+ \mathbf{c}_\theta\left[\frac{\partial S_{r\theta}}{\partial r} + \frac{1}{r}\frac{\partial S_{\theta\theta}}{\partial \theta} + \frac{\partial S_{\theta z}}{\partial z} + \frac{2}{r}S_{r\theta}\right]$$
$$+ \mathbf{c}_z\left[\frac{\partial S_{rz}}{\partial r} + \frac{1}{r}\frac{\partial S_{\theta z}}{\partial \theta} + \frac{\partial S_{zz}}{\partial z} + \frac{1}{r}S_{rz}\right]$$

$$|\mathscr{S}| = S_{rr} + S_{\theta\theta} + S_{zz}$$

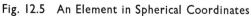

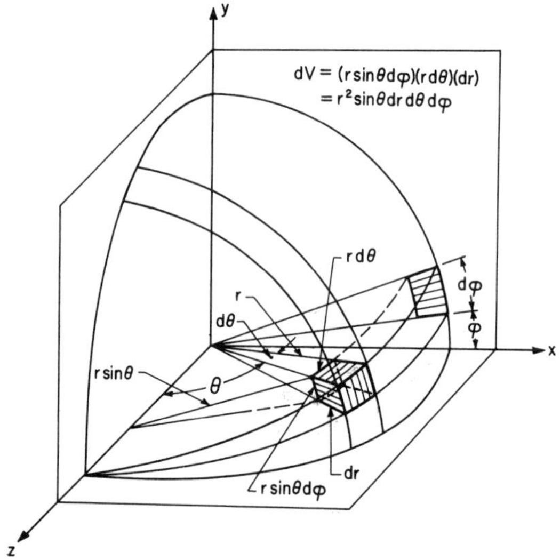

Fig. 12.5 An Element in Spherical Coordinates

For spherical coordinates,

$$\alpha_1 = r \qquad \alpha_2 = \theta \qquad \alpha_3 = \varphi \qquad h_1 = 1 \qquad h_2 = r \qquad h_3 = r\sin\theta$$
$$\mathbf{c}_1 = \mathbf{c}_r \qquad \mathbf{c}_2 = \mathbf{c}_\theta \qquad \mathbf{c}_3 = \mathbf{c}_\varphi$$
$$\mathbf{A} = \mathbf{c}_r A_r + \mathbf{c}_\theta A_\theta + \mathbf{c}_\varphi A_\varphi$$

Thus,

$$\frac{\partial \mathbf{c}_m}{\partial \alpha_n} = \begin{pmatrix} 0 & \mathbf{c}_\theta & \mathbf{c}_\varphi \sin\theta \\ 0 & -\mathbf{c}_r & \mathbf{c}_\varphi \cos\theta \\ 0 & 0 & -\mathbf{c}_r \sin\theta - \mathbf{c}_\theta \cos\theta \end{pmatrix}$$

$$\operatorname{grad} \phi = \mathbf{c}_r \frac{\partial \phi}{\partial r} + \frac{\mathbf{c}_\theta}{r}\frac{\partial \phi}{\partial \theta} + \frac{\mathbf{c}_\varphi}{r\sin\theta}\frac{\partial \phi}{\partial \varphi}$$

$$\operatorname{div} \mathbf{A} = \frac{1}{r^2}\frac{\partial}{\partial r}(r^2 A_r) + \frac{1}{r\sin\theta}\frac{\partial}{\partial \theta}(A_\theta \sin\theta) + \frac{1}{r\sin\theta}\frac{\partial A_\varphi}{\partial \varphi}$$

$$\operatorname{curl} \mathbf{A} = \frac{\mathbf{c}_r}{r\sin\theta}\left[\frac{\partial}{\partial \theta}(A_\varphi \sin\theta) - \frac{\partial A_\theta}{\partial \varphi}\right] + \frac{\mathbf{c}_\theta}{r}\left[\frac{1}{\sin\theta}\frac{\partial A_r}{\partial \varphi} - \frac{\partial}{\partial r}(rA_\varphi)\right] + \frac{\mathbf{c}_\varphi}{r}\left[\frac{\partial}{\partial r}(rA_\theta) - \frac{\partial A_r}{\partial \theta}\right]$$

$$\nabla^2 \phi = \frac{1}{r^2}\frac{\partial}{\partial r}\left(r^2 \frac{\partial \phi}{\partial r}\right) + \frac{1}{r^2\sin\theta}\frac{\partial}{\partial \theta}\left(\frac{\partial \phi}{\partial \theta}\sin\theta\right) + \frac{1}{r^2\sin^2\theta}\frac{\partial^2 \phi}{\partial \varphi^2}$$

$$\begin{aligned}\nabla^2 \mathbf{A} = {} & \mathbf{c}_r\left[\nabla^2 A_r - \frac{2}{r^2}A_r - \frac{2}{r^2\sin\theta}\frac{\partial}{\partial \theta}(A_\theta \sin\theta) - \frac{2}{r^2\sin\theta}\frac{\partial A_\varphi}{\partial \varphi}\right] \\ & + \mathbf{c}_\theta\left[\nabla^2 A_\theta - \frac{A_\theta}{r^2\sin^2\theta} + \frac{2}{r^2}\frac{\partial A_r}{\partial \theta} - \frac{2\cos\theta}{r^2\sin^2\theta}\frac{\partial A_\varphi}{\partial \varphi}\right] \\ & + \mathbf{c}_\varphi\left[\nabla^2 A_\varphi - \frac{A_\varphi}{r^2\sin^2\theta} + \frac{2}{r^2\sin\theta}\frac{\partial A_r}{\partial \varphi} + \frac{2\cos\theta}{r^2\sin^2\theta}\frac{\partial A_\theta}{\partial \varphi}\right]\end{aligned}$$

$$\begin{aligned}\nabla\mathbf{A} = {} & \frac{\partial A_r}{\partial r}\mathbf{c}_r\mathbf{c}_r + \frac{1}{r}\left(\frac{\partial A_\theta}{\partial \theta} + A_r\right)\mathbf{c}_\theta\mathbf{c}_\theta \\ & + \frac{1}{r\sin\theta}\left(\frac{\partial A_\varphi}{\partial \varphi} + A_r\sin\theta + A_\theta\cos\theta\right)\mathbf{c}_\varphi\mathbf{c}_\varphi + \frac{\partial A_\theta}{\partial r}\mathbf{c}_r\mathbf{c}_\theta \\ & + \frac{1}{r}\left(\frac{\partial A_r}{\partial \theta} - A_\theta\right)\mathbf{c}_\theta\mathbf{c}_r + \frac{\partial A_\varphi}{\partial r}\mathbf{c}_r\mathbf{c}_\varphi + \frac{1}{r\sin\theta}\left(\frac{\partial A_r}{\partial \varphi} - A_\varphi\sin\theta\right)\mathbf{c}_\varphi\mathbf{c}_r \\ & + \frac{1}{r}\frac{\partial A_\varphi}{\partial \theta}\mathbf{c}_\theta\mathbf{c}_\varphi + \frac{1}{r\sin\theta}\left(\frac{\partial A_\theta}{\partial \varphi} - A_\varphi\cos\theta\right)\mathbf{c}_\varphi\mathbf{c}_\theta\end{aligned}$$

$$\begin{aligned}\mathbf{A}\nabla = {} & \frac{\partial A_r}{\partial r}\mathbf{c}_r\mathbf{c}_r + \frac{1}{r}\left(\frac{\partial A_\theta}{\partial \theta} + A_r\right)\mathbf{c}_\theta\mathbf{c}_\theta \\ & + \frac{1}{r\sin\theta}\left(\frac{\partial A_\varphi}{\partial \varphi} + A_r\sin\theta + A_\theta\cos\theta\right)\mathbf{c}_\varphi\mathbf{c}_\varphi + \frac{\partial A_\theta}{\partial r}\mathbf{c}_\theta\mathbf{c}_r \\ & + \frac{1}{r}\left(\frac{\partial A_r}{\partial \theta} - A_\theta\right)\mathbf{c}_r\mathbf{c}_\theta + \frac{\partial A_\varphi}{\partial r}\mathbf{c}_\varphi\mathbf{c}_r + \frac{1}{r\sin\theta}\left(\frac{\partial A_r}{\partial \varphi} - A_\varphi\sin\theta\right)\mathbf{c}_r\mathbf{c}_\varphi \\ & + \frac{1}{r}\frac{\partial A_\varphi}{\partial \theta}\mathbf{c}_\varphi\mathbf{c}_\theta + \frac{1}{r\sin\theta}\left(\frac{\partial A_\theta}{\partial \varphi} - A_\varphi\cos\theta\right)\mathbf{c}_\theta\mathbf{c}_\varphi\end{aligned}$$

$$\operatorname{div}\mathscr{S}=\mathbf{c}_r\left[\frac{\partial S_{rr}}{\partial r}+\frac{1}{r\sin\theta}\frac{\partial S_{r\varphi}}{\partial\varphi}+\frac{1}{r}\frac{\partial S_{r\theta}}{\partial\theta}+2\,S_{rr}-S_{\varphi\varphi}-S_{\theta\theta}+S_{r\theta}\cot\theta\right]$$

$$+\mathbf{c}_\theta\left[\frac{\partial S_{r\theta}}{\partial r}+\frac{1}{r\sin\theta}\frac{\partial S_{\varphi\theta}}{\partial\varphi}+\frac{1}{r}\frac{\partial S_{\theta\theta}}{\partial\theta}+\frac{3S_{r\theta}+(S_{\theta\theta}-S_{\varphi\varphi})\cot\theta}{r}\right]$$

$$+\mathbf{c}_\varphi\left[\frac{\partial S_{r\varphi}}{\partial r}+\frac{1}{r\sin\theta}\frac{\partial S_{\varphi\varphi}}{\partial\varphi}+\frac{1}{r}\frac{\partial S_{\theta\varphi}}{\partial\theta}+\frac{3S_{r\varphi}+2S_{\varphi\theta}\cot\theta}{r}\right]$$

$$|\mathscr{S}|=S_{rr}+S_{\theta\theta}+S_{\varphi\varphi}$$

PROBLEMS

12-1 Show that the Jacobian of x_1, x_2, x_3 with respect to α_1, α_2, α_3, is equal to $h_1 h_2 h_3$. Hint: Use Eqs. (12.4) and (12.6) and the fact that the determinant (8.48) is unity.

12-2 Derive the expression for $\partial\mathbf{c}_1/\partial\alpha_2$ in Eqs. (12.13). Hint: Utilize the relation

$$2\sum_n\frac{\partial^2 x_n}{\partial\alpha_1\,\partial\alpha_2}\frac{\partial x_n}{\partial\alpha_3}=\sum_n\left[\frac{\partial}{\partial\alpha_2}\left(\frac{\partial x_n}{\partial\alpha_1}\frac{\partial x_n}{\partial\alpha_3}\right)-\frac{\partial}{\partial\alpha_3}\left(\frac{\partial x_n}{\partial\alpha_1}\frac{\partial x_n}{\partial\alpha_2}\right)+\frac{\partial}{\partial\alpha_1}\left(\frac{\partial x_n}{\partial\alpha_2}\frac{\partial x_n}{\partial\alpha_3}\right)\right]$$
$$=0$$

12-3 Show that the elementary volume of a rectangular parallelepiped with sides $h_1\,d\alpha_1$, $h_2\,d\alpha_2$, and $h_3\,d\alpha_3$ is

$$dV=\mathscr{J}\left(\frac{x_1, x_2, x_3}{\alpha_1, \alpha_2, \alpha_3}\right)d\alpha_1\,d\alpha_2\,d\alpha_3$$

12-4 Verify Eq. (12.14) for the expression for the radius of curvature of the α_2 coordinate line.

12-5 Derive Eq. (12.23). Keep in mind that the unit vectors $\mathbf{c}_1$, $\mathbf{c}_2$, and $\mathbf{c}_3$ are orthogonal and that the following vector formulas are true in any coordinate system:

$$\nabla\cdot(\phi\mathbf{A})=\phi(\nabla\cdot\mathbf{A})+(\nabla\phi)\cdot\mathbf{A}$$

$$\nabla\cdot(\mathbf{A}\times\mathbf{B})=\mathbf{B}\cdot(\nabla\times\mathbf{A})-\mathbf{A}\cdot(\nabla\times\mathbf{B})$$

$$\nabla\times(\nabla\phi)=0$$

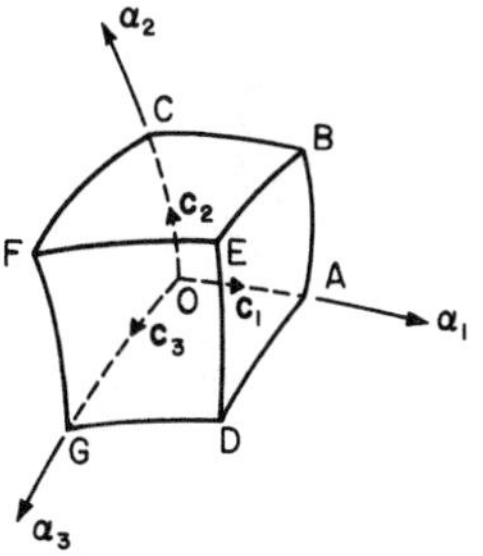

Problem 12-6

12-6 Derive the expression for div **A**, Eq. (12.24), by considering an infinitesimal element as shown and the relations

$$\int_V \operatorname{div} \mathbf{A}\, dV = \int_S \mathbf{A} \cdot \boldsymbol{\nu}\, dS \qquad \operatorname{div} \mathbf{A} \neq \lim_{V \to 0} \frac{\int_S \mathbf{A} \cdot \boldsymbol{\nu}\, dS}{V}$$

$$OA = h_1\, d\alpha_1$$

$$OC = h_2\, d\alpha_2$$

$$OG = h_3\, d\alpha_3$$

On $OABC$

$$\int \mathbf{A} \cdot \boldsymbol{\nu}\, dS = -A_3 h_1 h_2\, d\alpha_1\, d\alpha_2$$

On $DEFG$

$$\int \mathbf{A} \cdot \boldsymbol{\nu}\, dS = A_3 h_1 h_2\, d\alpha_1\, d\alpha_2 + \frac{\partial(A_3 h_1 h_2)}{\partial \alpha_3}\, d\alpha_1\, d\alpha_2\, d\alpha_3$$

12-7 Derive Eq. (12.26). The following vector formulas may be used

$$\nabla \times (\phi \mathbf{A}) = \phi(\nabla \times \mathbf{A}) + (\nabla\phi \times \mathbf{A})$$

$$\nabla \times (\nabla\phi) = 0$$

12-8 Derive the expression for $\nabla^2 \mathbf{A}$ in cylindrical coordinates by using (a) Eq. (12.29); (b) Eq. (12.30).

12-9 Prove that

$$\mathscr{A} \cdot \mathbf{B} = \mathbf{B} \cdot \mathscr{A}_c$$

$$(\mathscr{A} \cdot \mathscr{B})_c = (\mathscr{B}_c \cdot \mathscr{A}_c)$$

12-10 Derive all of the governing equations for plane strain and plane stress problems in vector and dyadic notation.

13

Displacement Functions and Stress Functions

13.1. Introduction

In Chapter 4 we demonstrated how the governing equations in elasticity may be reduced to Navier's equations in the displacement formulation, and to the equilibrium equations and Beltrami-Michell compatibility equations in the stress formulation. In Chapters 10 and 11, the same analyses were presented in Cartesian tensor and dyadic notations, respectively. In Chapter 5, it was also shown that in two-dimensional problems, the system of equilibrium and compatibility equations can be further simplified by introducing Airy's stress function. In this chapter, we shall show that Airy's stress function may be considered as a special case of a general set of stress functions. The stress function in torsion problems is another special case. In the displacement formulation, there are many displacement functions, the derivatives of which may be combined to give the displacement vector. In the following sections we shall first discuss the displacement functions and then the stress functions. The main purpose of this chapter is to point out many of the classical formulations in the theory of elasticity, and their interrelation. In some cases details are omitted, but references are supplied for those who wish to investigate these matters in more depth. These references are given at the end of the chapter, and are referred to in the text, where they are appropriate, by numbers in brackets, e.g., [1].

All equations which are applicable to any coordinate system will be written in vector and dyadic notation. The corresponding Cartesian tensor form of these equations can be deduced from these easily. It may be mentioned also that in the displacement formulation, only vector quantities are involved, and thus, dyadic notation is not required.

The relations in the various stress and displacement formulations are summarized in Tables 13.1 and 13.2. These tables may be helpful when reading the detailed discussion given below.

We shall only consider the case of zero body force. This is not a serious restriction because, in most practical problems, the body force is either a gravitational or centrifugal force (or zero), and it is not difficult to find a particular solution for a given body force **F** with no boundary conditions imposed. The final solution may be obtained by combining this particular solution with the solution of the homogeneous governing equations (no body force) which satisfies the prescribed boundary stress (or displacement) minus the boundary stress (or displacement) from the particular solution.

In our derivations, the following vector formulas will be used ([1], p. 18; [2], Chapter 4).

$$\begin{aligned}
&\nabla \cdot \nabla \times \mathbf{A} = \nabla \cdot (\nabla \times \mathbf{A}) = 0 && \text{(a)}\\
&\nabla \times \nabla\phi = \nabla \times (\nabla\phi) = 0 && \text{(b)}\\
&\nabla \times (\nabla \times \mathbf{A}) = \nabla(\nabla \cdot \mathbf{A}) - \nabla^2\mathbf{A} && \text{(c)}\\
&\nabla \cdot \nabla\phi = \nabla \cdot (\nabla\phi) = \nabla^2\phi && \text{(d)}\\
&\nabla \cdot (\nabla^2\mathbf{A}) = \nabla^2(\nabla \cdot \mathbf{A}) && \text{(e)}\\
&\nabla^2(\nabla B) = \nabla(\nabla^2 B) && \text{(f)}\\
&\nabla^2(\mathbf{A} \cdot \mathbf{s}) = 2\nabla \cdot \mathbf{A} + \mathbf{s} \cdot \nabla^2\mathbf{A} && \text{(g)}\\
&\qquad (\mathbf{s} = \text{position vector}) && \\
&\nabla \cdot (A\mathbf{B}) = A\nabla \cdot \mathbf{B} + (\nabla A) \cdot \mathbf{B} && \text{(h)}\\
&\nabla^2(B\mathbf{s}) = 2\nabla B + \mathbf{s}\nabla^2 B && \text{(i)}
\end{aligned} \qquad (13.1)$$

The following dyadic formulas are also included for easy reference.

$$\begin{aligned}
&\nabla\mathbf{s} = \mathscr{L} \qquad \text{where} \quad \mathscr{L} = \mathbf{ii} + \mathbf{jj} + \mathbf{kk} && \text{(a)}\\
&(\boldsymbol{\phi} \cdot \nabla)\mathbf{s} = \boldsymbol{\phi} \cdot \nabla\mathbf{s} = \boldsymbol{\phi} \cdot \mathscr{L} = \boldsymbol{\phi} && \text{(b)}\\
&\nabla(\mathbf{A} \cdot \mathbf{B}) = (\nabla\mathbf{A}) \cdot \mathbf{B} + (\nabla\mathbf{B}) \cdot \mathbf{A} && \text{(c)}\\
&\nabla(\mathbf{A} \cdot \mathbf{s}) = \mathbf{A} + (\nabla\mathbf{A}) \cdot \mathbf{s} && \text{(d)}
\end{aligned} \qquad (13.2)$$

13.2. Displacement Functions

Inside an elastic body, in the absence of body force, the displacement vector must satisfy Navier's equation

$$G\nabla^2\mathbf{u} + (\lambda + G)\nabla\nabla \cdot \mathbf{u} = 0 \qquad (13.3)$$

Table 13.1

DISPLACEMENT FUNCTIONS IN LINEAR ELASTICITY WITH NO BODY FORCE

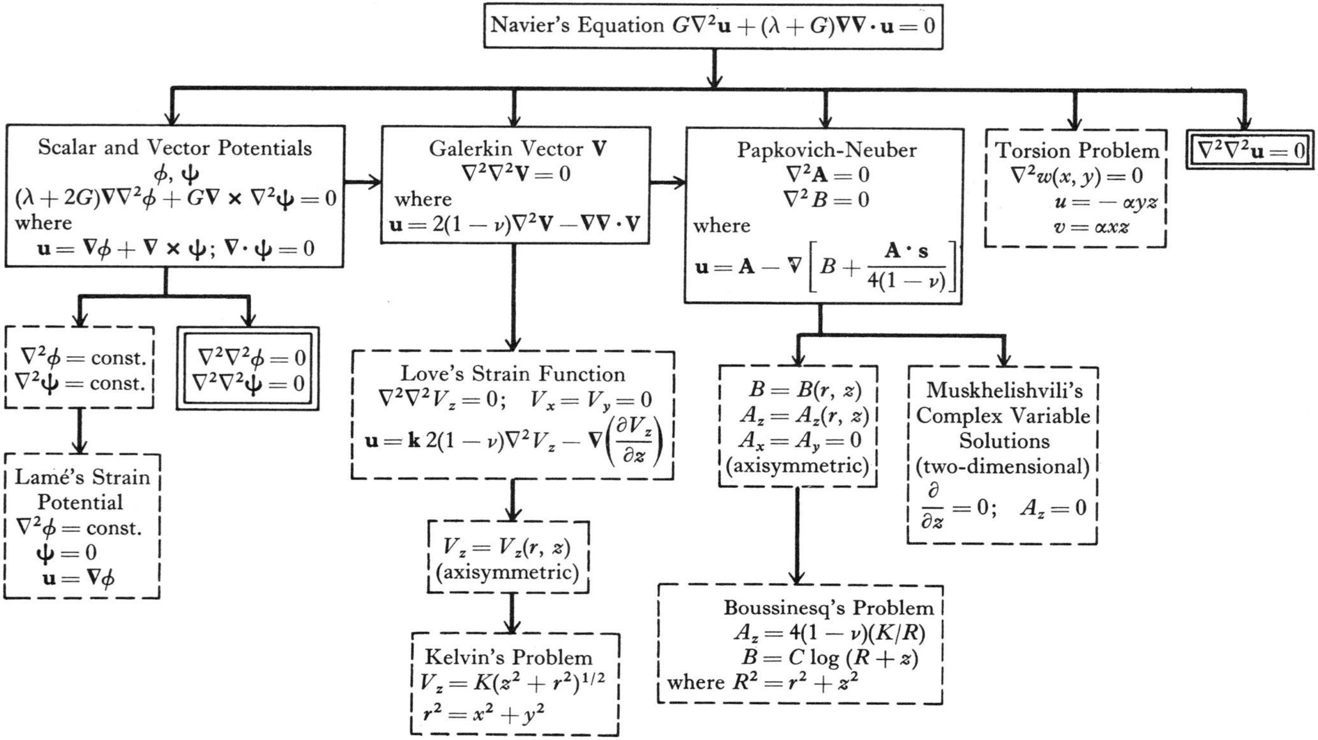

Table 13.2

STRESS FUNCTIONS IN LINEAR ELASTICITY WITH NO BODY FORCE

Equilibrium $\nabla \cdot \mathscr{T} = 0$

Beltrami-Michell Compatibility $\nabla^2 \mathscr{T} + \dfrac{1}{1+\nu} \nabla\nabla|\mathscr{T}| = 0$

where $|\mathscr{T}| = \sigma_x + \sigma_y + \sigma_z$

Dyadic Stress Function Φ

$\nabla^2(\nabla \times \Phi \times \nabla) + \dfrac{1}{1+\nu}\nabla\nabla(\nabla\cdot\nabla\cdot\Phi - \nabla^2|\Phi|) = 0$

where $\mathscr{T} = \nabla \times \Phi \times \nabla$; $\Phi = \Phi_c$

Morera

$\Phi = (\mathbf{ij}+\mathbf{ji})\phi_{xy} + (\mathbf{jk}+\mathbf{kj})\phi_{yz} + (\mathbf{ki}+\mathbf{ik})\phi_{xz}$

$\phi_{xx} = \phi_{yy} = \phi_{zz} = 0$

Maxwell

$\Phi = \mathbf{ii}\phi_{xx} + \mathbf{jj}\phi_{yy} + \mathbf{kk}\phi_{zz}$

$\phi_{xy} = \phi_{yz} = \phi_{zx} = 0$

Prandtl's Stress Function for Torsion ψ $\dfrac{\partial \psi}{\partial z} = 0$

$\nabla^2 \psi = -2G\alpha;$

$\phi_{xx,y} = z\psi_{,x}$

$\phi_{yy,x} = -z\psi_{,y}$

$\phi_{zz} = 0$

Airy's Stress Function ϕ°

$\nabla^2\nabla^2\phi^\circ = 0;$ $\dfrac{\partial \phi^\circ}{\partial z} = 0$

$\phi_{zx} = z\phi^\circ_{,x}$

$\phi_{yz} = z\phi^\circ_{,y}$

Plane Stress (approx.) $\phi_{xy} = 0$

Plane Strain $\phi_{xy,xy} = \nu\nabla^2\phi^\circ$

Prandtl's Stress Function for Torsion ψ

$\nabla^2 \psi = -2G\alpha$

$\psi = \phi_{yz,x};$ $\dfrac{\partial \phi_{yz}}{\partial z} = 0$

$\phi_{xy} = \phi_{xz} = 0$

Airy's Stress Function ϕ_{zz}

$\nabla^2\nabla^2\phi_{zz} = 0;$ $\dfrac{\partial \phi_{zz}}{\partial z} = 0$

Plane Stress (approximate) $\phi_{xx} = \phi_{yy} = 0$

Plane Strain $\phi_{xx} = \phi_{yy} = \nu\phi_{zz}$

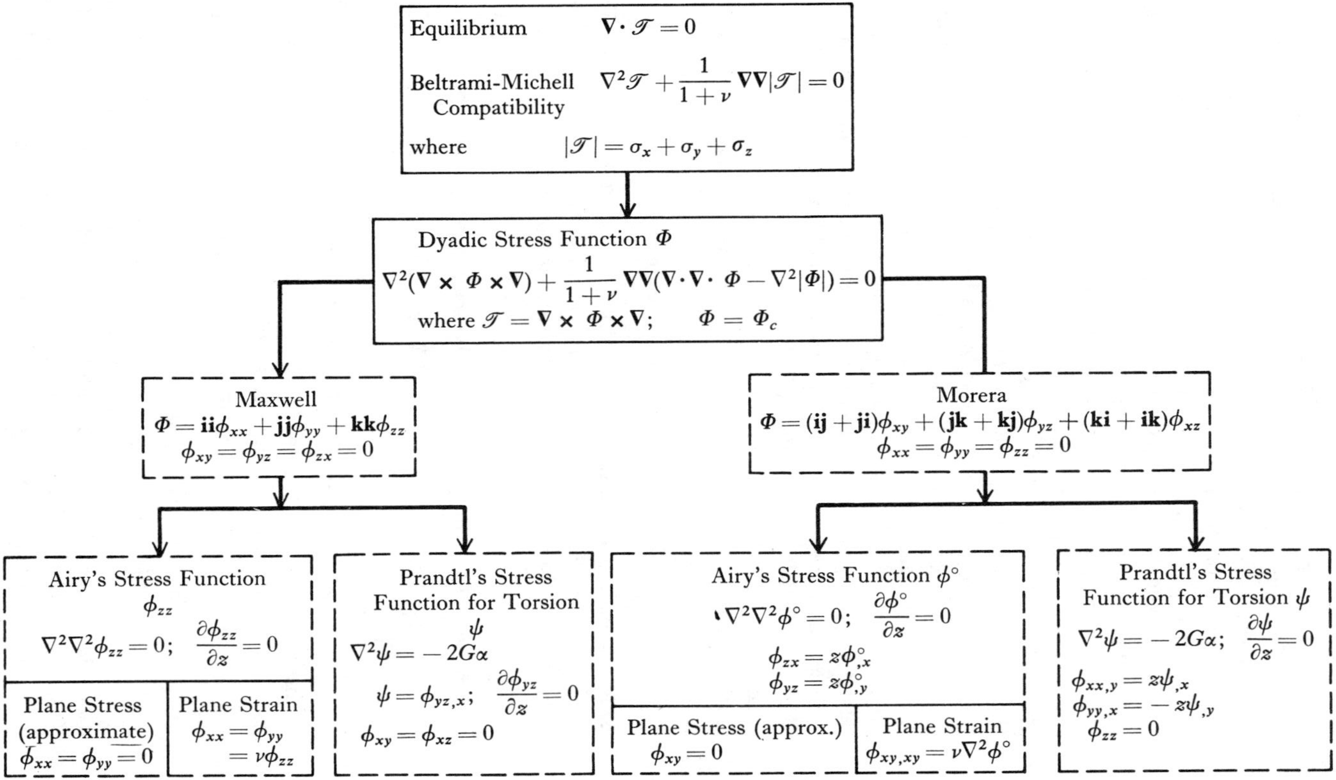

In addition, on the surface, the boundary conditions, either $\mathbf{u} = \mathbf{u}_b$ or $\mathbf{T}^\mu = \mathscr{T} \cdot \boldsymbol{\mu}$, must be satisfied. We shall be concerned with the field equation (Navier's equation) only, and introduce some potentials or displacement functions such that the displacement vector results from a combination of certain derivatives of these functions. These displacement functions are governed by equations, such as the Laplace equation or biharmonic equation, which have been analyzed thoroughly by mathematicians. Therefore, it is often easier to find the solutions of Laplace or biharmonic equations for the displacement potential than to find solutions for Navier's equation directly. It must be noted, however, that although the governing equations are simplified, the expressions for the boundary conditions become more complicated.

Let us first consider the scalar and vector potentials of the displacement vector. According to the well-known Helmholtz theorem, every vector field which is finite, uniform, and continuous, and which vanishes at infinity, may be resolved into the sum of an irrotational field and a solenoidal field ([3], p. 52; [4], p. 184). A field **A** is irrotational if $\nabla \times \mathbf{A} = 0$; a field **B** is solenoidal if $\nabla \cdot \mathbf{B} = 0$. Consequently, for every displacement vector field **u** satisfying the above conditions, there exist a scalar potential field $\phi(x_1, x_2, x_3)$ and a vector potential field $\boldsymbol{\psi}(x_1, x_2, x_3)$, such that

$$\mathbf{u} = \nabla\phi + \nabla \times \boldsymbol{\psi} \tag{13.4}$$

since according to Eqs. (13.1), $\nabla \cdot \nabla \times \boldsymbol{\psi} = 0$ and $\nabla \times \nabla\phi = 0$. Since **u** has three scalar components while ϕ and $\boldsymbol{\psi}$ together represent four scalar fields, no generality is lost by further imposing the condition

$$\nabla \cdot \boldsymbol{\psi} = 0 \tag{13.5}$$

From Eqs. (13.4) and (13.5) it can be seen that

$$\epsilon = \nabla \cdot \mathbf{u} = \nabla \cdot \nabla\phi = \nabla^2\phi \tag{13.6}$$

and

$$2\boldsymbol{\omega} = \nabla \times \mathbf{u} = \nabla \times (\nabla \times \boldsymbol{\psi}) \tag{13.7}$$

Substituting Eq. (13.4) into Eq. (13.3) and after some vector manipulation, we obtain the governing equation for ϕ and $\boldsymbol{\psi}$,

$$(\lambda + 2G)\nabla\nabla^2\phi + G\nabla \times \nabla^2\boldsymbol{\psi} = 0 \tag{13.8}$$

Any set of functions ϕ and $\boldsymbol{\psi}$ which satisfy this equation will produce, when substituted into Eq. (13.4), a displacement field that satisfies Navier's equation. Conversely, for every **u** that satisfies Navier's equation, there exists at least one set of ϕ and $\boldsymbol{\psi}$ which satisfies Eqs. (13.8), (13.4), and (13.5). Since **u** is equated to a combination of *first derivatives* of ϕ and $\boldsymbol{\psi}$, for a given **u**, ϕ and $\boldsymbol{\psi}$ will not be uniquely determined. But we are only interested in the complete determination of **u**, thus any set of ϕ and $\boldsymbol{\psi}$ that satisfies Eq. (13.8) may be used. In Table 13.1 the sides of the box containing Navier's equation and those of the box containing the scalar and vector potentials ϕ and $\boldsymbol{\psi}$ are both made of single solid lines. This indicates that every solution of the equations in one box is also a solution of those in the other.

Some particular solutions (not a general solution) of Eqs. (13.8) are functions which satisfy the following equations:

$$\begin{aligned} \nabla^2\phi &= \text{const.} \\ \nabla^2\boldsymbol{\psi} &= \text{const.} \end{aligned} \tag{13.9}$$

These two equations are enclosed by a box with dotted lines, indicating that solutions of these equations are also solutions of Eqs. (13.8); but a solution of Eqs. (13.8) is not necessarily a solution of Eqs. (13.9).

It may be pointed out here that if ϕ and $\boldsymbol{\psi}$ are allowed to be functions of time, then for every (static) solution of Eq. (13.8) there exists a set of ϕ and $\boldsymbol{\psi}$ related to **u** by Eq. (13.4), which satisfies the equations

$$\nabla^2\phi = \frac{1}{c_1^2}\frac{\partial^2\phi}{\partial t^2} \qquad \nabla^2\boldsymbol{\psi} = \frac{1}{c_2^2}\frac{\partial^2\boldsymbol{\psi}}{\partial t^2}$$

where $c_1^2 = (\lambda + 2G)/\rho$, and $c_2^2 = G/\rho$ and ρ is the density. Upon combining, the time-dependent parts of ϕ and $\boldsymbol{\psi}$ cancel out to yield a static displacement field [5].

Among the solutions of Eqs. (13.9), if we choose the following,

$$\begin{aligned} \boldsymbol{\psi} &= 0 \\ \nabla^2\phi &= \text{const.} \end{aligned} \tag{13.10}$$

then the function ϕ is the well-known Lamé strain potential. It can be seen from Eq. (13.10) that any function satisfying Poisson's equation (also any harmonic function) may be used as ϕ and the resulting displacement, $\mathbf{u} = \nabla\phi$, will satisfy Navier's equation. Many useful solutions satisfying practical boundary conditions in cylindrical and spherical coordinates may be generated from this potential ([6], Chapter 5). For instance, for the class of plane strain axisymmetric problems (polar cylindrical coordinates) in which ϕ is independent of θ, the second of Eqs. (13.10) becomes (see Section 12.7)

$$\frac{d^2\phi}{dr^2} + \frac{1}{r}\frac{d\phi}{dr} = \text{const.}$$

for which the general solution is

$$\phi = Ar^2 + B\log r + C$$

where A, B, and C are arbitrary constants. Since $\mathbf{u} = \nabla\phi$, we have

$$\mathbf{u} = (2Ar + B/r)\mathbf{c}_r$$

where the constants are evaluated as explained in Section 5.5.

If we take the divergence of Eqs. (13.8), and keeping in mind that the divergence of a curl vanishes, Eq. (13.1a), we obtain

$$\nabla^2\nabla^2\phi = 0 \tag{13.11}$$

By taking the curl of Eqs. (13.8) and utilizing Eqs. (13.1b), (13.1c), and (13.5), we have

$$\nabla^2\nabla^2\boldsymbol{\psi} = 0 \tag{13.12}$$

These two equations show that both ϕ and $\boldsymbol{\psi}$ satisfy the biharmonic equation. Each of the three scalar functions ψ_x, ψ_y, and ψ_z, i.e., the Cartesian components of $\boldsymbol{\psi}$, is biharmonic; but the functions which are the components of $\boldsymbol{\psi}$ in curvilinear coordinates do not necessarily satisfy the biharmonic equation. Since Eq. (13.8) is a third-order partial differential equation for ϕ and $\boldsymbol{\psi}$, while Eqs. (13.11) and (13.12) are fourth-order equations, the latter are necessary, but not sufficient, for the satisfaction of Eq. (13.8). In Table 13.1, the box containing $\nabla^2\nabla^2\phi = 0$ and $\nabla^2\nabla^2\boldsymbol{\psi} = 0$ is drawn with double solid lines, indicating the fact that a solution of Eq. (13.8) must also be a solution of Eqs. (13.11) and (13.12); but a solution of the latter may not satisfy Eq. (13.8). An example of this is shown in Problem 13-1.

13.3. The Galerkin Vector

In the preceding section, the displacement vector was represented in terms of the first derivatives of a scalar function and a vector function. It is also convenient to represent **u** by a combination of the second derivatives of a *single* vector function. Let us consider a vector function **V**, which is related to the displacement vector by

$$\mathbf{u} = 2(1 - \nu)\nabla^2\mathbf{V} - \nabla\nabla\cdot\mathbf{V} \tag{13.13}$$

This vector, **V**, is known as the Galerkin vector. It can be proved ([6], Section 69; see also the references cited in [7]) that for any vector function **u**, there exists a vector function **V** which satisfies Eq. (13.13), and therefore the Galerkin vector supplies the general solution. Substituting Eq. (13.13) into Navier's equation, and utilizing Eqs. (13.1d), (13.1e), and (13.1f) and the relation $\lambda = 2\nu G/(1 - 2\nu)$, we obtain

$$\nabla^2\nabla^2\mathbf{V} = 0 \tag{13.14}$$

This shows that any biharmonic vector function may be used as the Galerkin vector, and the resulting displacement vector, according to

Eq. (13.13), always satisfies Navier's equation. For a given **u** distribution, **V** contains certain arbitrary functions since **u** involves second derivatives of **V**, but we are interested only in **u**; therefore the system of equations (13.13) and (13.14) is equivalent to Navier's equation. A solution (in **u**) of one is also a solution of the other, as indicated by the single solid line box surrounding the Galerkin vector in Table 13.1.

For the historic background on how Galerkin first introduced the three component functions of **V** and how Papkovich treated **V** as a vector, the reader is referred to Westergaard's book [6] and the article by Sternberg [7]. The Galerkin vector is related to the scalar potential ϕ and vector potential $\boldsymbol{\psi}$, as defined in Eqs. (13.4) and (13.8), by the following equations:

$$\begin{aligned} \phi &= -\nabla \cdot \mathbf{V} \\ \nabla \times \boldsymbol{\psi} &= 2(1-\nu)\nabla^2 \mathbf{V} \end{aligned} \tag{13.15}$$

If we restrict the **V** to be not only biharmonic, but also harmonic, i.e., $\nabla^2 \mathbf{V} = 0$, then $\nabla \times \boldsymbol{\psi} = 0$ and ϕ satisfies the equation

$$\nabla^2 \phi = -\nabla \cdot (\nabla^2 \mathbf{V}) = 0 \tag{13.16}$$

This ϕ is called Lamé's strain potential. Thus Lamé's strain potential can be derived from either ϕ and $\boldsymbol{\psi}$, or from **V**. The flow chart shown in Table 13.1 is by no means the only way to derive the relations between the various functions. For the sake of clarity, many lines joining boxes in two different columns are omitted (e.g., a line may be drawn between the Galerkin vector and Lamé's potential).

A few special applications of the Galerkin vector will now be discussed. If only the z component of **V** is nonvanishing, we have Love's strain function

$$\mathbf{V} = \mathbf{k} V_z \qquad V_x = V_y = 0 \tag{13.17}$$

The governing equation is reduced to the scalar biharmonic equation

$$\nabla^2 \nabla^2 V_z = 0 \tag{13.18}$$

and the expression for **u** is

$$\mathbf{u} = 2(1-\nu)\mathbf{k}\nabla^2 V_z - \nabla\left(\frac{\partial V_z}{\partial z}\right) \tag{13.19}$$

Notice that the z coordinate is rectilinear. Therefore Love's strain function is only applicable in cylindrical coordinates. The cylindrical coordinates could be circular cylindrical, elliptical cylindrical, or parabolic cylindrical, as long as one of the coordinates is rectilinear. It is customary to use the terms "cylindrical," "circular cylindrical," and "polar cylindrical" as being synonymous, as we did in Chapters 5 and 11. An example of the problems that can be solved by this function is

$$V_z = (A + Bcz)\psi$$

where

$$\psi = \cos ax \cos by \, e^{-cz}$$

$$c = \sqrt{a^2 + b^2}$$

which is the solution for the problem of a vertical pressure $p = p_0 \cos(ax)\cos(by)$ applied on the horizontal surface of the semi-infinite solid $z \geq 0$ (see Problem 13-4 and [6], p. 131).

In another class of problems, we may use circular cylindrical coordinates and restrict V_z to be a function of r and z only, i.e., independent of θ. One application of this is the problem of a single concentrated force acting in the interior of an infinite body. This problem is known as Kelvin's problem. In this case (see Problem 13-5),

$$V_z = K(z^2 + r^2)^{1/2}$$

$$K = \frac{P}{4\pi(1 - \nu)} \tag{13.20}$$

where $r^2 = x^2 + y^2$ and $4GP$ is the magnitude of the force, which acts in the z direction.

13.4. The Solution of Papkovich-Neuber

The potentials ϕ and ψ must satisfy Eq. (13.8), which is a third-order equation. The biharmonic equation governing the Galerkin vector is of the fourth order. It is desirable to find a system of second-order equations which is equivalent to Navier's equation. (Remember that harmonic functions can be used for ϕ and ψ, but they constitute some particular solutions, not the general solution.) The approach used by Papkovich and Neuber is to use a combination of harmonic functions to represent the displacement vector **u**. Let us introduce

$$\mathbf{u} = \mathbf{A} - \nabla\left[B + \frac{\mathbf{A}\cdot\mathbf{s}}{4(1-\nu)}\right] \tag{13.21}$$

where **A** is a vector field, B is a scalar field, and **s** is the position vector. Substituting Eq. (13.21) into Eq. (13.3), and keeping in mind formulas (13.1f), (13.1d), and (13.1g) and the relation $\lambda = 2\nu G/(1 - 2\nu)$, we obtain

$$G\nabla^2\mathbf{A} - (\lambda + 2G)\nabla\nabla^2 B - \left(\frac{\lambda + G}{2}\right)\nabla(\mathbf{s}\cdot\nabla^2\mathbf{A}) = 0 \tag{13.22}$$

This equation is satisfied if

$$\nabla^2 \mathbf{A} = 0$$
$$\nabla^2 B = 0 \tag{13.23}$$

In other words, any four harmonic functions, A_x, A_y, A_z, and B, can be substituted into Eq. (13.21), and the resulting $\mathbf{u}$ always satisfies Navier's equation. The functions $\mathbf{A}$ and B may also be derived from the Galerkin vector $\mathbf{V}$. If we write

$$\mathbf{A} = 2(1-\nu)\nabla^2 \mathbf{V}$$
$$B = \nabla \cdot \mathbf{V} - \frac{\mathbf{A} \cdot \mathbf{s}}{4(1-\nu)} \tag{13.24}$$

it can be seen that Eq. (13.21) becomes Eq. (13.13). This connection between $\mathbf{A}$, B, and $\mathbf{V}$ was observed by Mindlin ([8], p. 373). He also proved that the Papkovich-Neuber solution is a complete, or general, solution, by using the Stokes-Helmholtz theory on resolution of vector fields.

The four scalar functions $\mathbf{A}$ and B are not completely independent. It can be proved ([7] and [9]) that for an arbitrary three-dimensional convex domain, the number of independent functions is reducible from four to three.

A special form of the solution based on Eq. (13.21) for axisymmetric problems is

$$A_x = A_y = 0$$
$$A_z = A_z(r, z) \tag{13.25}$$
$$B = B(r, z)$$

where $r^2 = x^2 + y^2$. Boussinesq's problem of a force P acting in the z direction at the origin of coordinates on the semi-infinite solid occupying the space $z \geq 0$, has the solution

$$A_z = 4(1-\nu)\frac{K}{R} \tag{13.26}$$
$$B = C \log(R + z)$$

where

$$R^2 = r^2 + z^2 \qquad K = \frac{P}{4G\pi} \qquad C = -\frac{(1-2\nu)P}{4G\pi}$$

More details on Boussinesq's problem may be found in Chapter 6 of [6] and Chapter 8 of [4].

The solution of Papkovich-Neuber may be specialized, in two-dimensional problems, to the complex variable solutions developed by Kolossoff and Muskhelishvili. On carrying out the indicated differentiation in Eq. (13.21) and simplifying, we obtain

$$\mathbf{u} = \frac{3 - 4\nu}{4(1-\nu)} \mathbf{A} - \nabla B - \frac{1}{4(1-\nu)} (\nabla \mathbf{A}) \cdot \mathbf{s} \qquad (13.27)$$

where Eq. (13.2d) was used. Limiting ourself to two-dimensional problems with $w = A_z = \partial/\partial z = 0$, and introducing the harmonic functions ϕ_1, ϕ_2, ψ_1, and ψ_2 as follows:

$$\begin{aligned} \phi_\alpha(x_1, x_2) &= \frac{G}{2(1-\nu)} A_\alpha \qquad (\alpha = 1, 2) \\ \psi_1(x_1, x_2) &= 2GB_{,1} \\ \psi_2(x_1, x_2) &= -2GB_{,2} \end{aligned} \qquad (13.28)$$

where indicial notation is adapted, i.e., $A_x = A_1$, $x = x_1$, $y = x_2$, etc. By substituting Eqs. (13.28) into (13.27), we find that

$$2Gu_\alpha = (3 - 4\nu)\phi_\alpha \mp \psi_\alpha - x_\beta \phi_{\beta,\alpha} \qquad (\alpha = 1, 2) \qquad (13.29)$$

where the upper sign is for $\alpha = 1$ and lower sign for $\alpha = 2$. Let us construct an analytic function $\psi(z)$ of the complex variable $z = x_1 + ix_2$, where

$$\psi(z) = \psi_1(x_1, x_2) + i\psi_2(x_1, x_2) \qquad (13.30)$$

Since B is harmonic, it can be seen from Eqs. (13.28) that the Cauchy-Riemann conditions for an analytic function, $\psi_{1,1} = \psi_{2,2}$, $\psi_{1,2} = -\psi_{2,1}$, are satisfied. If we impose further restrictions on ϕ_1 and ϕ_2, such that

$$\phi_{1,1} = \phi_{2,2} \qquad \phi_{1,2} = -\phi_{2,1}$$

then

$$\phi(z) = \phi_1(x_1, x_2) + i\phi_2(x_1, x_2) \qquad (13.31)$$

is an analytic function. The complex displacement function, $u + iv$, is then

$$2G(u + iv) = (3 - 4v)\phi(z) - \psi^*(z^*) - z\phi^{*\prime}(z^*) \qquad (13.32)$$

where $\psi^*(z^*) = \psi_1 - i\psi_2$ is the conjugate of $\psi(z)$. Any analytic functions can be used as $\phi(z)$ and $\psi(z)$, and the resulting u and v will satisfy Navier's equation. By using theories in complex variables and conformal mapping, specific functions may be found for certain specified boundary conditions. Further details may be found in [10] and [11].

Before we leave the topic of displacement functions, it may be mentioned that any displacement vector field that satisfies Navier's equation must be biharmonic. This is true not only for the case of vanishing body force, but also for the case in which the body force vector field $\mathbf{F}$ is derived from a harmonic potential, or

$$\mathbf{F} = \nabla\phi \qquad \nabla\cdot\mathbf{F} = \nabla^2\phi = 0 \tag{13.33}$$

where ϕ is the body force potential field. Applying the divergence operator to the Navier equation (Table 11.1), and noting that $\nabla\cdot\nabla^2\mathbf{u} = \nabla^2(\nabla\cdot\mathbf{u})$, we get

$$\nabla^2(\nabla\cdot\mathbf{u}) = 0 \tag{13.34}$$

or, $\nabla\cdot\mathbf{u}$ is harmonic. Now, application of the Laplace operator to Navier's equation gives us

$$G\nabla^2\nabla^2\mathbf{u} + (\lambda + G)\nabla^2\nabla\nabla\cdot\mathbf{u} + \nabla^2\nabla\phi = 0$$

or

$$G\nabla^2\nabla^2\mathbf{u} + (\lambda + G)\nabla\nabla^2(\nabla\cdot\mathbf{u}) + \nabla\nabla^2\phi = 0$$

therefore,

$$\nabla^2\nabla^2\mathbf{u} = 0 \tag{13.35}$$

Again, it must be emphasized that, although the vector displacement $\mathbf{u}$, as well as its three Cartesian components u, v, and w, are all biharmonic, its components in any curvilinear coordinates are not necessarily biharmonic. Equation (13.35) is enclosed by double solid lines in Table 13.1, indicating that it is a necessary but not sufficient condition for the satisfaction of Navier's equation. For a particular example, see Problem 13-2.

The displacement components in the torsion problem of prismatic bars discussed in Chapter 6 also satisfy Navier's equation and may be considered as one class of its special solutions, as shown in Table 13.1.

13.5. Stress Functions

The stress tensor field is governed by the equilibrium equation and the Beltrami-Michell compatibility equation. In a manner analogous to the study of displacement functions, many authors have investigated the corresponding stress functions. Sternberg [7] presented a brief discussion on this topic and included a list of references. In this section, we shall first present the general three-dimensional stress functions and then specialize them into a few simple cases.

Since the dependent variable, stress, is a tensor quantity, vector notation alone is not sufficient in the symbolic approach, and dyadic notation must be used. In the absence of body force the equilibrium equation is

$$\nabla\cdot\mathscr{T} = 0 \tag{13.36}$$

and the Beltrami-Michell compatibility equation is

$$\nabla^2 \mathscr{T} + \frac{1}{1+\nu} \nabla\nabla |\mathscr{T}| = 0 \tag{13.37}$$

where

$$|\mathscr{T}| = \Theta = \sigma_x + \sigma_y + \sigma_z$$

is the trace of $\mathscr{T}$. (See Section 8.6 and Eq. (11.93).)

The equilibrium equation can be satisfied if we express $\mathscr{T}$ as the curl of some vector function, since the divergence of a curl vanishes. But $\mathscr{T}$ must also be symmetric, or $\mathscr{T} = \mathscr{T}_c$. To satisfy this symmetry condition, we shall introduce a symmetric dyadic function Φ, ($\Phi = \Phi_c$), such that

$$\mathscr{T} = \nabla \times \Phi \times \nabla \tag{13.38}$$

Following the definition of the curl operator on dyadics as given in Section 11.9, we find the unabridged form of (13.38) in Cartesian coordinates to be

$$\begin{aligned} \mathscr{T} &= \nabla \times \Phi \times \nabla \\ &= \mathbf{ii}(-\phi_{zz,yy} - \phi_{yy,zz} + 2\phi_{yz,yz}) + \mathbf{jj}(-\phi_{xx,zz} - \phi_{zz,xx} + 2\phi_{xz,xz}) \\ &\quad + \mathbf{kk}(-\phi_{xx,yy} - \phi_{yy,xx} + 2\phi_{xy,xy}) + (\mathbf{ij} + \mathbf{ji})(-\phi_{yz,zx} - \phi_{zx,yz} \\ &\quad + \phi_{zz,yx} + \phi_{yx,zz}) + (\mathbf{jk} + \mathbf{kj})(-\phi_{yx,xz} - \phi_{xz,yx} + \phi_{yz,xx} + \phi_{xx,yz}) \\ &\quad + (\mathbf{ki} + \mathbf{ik})(-\phi_{yz,xy} - \phi_{xy,yz} + \phi_{yy,xz} + \phi_{xz,yy}) \end{aligned} \tag{13.39}$$

This general stress function, in Cartesian notation, was first presented by Finzi [12]. The sum of the three normal stresses, Θ, or $|\mathscr{T}|$, is then

$$\Theta = |\mathscr{T}| = \nabla \cdot (\nabla \cdot \Phi) - \nabla^2 |\Phi| \tag{13.40}$$

where $|\Phi| = \phi_{xx} + \phi_{yy} + \phi_{zz}$. The compatibility equation becomes

$$\nabla^2(\nabla \times \Phi \times \nabla) + \frac{1}{1+\nu} \nabla\nabla[\nabla \cdot (\nabla \cdot \Phi) - \nabla^2 |\Phi|] = 0 \tag{13.41}$$

Not all of the six scalar functions, the six components of Φ, are independent. Two alternative ways of generating complete solutions from Φ are the Maxwell stress functions and the Morera stress functions (see Love [13]). If only the diagonal components of Φ are retained, we have the Maxwell stress functions, whereas if only the nondiagonal components are retained, we have the Morera stress functions. In Cartesian coordinates, the Maxwell stress functions are

$$\Phi = \mathbf{ii}\phi_{xx} + \mathbf{jj}\phi_{yy} + \mathbf{kk}\phi_{zz} \tag{13.42}$$

and the Morera stress functions are

$$\Phi = (\mathbf{ij} + \mathbf{ji})\phi_{xy} + (\mathbf{jk} + \mathbf{kj})\phi_{yz} + (\mathbf{ki} + \mathbf{ik})\phi_{zx} \tag{13.43}$$

Each of these two expressions is complete, i.e., for every stress distribution that satisfies the equilibrium equation there exists a set of Maxwell functions and a set of Morera stress functions (see [7] and [14]). In Table 13.2, the boxes enclosing the stress functions of Maxwell and Morera are drawn in dotted lines, not because they are incomplete, but because these equations are written in Cartesian coordinates and are not applicable in other coordinates.

Many authors have pointed out the fact that in the Maxwell representation, if only the ϕ_{zz} function is different from zero, it results in Airy's solution for two-dimensional problems. We may move one step further and make the distinction between plane stress and plane strain problems. If all components of Φ except ϕ_{zz} vanish, and with $\partial\phi_{zz}/\partial z = 0$, then from Eq. (13.39) we have,

$$\left.\begin{aligned} \sigma_x &= -\phi_{zz,yy} \qquad & \sigma_y &= -\phi_{zz,xx} \\ \tau_{xy} &= \phi_{zz,yx} \qquad & \tau_{zx} &= \tau_{zy} = 0 \end{aligned}\right\} \tag{13.44}$$

$$\sigma_z = 0 \tag{13.45}$$

Since σ_z vanishes, this is the case of plane stress. Three of the six scalar compatibility equations (13.41) are satisfied identically. The combination of the **ii** and **jj** components of Eq. (13.41) yields the biharmonic equation

$$\nabla^2 \nabla^2 \phi_{zz} = 0 \tag{13.46}$$

However, the **ii** and **jj** components individually, as well as the $(\mathbf{ij} + \mathbf{ji})$ component of Eq. (13.41), are not satisfied by $\Phi = \mathbf{kk}\phi_{zz}$, wherein lies the approximate nature of plane stress solutions as discussed in Section 5.3.

For plane strain problems, the normal stress in the axial direction, σ_z, does not vanish. It is related to σ_x and σ_y by the relation (see Eq. (4.9))

$$\sigma_z = \nu(\sigma_x + \sigma_y) \tag{13.47}$$

To satisfy this equation, we shall include ϕ_{xx} and ϕ_{yy} in addition to ϕ_{zz} within the Maxwell representation, but impose the condition

$$\phi_{xx} = \phi_{yy} = \nu\phi_{zz} \qquad (\phi_{xy} = \phi_{yz} = \phi_{zx} = 0) \tag{13.48}$$

and the condition that all these functions are independent of z. Substituting Eqs. (13.48) into Eqs. (13.39), we obtain the stress distribution for plane strain, Eqs. (13.44) and (13.47). Substitution of Eq. (13.48) into the compatibility equation (13.41) shows that each of the **ii**, **jj**, and **kk** components reduces to the biharmonic equation

$$\nabla^2 \nabla^2 \phi_{zz} = 0$$

while the other three scalar component equations are identically satisfied. The function ϕ_{zz}, in both plane stress and plane strain, is the negative of Airy's stress function.

In the Morera representation, if we choose the three functions as follows

$$\begin{aligned} \phi_{zx} &= z\phi^\circ_{,x} \\ \phi_{yz} &= z\phi^\circ_{,y} \\ \phi_{xy} &= 0 \end{aligned} \tag{13.49}$$

where ϕ° is independent of z, the stress components are

$$\left.\begin{aligned} \sigma_x = 2\phi_{yz,yz} = 2\phi^\circ_{,yy} \qquad \sigma_y = 2\phi_{xz,xz} = 2\phi^\circ_{,xx} \\ \tau_{xy} = -\phi_{yz,zx} - \phi_{zx,yz} = -2\phi^\circ_{,xy} \end{aligned}\right\} \tag{13.50}$$

$$\sigma_z = \tau_{xz} = \tau_{yz} = 0 \tag{13.51}$$

which represent a plane stress state. Again, the sum of the **ii** and **jj** components of the compatibility equation reduces to the biharmonic equation

$$\nabla^2 \nabla^2 \phi^\circ = 0$$

The **ii** and **jj** components individually, as well as the (**ij** + **ji**) component of the compatibility equation, are not satisfied.

In the Morera representation, plane strain problems are generated by taking

$$\begin{aligned} \phi_{zx} &= z\phi^\circ_{,x} \\ \phi_{yz} &= z\phi^\circ_{,y} \\ \phi_{xy,xy} &= \nu\nabla^2 \phi^\circ \end{aligned} \tag{13.52}$$

where ϕ° is independent of z. The stress components are, in addition to those given by Eqs. (13.50),

$$\begin{aligned} \sigma_z = 2\phi_{xy,xy} = 2\nu\nabla^2 \phi^\circ = \nu(\sigma_x + \sigma_y) \\ \tau_{xz} = \tau_{yz} = 0 \end{aligned} \tag{13.53}$$

which agree with the plane strain conditions. Each of the six scalar components of the compatibility equations is either satisfied identically or reduces to the biharmonic equation $\nabla^2 \nabla^2 \phi^\circ = 0$. The function ϕ° is equal to one-half of the Airy stress function.

Plane strain problems may also be produced by a combination of ϕ_{zz} and ϕ_{xy}. If we let

$$\Phi = (\mathbf{ij} + \mathbf{ji})\phi_{xy} + \mathbf{kk}\phi_{zz} \tag{13.54}$$

and require that

$$\phi_{xy,xy} = -\frac{\nu}{2}\nabla^2\phi_{zz} \tag{13.55}$$

we shall also obtain the plane strain distribution, and all the compatibility equations will be satisfied.

Prandtl's stress function for torsion may also be generated from Morera's expression. Let $\psi(x, y)$ be Prandtl's stress function (in Chapter 6, the notation $\phi(x, y)$ was used for the same function; see Section 6.1). If we choose the three Morera functions as

$$\begin{aligned} \phi_{yz} &= \phi_{yz}(x, y) \\ \phi_{yz,x} &= \psi \\ \phi_{xz} &= \phi_{xy} = 0 \end{aligned} \tag{13.56}$$

then from (13.39) we obtain the torsion stress distribution and from (13.41) we obtain the governing equations

$$\nabla^2 \frac{\partial}{\partial x}\psi = 0 \qquad \nabla^2 \frac{\partial}{\partial y}\psi = 0$$

which require that

$$\nabla^2\psi = \text{const.} = C \tag{13.57}$$

If we let $C = -2G\alpha$, this is exactly the governing equation for ψ as derived in Chapter 6, Eq. (6.7).

In the Maxwell representation, the torsion problem is obtained by the special restrictions,

$$\begin{aligned} \phi_{xx,y} &= z\psi_{,x} \\ \phi_{yy,x} &= -z\psi_{,y} \\ \phi_{zz} &= 0 \end{aligned} \tag{13.58}$$

where ψ is independent of z. The stress components are then

$$\begin{aligned} \tau_{yz} &= \phi_{xx,yz} = \psi_{,x} \\ \tau_{xz} &= \phi_{yy,xz} = -\psi_{,y} \\ \sigma_x &= \sigma_y = \sigma_z = \tau_{xy} = 0 \end{aligned} \tag{13.59}$$

From the compatibility equation (13.41), we have

$$\frac{\partial}{\partial y}(\nabla^2\psi) = 0 \qquad \frac{\partial}{\partial x}(\nabla^2\psi) = 0$$

or, we may let

$$\nabla^2\psi = -2G\alpha$$

Airy's stress function for two-dimensional problems and Prandtl's function for torsion in both the Maxwell and Morera representations are all summarized in Table 13.2.

PROBLEMS

13-1 Given a scalar and vector potential as follows:

$$\phi = x^3 \qquad \boldsymbol{\psi} = 0$$

Is the function ϕ biharmonic? Does the displacement field generated by these ϕ and $\boldsymbol{\psi}$ according to Eq. (13.4) satisfy Navier's equation?

13-2 Is the displacement field $\mathbf{u} = \mathbf{i}x^2$ biharmonic? Does it represent a possible displacement field in elasticity?

13-3 In a problem with spherical symmetry, all variables are functions of r only. What is the expanded form of $\nabla^2\phi = 0$ if spherical symmetry exists? What is the r component of the equation

$$\nabla^2\boldsymbol{\psi} = 0$$

where

$$\boldsymbol{\psi} = \mathbf{c}_r\psi_r + \mathbf{c}_\theta\psi_\theta + \mathbf{c}_\varphi\psi_\varphi$$

13-4 Show that when the only nonvanishing component of the Galerkin vector is

$$V_z = (A + Bcz)\psi$$

where

$$\psi = e^{-cz}\cos ax \cos by$$

$$c = \sqrt{a^2 + b^2}$$

V_z is biharmonic and satisfies the boundary condition of the vertical load described in Section 13.3 applied on the horizontal surface of a semi-infinite solid, $z \geq 0$.

13-5 Show that the solution of Kelvin's problem, Eq. (13.20), satisfies the governing equations and the corresponding boundary conditions.

13-6 Show that the expressions for the functions A_z and B in Eq. (13.26) give the solution of Boussinesq's problem. Refer to [6], p. 139.

13-7 Show that Boussinesq's problem may also be solved by the combination of a Galerkin vector $V_z = KR$ and a Lamé strain potential $\phi = C \log (R + z)$, where $R^2 = x^2 + y^2 + z^2$. (See Eq. 532, [6].)

13-8 In Chapter 10 (Table 10.1), Chapter 11 (Table 11.2), and this Chapter (Table 13.1), we have used the following form of Navier's equation (in the absence of body force),

$$G\nabla^2 \mathbf{u} + (\lambda + G)\nabla\nabla \cdot \mathbf{u} = 0$$

In Problem 11-1, the form

$$2(1 - \nu)\nabla\nabla \cdot \mathbf{u} - (1 - 2\nu)\nabla \times \nabla \times \mathbf{u} = 0$$

was used. Show that the following expressions are all equivalent to Navier's equation.

$$(\lambda + 2G)\nabla\nabla \cdot \mathbf{u} - G\nabla \times \nabla \times \mathbf{u} = 0$$

$$(\lambda + 2G)\nabla^2\mathbf{u} + (\lambda + G)\nabla \times \nabla \times \mathbf{u} = 0$$

$$2(1 - \nu)\nabla^2\mathbf{u} + \nabla \times \nabla \times \mathbf{u} = 0$$

$$\nabla^2[\mathbf{u} + \tfrac{1}{2}(1 + \lambda/G)\mathbf{s}(\nabla \cdot \mathbf{u})] = 0 \quad (\mathbf{s} = \text{position vector})$$

$$(1 - 2\nu)\nabla^2\mathbf{u} + \nabla\nabla \cdot \mathbf{u} = 0$$

13-9 Expand Eqs. (13.38) and (13.41) into circular cylindrical coordinates (r, θ, z). Select a special form of Φ, in these coordinates, which gives the solution of Kelvin's problem, and also select one for Boussinesq's problem.

Chapter 13

REFERENCES

1. Gerard Nadeau, *Introduction to Elasticity*, Holt, Rinehart, and Winston, Inc., New York, 1964.

2. Murray R. Spiegel, *Vector Analysis*, Schaum Publishing Company, New York, 1959.

3. Morse and Feshbach, *Methods of Theoretical Physics*, Part I, McGraw-Hill Book Company, Inc., New York, 1953.

4. Y. C. Fung, *Foundations of Solid Mechanics*, Prentice Hall, Inc., Englewood Cliffs, N. J., 1965.

5. E. Sternberg, "On the Integration of the Equations of Motion in the Classical Theory of Elasticity," *Arch. Rat. Mech. Anal.*, Vol. 6, pp. 34–50, 1960.

6. H. M. Westergaard, *Theory of Elasticity and Plasticity*, John Wiley & Sons, Inc., New York, 1952; also available from Dover Publications, New York, 1964.

7. E. Sternberg, "On Some Recent Developments in the Linear Theory of Elasticity," *Structural Mechanics*, Proceedings of the First Symposium on Naval Structural Mechanics, Pergamon Press, New York, 1960.

8. R. D. Mindlin, "Note on the Galerkin and Papkovich Stress Functions," *Bull. Amer. Math. Soc.*, Vol. 42, 1936.

9. Eubanks and Sternberg, "On the Completeness of the Boussinesq-Papkovich Stress Functions," *J. Rat. Mech. Anal.*, Vol. 5, p. 735, 1956.

10. I. S. Sokolnikoff, *Mathematical Theory of Elasticity*, Second Edition, McGraw-Hill Book Company, Inc., New York, 1956 (Chapter 7).

11. N. J. Muskhelishvili, *Some Basic Problems of the Mathematical Theory of Elasticity* (English translation by J. R. M. Radok), Noordhoff, Groningen, 1953.

12. B. Finzi, "Integrazione Delle Equazioni Indefinite Della Meccanica dei Sistemi Continui," *Atti. Accad. Nazl. Linei*, Ser. 6, Vol. 19, p. 578, 1934.

13. A. E. H. Love, *A Treatise on the Mathematical Theory of Elasticity*, Fourth Edition, Dover Publications, New York, p. 88, 1927.

14. R. V. Southwell, "Castigliano's Principle of Minimum Strain-Energy and the Conditions of Compatibility for Strain," *Timoshenko 60th Anniversary Volume*, Macmillan Co., New York, p. 211, 1938.

Index

A CATALOG OF SELECTED

DOVER BOOKS

IN SCIENCE AND MATHEMATICS

CATALOG OF DOVER BOOKS

ORDINARY DIFFERENTIAL EQUATIONS, Morris Tenenbaum and Harry Pollard. Exhaustive survey of ordinary differential equations for undergraduates in mathematics, engineering, science. Thorough analysis of theorems. Diagrams. Bibliography. Index. 818pp. 5⅜ × 8½. 64940-7 Pa. $16.95

STATISTICAL MECHANICS: Principles and Applications, Terrell L. Hill. Standard text covers fundamentals of statistical mechanics, applications to fluctuation theory, imperfect gases, distribution functions, more. 448pp. 5⅜ × 8½. 65390-0 Pa. $9.95

ORDINARY DIFFERENTIAL EQUATIONS AND STABILITY THEORY: An Introduction, David A. Sánchez. Brief, modern treatment. Linear equation, stability theory for autonomous and nonautonomous systems, etc. 164pp. 5⅜ × 8¼. 63828-6 Pa. $5.95

THIRTY YEARS THAT SHOOK PHYSICS: The Story of Quantum Theory, George Gamow. Lucid, accessible introduction to influential theory of energy and matter. Careful explanations of Dirac's anti-particles, Bohr's model of the atom, much more. 12 plates. Numerous drawings. 240pp. 5⅜ × 8½. 24895-X Pa. $6.95

THEORY OF MATRICES, Sam Perlis. Outstanding text covering rank, nonsingularity and inverses in connection with the development of canonical matrices under the relation of equivalence, and without the intervention of determinants. Includes exercises. 237pp. 5⅜ × 8½. 66810-X Pa. $7.95

GREAT EXPERIMENTS IN PHYSICS: Firsthand Accounts from Galileo to Einstein, edited by Morris H. Shamos. 25 crucial discoveries: Newton's laws of motion, Chadwick's study of the neutron, Hertz on electromagnetic waves, more. Original accounts clearly annotated. 370pp. 5⅜ × 8½. 25346-5 Pa. $10.95

INTRODUCTION TO PARTIAL DIFFERENTIAL EQUATIONS WITH APPLICATIONS, E.C. Zachmanoglou and Dale W. Thoe. Essentials of partial differential equations applied to common problems in engineering and the physical sciences. Problems and answers. 416pp. 5⅜ × 8½. 65251-3 Pa. $10.95

BURNHAM'S CELESTIAL HANDBOOK, Robert Burnham, Jr. Thorough guide to the stars beyond our solar system. Exhaustive treatment. Alphabetical by constellation: Andromeda to Cetus in Vol. 1; Chamaeleon to Orion in Vol. 2; and Pavo to Vulpecula in Vol. 3. Hundreds of illustrations. Index in Vol. 3. 2,000pp. 6⅛ × 9¼. 23567-X, 23568-8, 23673-0 Pa., Three-vol. set $41.85

CHEMICAL MAGIC, Leonard A. Ford. Second Edition, Revised by E. Winston Grundmeier. Over 100 unusual stunts demonstrating cold fire, dust explosions, much more. Text explains scientific principles and stresses safety precautions. 128pp. 5⅜ × 8½. 67628-5 Pa. $5.95

AMATEUR ASTRONOMER'S HANDBOOK, J.B. Sidgwick. Timeless, comprehensive coverage of telescopes, mirrors, lenses, mountings, telescope drives, micrometers, spectroscopes, more. 189 illustrations. 576pp. 5⅜ × 8¼. (Available in U.S. only) 24034-7 Pa. $9.95

ROTARY-WING AERODYNAMICS, W.Z. Stepniewski. Clear, concise text covers aerodynamic phenomena of the rotor and offers guidelines for helicopter performance evaluation. Originally prepared for NASA. 537 figures. 640pp. 6⅛ × 9¼. 64647-5 Pa. $15.95

DIFFERENTIAL GEOMETRY, Heinrich W. Guggenheimer. Local differential geometry as an application of advanced calculus and linear algebra. Curvature, transformation groups, surfaces, more. Exercises. 62 figures. 378pp. 5⅜ × 8½. 63433-7 Pa. $8.95

INTRODUCTION TO SPACE DYNAMICS, William Tyrrell Thomson. Comprehensive, classic introduction to space-flight engineering for advanced undergraduate and graduate students. Includes vector algebra, kinematics, transformation of coordinates. Bibliography. Index. 352pp. 5⅜ × 8½. 65113-4 Pa. $8.95

A SURVEY OF MINIMAL SURFACES, Robert Osserman. Up-to-date, in-depth discussion of the field for advanced students. Corrected and enlarged edition covers new developments. Includes numerous problems. 192pp. 5⅜ × 8½. 64998-9 Pa. $8.95

ANALYTICAL MECHANICS OF GEARS, Earle Buckingham. Indispensable reference for modern gear manufacture covers conjugate gear-tooth action, gear-tooth profiles of various gears, many other topics. 263 figures. 102 tables. 546pp. 5⅜ × 8½. 65712-4 Pa. $14.95

SET THEORY AND LOGIC, Robert R. Stoll. Lucid introduction to unified theory of mathematical concepts. Set theory and logic seen as tools for conceptual understanding of real number system. 496pp. 5⅜ × 8¼. 63829-4 Pa. $10.95

A HISTORY OF MECHANICS, René Dugas. Monumental study of mechanical principles from antiquity to quantum mechanics. Contributions of ancient Greeks, Galileo, Leonardo, Kepler, Lagrange, many others. 671pp. 5⅜ × 8½. 65632-2 Pa. $14.95

FAMOUS PROBLEMS OF GEOMETRY AND HOW TO SOLVE THEM, Benjamin Bold. Squaring the circle, trisecting the angle, duplicating the cube: learn their history, why they are impossible to solve, then solve them yourself. 128pp. 5⅜ × 8½. 24297-8 Pa. $4.95

MECHANICAL VIBRATIONS, J.P. Den Hartog. Classic textbook offers lucid explanations and illustrative models, applying theories of vibrations to a variety of practical industrial engineering problems. Numerous figures. 233 problems, solutions. Appendix. Index. Preface. 436pp. 5⅜ × 8½. 64785-4 Pa. $10.95

CURVATURE AND HOMOLOGY, Samuel I. Goldberg. Thorough treatment of specialized branch of differential geometry. Covers Riemannian manifolds, topology of differentiable manifolds, compact Lie groups, other topics. Exercises. 315pp. 5⅜ × 8½. 64314-X Pa. $8.95

HISTORY OF STRENGTH OF MATERIALS, Stephen P. Timoshenko. Excellent historical survey of the strength of materials with many references to the theories of elasticity and structure. 245 figures. 452pp. 5⅜ × 8½. 61187-6 Pa. $11.95

- Returns must be accompanied by the original receipt.
- Returns must be completed within 30 days.
- Merchandise must be in salable condition.
- Opened videos, discs and cassettes may be exchanged for replacement copies of the original items only.
- Periodicals and newspapers may not be returned.
- Items purchased by check may be returned for cash after 10 business days.
- All returned checks will incur a $15 service charge.
- All other refunds will be granted in the form of the original payment.

BORDERS®

- Returns must be accompanied by the original receipt.
- Returns must be completed within 30 days.
- Merchandise must be in salable condition.
- Opened videos, discs and cassettes may be exchanged for replacement copies of the original items only.
- Periodicals and newspapers may not be returned.
- Items purchased by check may be returned for cash after 10 business days.
- All returned checks will incur a $15 service charge.
- All other refunds will be granted in the form of the original payment.

BORDERS®

- Returns must be accompanied by the original receipt.
- Returns must be completed within 30 days.
- Merchandise must be in salable condition.

BORDERS
BOOKS AND MUSIC
300 ROUTE 18
EAST BRUNSWICK NJ 08816
(732) 238-7000

STORE: 0042 REG: 02/86 TRAN#: 8705
SALE 07/28/2001 EMP: 00125

ELASTICITY
0551624 QP T 9.95

	Subtotal	9.95
	NJ 6%	.60
1 Item	Total	10.55
	CASH	20.00
	CASH	9.45-

07/28/2001 08:47PM

THANK YOU FOR SHOPPING AT BORDERS
PLEASE ASK ABOUT OUR SPECIAL EVENTS